Food Proteins and Bioactive Peptides

Special Issue Editor

Maria Hayes

MDPI • Basel • Beijing • Wuhan • Barcelona • Belgrade

Special Issue Editor
Maria Hayes
Scientific Research Officer, Teagasc Food Research Centre
Ireland

Editorial Office
MDPI
St. Alban-Anlage 66
Basel, Switzerland

This edition is a reprint of the Special Issue published online in the open access journal *Foods* (ISSN 2304-8158) from 2017–2018 (available at: http://www.mdpi.com/journal/foods/special_issues/bioactive_peptides).

For citation purposes, cite each article independently as indicated on the article page online and as indicated below:

Lastname, F.M.; Lastname, F.M. Article title. *Journal Name* **Year**, *Article number*, page range.

First Editon 2018

ISBN 978-3-03842-863-3 (Pbk)
ISBN 978-3-03842-864-0 (PDF)

Table of Contents

About the Special Issue Editor

Maria Hayes, BSc., PhD, is a microbiologist and chemist based at Teagasc Food Research Centre, Ashtown, with a specific focus on by-product utilisation and bioeconomy and experience in the isolation and characterisation of bioactive peptides from protein sources using microbial strains and enzymes through controlled fermentations and hydrolysis. Maria previously was the Scientific Project Manager of the Marine Functional Foods Research Initiative (NutraMara Project), which encompasses five Irish Universities, namely, University College Dublin (UCD), University of Limerick (UL), University of Ulster Coleraine (UU), University College Galway (NUIG), and University College Cork (UCC), together with Teagasc. She has published over 90 papers in high-impact-factor journals and specialises in peptides and proteins for health and product development.

foods

MDPI

Editorial

Food Proteins and Bioactive Peptides: New and Novel Sources, Characterisation Strategies and Applications

Maria Hayes

The Food BioSciences Department, Teagasc Food Research Centre, Ashtown, Dublin 15, Ireland; Maria.Hayes@teagasc.ie; Tel.: +353-1-8059957

Received: 7 March 2018; Accepted: 12 March 2018; Published: 14 March 2018

Abstract: By 2050, the world population is estimated to reach 9.6 billion, and this growth continues to require more food, particularly proteins. Moreover, the Westernisation of society has led to consumer demand for protein products that taste good and are convenient to consume, but additionally have nutritional and health maintenance and well-being benefits. Proteins provide energy, but additionally have a wide range of functions from enzymatic activities in the body to bioactivities including those associated with heart health, diabetes-type 2-prevention and mental health maintenance; stress relief as well as a plethora of other health beneficial attributes. Furthermore, proteins play an important role in food manufacture and often provide the binding, water- or oil-holding, emulsifying, foaming or other functional attributes required to ensure optimum sensory and taste benefits for the consumer. The purpose of this issue is to highlight current and new protein sources and their associated functional, nutritional and health benefits as well as best practices for quantifying proteins and bioactive peptides in both a laboratory and industry setting. The bioaccessibility, bioavailability and bioactivities of proteins from dairy, cereal and novel sources including seaweeds and insect protein and how they are measured and the relevance of protein quality measurement methods including the Protein Digestibility Amino Acid Score (PDCAAS) and Digestible Indispensable Amino Acid Score (DIAAS) are highlighted. In addition, predicted future protein consumption trends and new markets for protein and peptide products are discussed.

Keywords: protein; bioactive peptides; Protein Digestibility Amino Acid Score (PDCAAS); Digestible Indispensable Amino Acid Score (DIAAS); functional activities; heart health; diabetes; mental health

---

The increasing global population and the need for reduction of or alternatives to animal protein consumption are two factors currently driving the development of new and alternative protein research and product development along with increased consumer awareness of the need to reduce carbon emissions. In addition, it has been identified that plant-based proteins from the likes of lentils, fava beans, peas and dry beans and soy are beneficial for health including the cardiovascular system, in glycaemic control, for the provision of minerals, vitamins, and fibre.

In this edition, Maehre and colleagues review the best method for use in the determination of protein quantities and qualities in food products and ingredients and conclude that the amino acid analysis method is the only one currently available where interference doesn't occur from some other substance, such as carbohydrates, lectins and anti-nutritional factors [1]. Inaccuracies were linked to indirect measurements, i.e., nitrogen determination and subsequent conversion to protein. In addition, novel protein processing methods were reviewed and implemented on soy protein. Vagadia and colleagues looked at conventional and microwave treatment methods for the inactivation of Trypsin inhibitors in order to improve in vitro digestibility of this plant protein source. Both microwave and conventional treatments were found to increase the digestibility of soymilk by 7 and 11%, respectively, and also reduced Trypsin inhibitor activities by 1 and 3%, respectively. This finding is important as, if implemented commercially it could increase the nutritional quality of soymilk by removal of anti-nutritional factors thereby improving the nutritional health of consumer groups including vegans and vegetarians as well as those that choose soymilk instead of traditional cow's milk due to lactose intolerance [2].

The generation of protein hydrolysates and ingredients is a strategy used by industry to improve the solubility of proteins. In addition to this functional benefit, hydrolysis is a strategy that may be employed to generate health benefits to protein ingredients as it results in the creation of bioactive peptides that often have hormone-like beneficial health effects once consumed. Bioactive peptides can be generated from any protein source using hydrolysis with commercially available enzymes or fermentation with proteolytic lactic acid bacteria (LAB). Importantly, they are inactive within the parent protein sequence and must be released before becoming active. In addition, they must survive the conditions within the gastrointestinal tract and reach their targets intact and in sufficient concentrations to provide an in vivo, beneficial health effect [3,4]. Daliri et al. [5] provide a detailed overview of the potential health benefits of bioactive peptides as well as a number of the challenges that exist concerning their development. Albenzio reviewed the use of bioactive peptides in animal food products [6]. Bleakley and colleagues look at the potential health benefits of cereal proteins using a novel strategy of combining an in silico computer-based screening approach with traditional protein extraction using salting out buffers followed by hydrolysis and peptide characterisation and synthesis for the generation and identification of potential application of bioactive peptides, specifically those related to the Renin-Angiotensin-Aldosterone System (RAAS) [7]. Nine Angiotensin-I-converting enzyme (ACE-I) inhibitory peptides with the sequences FFG, FFFL, PFL, WWK, WCY, FPIL, CPA, FLLA, FEPL and were found to inhibit ACE-I by 96.5% compared to the positive and commercially available drug Captopril® when assayed in vitro at a concentration of 1 mg protein/mL. In addition to this paper, Feeney and colleagues looked at the potential health benefits of glycosylated food proteins, specifically Glycomacropeptide (GMP), which is the terminal end kappa casein and is released from whey during the manufacture of cheese [8]. Using cell assays with HT-29 and Caco-2 cells, they assessed the ability of GMP to prevent adhesion of pathogenic *Escherichia coli* to the cell epithelium and concluded that GMP could be used as a gut health ingredient to protect the gut lining from damage [8]. In addition, Vijayakumar and Muriana discuss the potential use of the bioactive peptides produced from bacteria known as bacteriocins as a food protection agent [9]. Bacteriocins may be defined as ribosomal synthesized antimicrobial peptides produced by bacteria, which can kill or inhibit bacterial strains closely-related or non-related to produced bacteria, but will not harm the bacteria themselves by specific immunity proteins [10].

Novel proteins from seaweeds and macroalgae are reviewed by Bleakley and colleagues [11] and this review includes extraction and characterisation methods for these novel proteins including filtration and mass spectrometry methods. It also details the challenges faced in the development of this new protein resource including access, processing, characterisation, formulation and market challenges including legislation and consumer challenges such as sensory and taste. Total utilisation of by-products or co-products from food processing is important, not just in terms of the bio-economy, but it can also reduce environmental and economic costs for the processors while at the same time developing new products and markets [12]. In this special issue, Ofori and Hsieh examined the use of a monoclonal antibody to detect an antigenic protein in the red blood cells of porcine blood [13]. This development has potential for use by industry to protect consumers from eating porcine blood in food products and could be used for the vegan, vegetarian, Halal and Kosher product development markets. Finally, the review by Henchion and colleagues details potential future protein markets and consumption trends and provides an overview of the positive and negative attributes associated with meat, vegetable, marine and novel protein sources including insect, algal and others [14].

Conflicts of Interest: The author declares no conflict of interest.

References

1. Maehre, H.K.; Dalheim, L.; Edvinsen, G.K.; Elvevoll, E.O.; Jensen, I.-J. Protein determination—Methods matter. *Foods* **2017**, *7*, 5. [CrossRef] [PubMed]
2. Vagadia, B.H.; Vanga, S.K.; Singh, A.; Gariepy, Y.; Raghavan, V. Comparison of conventional and microwave treatment on soymilk for inactivation of Trypsin inhibitors and in vitro protein digestibility. *Foods* **2018**, *7*, 6. [CrossRef] [PubMed]
3. Hayes, M.; Stanton, C.; Fitzgerald, G.F.; Ross, R.P. Putting microbes to work: Dairy fermentation, cell factories and bioactive peptides: Part II: Bioactive peptide functions. *Biotechnol. J.* **2007**, *2*, 435–449. [CrossRef] [PubMed]
4. Pihlanto, A.; Mäkinen, S. Chapter 13: The function of Renin and the role of food-derived peptides as direct renin inhibitors. In *Renin-Angiotensin System—Past, Present and Future*; InTech: Rijeka, Croatia, 2017.
5. Daliri, E.B.-M.; Oh, D.H.; Lee, B.H. Bioactive peptides. *Foods* **2017**, *6*, 32. [CrossRef] [PubMed]
6. Albenzio, M.; Santillo, A.; Caroprese, M.; Malva, A.; Marino, R. Bioactive peptides in animal food products. *Foods* **2017**, *6*, 35. [CrossRef] [PubMed]
7. Bleakley, S.; Hayes, M.; O'Shea, N.; Gallagher, E.; Lafarga, T. Predicted release and analysis of novel ACE-I, Renin and DPP-IV inhibitory peptides from common oat (*Avena sativa*) protein hydrolysates using in silico analysis. *Foods* **2017**, *6*, 108. [CrossRef] [PubMed]
8. Feeney, S.; Ryan, J.T.; Kilcoyne, M.; Joshi, L.; Hickey, R. Glycomacropeptide reduces intestinal epithelial cell barrier dysfunction and adhesion of Entero-hemorrhagic and entero-pathogenic *Escherichia coli* in vitro. *Foods* **2017**, *6*, 93. [CrossRef] [PubMed]
9. Vijayakumar, P.P.; Muriana, P.M. Inhibition of *Listeria monocytogenes* on ready to eat meats using Bacteriocin mixtures based on mode of action. *Foods* **2017**, *6*, 22. [CrossRef] [PubMed]
10. Yang, S.-C.; Lin, C.-H.; Sung, C.T.; Fang, J.-Y. Antibacterial activities of bacteriocins: Application in foods and pharmaceuticals. *Front. Microbiol.* **2014**, *5*, 241. [CrossRef] [PubMed]
11. Bleakley, S.; Hayes, M. Algal proteins: Extraction, application and challenges concerning production. *Foods* **2017**, *6*, 33. [CrossRef] [PubMed]
12. Lafarga, T.; Hayes, M. Bioactive peptides from meat muscle and by-products: Generation, functionality and application as functional ingredients. *Meat Sci.* **2014**, *98*, 227–239. [CrossRef] [PubMed]
13. Ofori, J.A.; Hsieh, Y.-H.P. Immunodetection of porcine red blood cell containing food ingredients using a porcine hemoglobin-specific monoclonal antibody. *Foods* **2017**, *6*, 101. [CrossRef] [PubMed]
14. Henchion, M.; Hayes, M.; Mullen, A.M.; Fenelon, M.; Tiwari, B. Future protein supply and demand: Strategies and factors influencing a sustainable equilibrium. *Foods* **2017**, *6*, 53. [CrossRef] [PubMed]

foods

MDPI

Article

Protein Determination—Method Matters

Hanne K. Mæhre *, Lars Dalheim, Guro K. Edvinsen, Edel O. Elvevoll and Ida-Johanne Jensen

Norwegian College of Fishery Science, Faculty of Biosciences, Fisheries and Economics, UIT The Arctic University of Norway, N-9037 Tromsø, Norway; lars.dalheim@uit.no (L.D.); guro.k.edvinsen@uit.no (G.K.E.); edel.elvevoll@uit.no (E.O.E.); ida-johanne.jensen@uit.no (I.-J.J.)

* Correspondence: hanne.maehre@uit.no; Tel.: +47-7764-6793

Received: 15 November 2017; Accepted: 28 December 2017; Published: 1 January 2018

Abstract: The reported protein content of foods depends on the analytical method used for determination, making a direct comparison between studies difficult. The aim of this study was to examine and compare protein analytical methods. Some of these methods require extraction preceding analysis. The efficacy of protein extraction differs depending on food matrices and thus extraction yield was determined. Overall, most analytical methods overestimated the protein contents. The inaccuracies were linked to indirect measurements, i.e., nitrogen determination and subsequent conversion to protein, or interference from other chemical substances. Amino acid analysis is the only protein analysis method where interfering substances do not affect the results. Although there is potential for improvement in regards to the hydrolysis method, we recommend that this method should be the preferred for food protein determination.

Keywords: proteins; amino acids; analytical methods; extraction methods; Kjeldahl; Bradford; Lowry

1. Introduction

Proteins have a major role in the growth and maintenance of the human body and are, along with carbohydrates and lipids, the energy giving nutrients in the diet. In addition, proteins also pose a wide range of other functions in the body, such as enzymatic activity and transport of nutrients and other biochemical compounds across cellular membranes [1]. In order to maintain these important functions, it is essential to provide the body with good quality proteins through diet. Inadequate intake of dietary proteins containing essential amino acids results in increased turnover of muscular proteins, leading to reduced growth and loss of muscle mass. Impaired immunity, as well as reduced hormonal and enzymatic activity may subsequently follow [2]. Being such important constituents of human diet it is crucial to know the protein content in foods and thus it is important to have reliable analytical methods.

Food protein analysis is not necessarily a straightforward procedure. This is partly due to foods being heterogenic materials, comprised of a range of different nutrients, such as lipids, carbohydrates and a variety of micronutrients. Composition, food structure, or matrix, and interactions between the different nutrients may reduce the accessibility of the proteins leading to underestimation of the protein content. In addition, different methods are based on different analytical principles, determining protein content either directly or indirectly. Direct protein determination is when protein content is calculated based on the analysis of amino acid residues. Indirect protein determination can for instance be inferred following the determination of the nitrogen content, or after chemical reactions with functional groups within the protein. An additional factor that can contribute to inaccuracies in the determination of protein content is protein extraction. Some methods require some degree of protein extraction prior to analysis and thus, extraction yields can affect the results [3].

Varieties of different analytical methods have been developed throughout the years. Only a few of these however, are frequently used, and the reason for the choice of method used in many studies is seldom described. This may be due to a variety of factors, for instance tradition (using established analytical procedures in laboratories), lack of analytical infrastructure or high economic costs associated with certain methods. For instance, in a recent review by Angell et al. [4] it was shown that 52% of all studies on the protein content of seaweeds used nitrogen determination with subsequent conversion using a nitrogen-to-protein conversion factor of 6.25. This is in spite of many studies documenting that this factor leads to an overestimation of the protein content in most foods and, in particular, plant foods [5–7].

All of these variables make performing traditional reviews of studies concerning protein content of foods quite difficult. Thus, the main objective of this study was to document how the determined protein content in several common foods with different matrix compositions varies as a result of the choice of extraction and analytical method. The chosen foods were one lean and one fatty fish (cod and salmon), one crustacean (shrimp), one marine plant (dulse, red seaweed), as well as whole and white wheat flour. The chosen analytical methods were amino acid analysis, Kjeldahl's method, Bradford's method and a modified version of the Lowry method. The two extraction solutions were one consisting of a combination of salt and alkaline solutions, and one using a so-called Good's buffer [8].

2. Materials and Methods

2.1. Raw Materials

Salmon (*Salmo salar*) loins (n = 5) were purchased frozen in local supermarkets in Tromsø (Norway) and kept at −20 °C. Cod (*Gadus morhua*) (n = 5) was caught with a fishing rod off the coast of the Lofoten islands (Norway), gutted, put on ice and *post rigor* filleted. Loins were frozen at −20 °C prior to analysis. Batches (100 g) of peeled shrimp (*Pandalus borealis*) (n = 5) were obtained frozen from Marealis AS® (Tromsø, Norway) and kept at −20 °C. Prior to analysis, salmon, cod and shrimp samples were thawed and homogenized. White flour (n = 5) and whole flour (n = 5) of wheat were purchased at local supermarkets. Dehydrated samples of the red seaweed dulse (*Palmaria palmata*) (n = 5), harvested at the south coast of Iceland, were purchased from "The Northern Company" (Oslo, Norway). The flour and seaweed samples were kept dark at room temperature until analyses.

2.2. Protein Extraction Methods

2.2.1. The Salt/Alkaline Extraction Method

The salt/alkaline extraction was performed as described by Mæhre et al. [9], with minor modifications. Briefly, 0.5 g of raw material was homogenized with 30 mL of 0.1 M sodium hydroxide (NaOH) in 3.5% sodium chloride (NaCl) using an UltraTurrax homogenizer (IKA Werke GmbH, Staufen, Germany). The homogenates were incubated at 60 °C for 90 min before centrifugation at 4000× g for 30 min at 4 °C. The supernatants were frozen and kept at −20 °C until analyses.

2.2.2. The Good's Buffer Extraction

This extraction was performed as described by Alhamdani et al. [10] with minor modifications. Raw materials were homogenized 1:6 (*w*:*v*) with a buffer consisting of 20 mM 4-(2-hydroxyethyl) piperazine-1-ethanesulfonic acid (HEPES), 1 mg magnesium chloride ($MgCl_2$) and 0.5% 3-((3-holamidopropyl)dimethylammonio)-1-propanesulfonate (CHAPS), pH 7.9 using an UltraTurrax homogenizer. The homogenates were incubated on ice for 30 min with occasional mixing, before centrifugation at 4000× g, for 20 min at 4 °C. The supernatants were frozen and kept at −20 °C until analyses.

2.3. *Direct Protein Determination*

Amino Acid Analysis

Sample preparations for analysis of total amino acids were performed as described by Mæhre et al. [11]. For the raw material samples, approximately 200 mg of fish and shrimp samples and approximately 50 mg of flour and dulse samples, were dissolved in 0.7 mL distilled H_2O and 0.5 mL 20 mM norleucine (internal standard). For the protein extract samples, 500 µL extract was mixed with 50 µL 20 mM norleucine. Subsequently, for all samples, concentrated hydrochloric acid (HCl, 12 M) was added, to a final concentration of 6 M. The sample mixtures were flushed with nitrogen gas for 15 s in order to minimize oxidation, before hydrolysis at 110 °C for 24 h according to Moore and Stein [12]. Following hydrolysis, 100 µL aliquots of the hydrolysates were evaporated under nitrogen gas until complete dryness and re-dissolved to a suitable concentration in lithium citrate buffer at pH 2.2. All amino acids were analyzed chromatographically using an ion exchange column followed by ninhydrin post column derivatization on a Biochrom 30 amino acid analyzer (Biochrom Co., Cambridge, UK). Amino acid residues were identified using the A9906 physiological amino acids standard (Sigma Chemical Co., St. Louis, MO, USA) as described previously [13]. Protein content was calculated as the sum of individual amino acid residues (the molecular weight of each amino acid after subtraction of the molecular weight of H_2O).

2.4. *Indirect Protein Determinations*

2.4.1. The Kjeldahl Method

The Kjeldahl method was performed according to method 981.10 of the AOAC International [14]. Approximately 1 g of raw material was hydrolyzed with 15 mL concentrated sulfuric acid (H_2SO_4) containing two copper catalyst tablets in a heat block (Kjeltec system 2020 digestor, Tecator Inc., Herndon, VA, USA) at 420 °C for 2 h. After cooling, H_2O was added to the hydrolysates before neutralization and titration. The amount of total nitrogen in the raw materials were multiplied with both the traditional conversion factor of 6.25 [15] and species-specific conversion factors [7,16] in order to determine total protein content. The species-specific conversion factors were 5.6 for fish and shrimp, 5.4 for flours and 4.59 for seaweed, respectively.

2.4.2. The Modified Lowry Method

The modified Lowry protein measurement was conducted according to the method described by Hartree [17]. The assay was carried out by diluting the extracts to 1 mL with H_2O and adding 0.9 mL of solution A (2 g L^{-1} potassium sodium tartrate ($KNaC_4H_4O_6 \cdot 4H_2O$) and 100 g L^{-1} sodium carbonate (Na_2CO_3) in 0.5 M NaOH) before incubation for 10 min at 50 °C. Following this, the samples were cooled down to room temperature, added 1 mL of solution B (0.2 g L^{-1} $KNaC_4H_4O_6 \cdot 4H_2O$ and 0.1 g L^{-1} copper sulfate pentahydrate ($CuSO_4 \cdot 5H_2O$) in 0.1 M NaOH) and left for 10 min. Finally, 3 mL of solution C (Folin–Ciocalteu phenol reagent in H_2O (1:16 v/v)) was added before incubation for 10 min at 50 °C. A standard curve was made of bovine serum albumin (BSA; 0, 0.0625, 0.125, 0.25, 0.5 and 1 g L^{-1}) and absorbance was read at 650 nm.

2.4.3. The Bradford Method

The Bradford assay was conducted according to the method described by Bradford [18]. Briefly, 100 mg Coomassie Brilliant Blue G-250 was dissolved in 50 mL 95% ethanol (C_2H_5OH). Thereafter, 100 mL of 85% phosphoric acid (H_3PO_4) was carefully added under stirring, before H_2O was added to a total volume of 1 L. The solution was filtered and kept at 4 °C. For the measurements, 100 µL extract and 5 mL Bradford solution were mixed and incubated for 5 min. A standard curve was made of BSA (0, 0.0625, 0.125, 0.25, 0.5 and 1 g L^{-1}) and absorbance was read at 595 nm.

2.5. Statistics Description

All results are presented as arithmetic mean of 5 parallels ± standard deviation (SD). Statistical package for the social sciences 23 (SPSS Inc., Chicago, IL, USA) was used to perform statistical analyses. Shapiro-Wilk's test for normality and Levene's test for homogeneity of variance were performed and for samples returning normal distribution, one-way analysis of variance (ANOVA) was performed. For non-normal distributions, the non-parametric Mann-Whitney U test was used. For evaluation of statistics, Tukey and Dunnett's T3 post-hoc tests were run for equal and un-equal variances, respectively. Means were considered significantly different at $p < 0.05$.

3. Results and Discussion

3.1. Direct Protein Determination

Amino Acid Analysis

Amino acid analysis is one of the analytical principles for protein determination. Here, the proteins are broken down into their constituent amino acids by hydrolysis of the peptide bonds. The liberated amino acid residues are then determined, most often chromatographically, and protein content is calculated as the sum of individual amino acid residues after subtraction of the molecular mass of H_2O. One major drawback of this method, however, is the protein hydrolysis prior to analysis. The commonest method is hydrolysis in 6 M HCl at 110 °C for 24 h, as described by Moore and Stein [12]. This procedure efficiently hydrolyzes most of the peptide bonds, but at the same time the content of some amino acids are reduced or even destroyed completely [19], and thus, the protein content analyzed by this method may be underestimated. Another drawback of this method is the costs related to investment and analysis. This makes the method unavailable for many food science laboratories. Due to amino acid analysis being the only protein analysis method determining protein contents directly (based solely on amino acid residues and where interfering substances do not affect the results), it was decided to use this method as the reference method in this study and compare all other analyses with the results from this. This is also in accordance with the recommendations given by the Food and Agricultural Organization of the United Nations (FAO) regarding determination of food proteins [20].

3.2. Indirect Protein Determination

3.2.1. Protein Content Based on Nitrogen Determination

Some of the most frequently used methods for food protein determination are based on analysis of the total nitrogen content in the samples. Examples of such methods are the Dumas method [21] and the Kjeldahl method [15]. In both methods, the total nitrogen in the sample is liberated at high temperature. In the Kjeldahl method, the nitrogen is released into a strong acid and the content is measured after neutralization and titration. In the Dumas method, the nitrogen is liberated in a gaseous form and is determined with a thermal conductivity detector, after removal of carbon dioxide and water aerosols. The Kjeldahl method was chosen as an example of this analytical principle in this study as it is still recognized as the official method for food protein determination by the AOAC International [14]. Following the nitrogen determination, crude protein content is calculated using a conversion factor. The original, and still frequently used, conversion factor 6.25 is based on an assumption that the general nitrogen content in food proteins is 16% and that all nitrogen in foods is protein-bound. These are, however, quite rough assumptions as the relative nitrogen content varies between amino acids and amino acid composition varies between food proteins [22]. In addition, a wide range of other compounds, such as nitrate, ammonia, urea, nucleic acids, free amino acids, chlorophylls and alkaloids contain nitrogen. These compounds are called non-protein nitrogen and their relative contents are often higher in vegetables than in foods of animal origin [5]. Throughout the years, it has been proven that the conversion factor of 6.25 in most cases overestimates the protein content and in order to adjust for these variations, several species-specific conversion factors have been suggested [6,7,16,22], making the conversion from nitrogen to protein more precise.

3.2.2. Comparison between Amino Acid Analysis and the Kjeldahl Method

In Table 1 the results of the amino acid analysis and the Kjeldahl analysis of the raw materials are shown. Both the traditional conversion factor of 6.25 and the respective species-specific conversion factors are presented. For fish, shrimp and flour, the species-specific factors used in these calculations were the average conversion factors suggested by Mariotti et al. [7], namely 5.6 for fish and shrimp and 5.4 for cereals. The conversion factor used for dulse is the factor suggested by Lourenço et al. [16] as an average for red algae, namely 4.59.

Table 1. Protein content from amino acid analysis and Kjeldahl nitrogen analysis of cod, salmon, shrimp, white and whole wheat flour and dulse (red seaweed).

	Raw Materials		
Species	**Amino Acid Analysis**	**Kjeldahl (Factor 6.25)**	**Kjeldahl (Species-Specific Factors)**
Cod	111.6 ± 9.1 [a]	185.7 ± 15.0 [b]	166.4 ± 13.4 [b] (5.6)
Salmon	121.4 ± 6.3 [a]	208.1 ± 7.3 [c]	186.5 ± 6.6 [b] (5.6)
Shrimp	83.8 ± 8.6 [a]	132.7 ± 4.7 [c]	117.8 ± 4.4 [b] (5.6)
White flour (wheat)	77.9 ± 6.5 [a]	117.6 ± 7.3 [c]	101.6 ± 6.3 [b] (5.4)
Whole flour (wheat)	88.2 ± 3.8 [a]	133.4 ± 7.8 [c]	115.3 ± 6.8 [b] (5.4)
Dulse (red seaweed)	105.3 ± 9.1 [a]	152.1 ± 10.0 [b]	111.7 ± 7.3 [a] (4.59)

The conversion factors used for the Kjeldahl analysis are the commonly used conversion factor 6.25 and species-specific conversion factors (in parentheses) 5.4 for the flours, 5.6 for fish and shrimp [7] and 4.59 for seaweed [15]. Values are reported as mean value ± standard deviation (SD) (n = 5) and in g protein kg^{-1} raw material. Different letters within the same row indicate significant differences between analyses within each species.

As expected, all of the Kjeldahl results using the traditional conversion factor were significantly higher, ranging from 44% to 71% higher, than the corresponding amino acid analysis results. More surprisingly, and except for the red algae, the species-specific conversion factors also gave significantly higher protein content than the amino acid analysis. One possible explanation may be that the concentration of some of the amino acids in the sample was reduced as a result of the hydrolysis process prior to the amino acid analysis and that the protein concentration calculated from this analysis was in fact underestimated. Another possible explanation could be that the "real" conversion factors for these species should in fact be even lower than the average values reported by Mariotti et al. [7]. Calculating conversion factors based on the results from this study gave factors of 4.9 for fish and shrimp and 4.7 for flours, respectively.

There are risks associated with calculating protein from nitrogen and the resulting overestimation of protein content. One is the possibility of food adulteration, as a high protein content often raises the economic value of a product [23]. There have been some cases where the producer in order to increase the apparent protein content [24], and subsequently the economic value of the food product have added non-protein nitrogen, such as melamine. This may compromise food safety for consumers and hence, it is important that such food adulteration is rendered impossible. Another risk is that the utilization potential and economic feasibility of "new" raw materials could be overestimated, as protein is one of the constituents with value-added potential in a product. Overestimation could thus give false premises for the establishment of new industries. For instance, the interest of increased industrial utilization of seaweeds has increased greatly the last decades. Several studies have presented a protein content of some red seaweed species of 30–45% [25,26], while when analyzed with amino acid analysis, it commonly ranges between 10% and 20% [11,27,28]. Such a difference could be crucial for the economy of small scale industries.

3.2.3. Spectrophotometric Methods and Protein Extraction

The third common analytical technique for protein analysis is spectrophotometry. Here, the principle is that functional groups or regions within the protein absorb light in the ultraviolet or visible range of the electromagnetic spectrum (200–800 nm). This absorbance is read and compared with known protein standards. Examples of such functional groups or regions are basic groups, aromatic groups, peptide bonds or aggregated proteins.

While both nitrogen analysis and amino acid analysis may be performed without any pre-treatment of the raw material, extraction of proteins is a prerequisite before submitting the material to spectrophotometric analysis. Also for protein extraction, the available methods are many. Most common protein extraction protocols are based on exposure of tissue to weak buffers or water, leading to collapse of cells with subsequent release of intracellular proteins as a response to the hypotonic shock that arises. This is very efficient for tissues containing cells without cell walls (animal cells), but not as efficient for cells with cell walls (plant cells). The latter being due to the cell walls protecting the cell against collapse [29]. In this study both protein sources of animal origin and protein sources of plant origin were included in order to evaluate these differences.

Of the commonest protein extraction methods, especially prior to electrophoresis, is the use of so-called Good's buffers, a range of buffers first described in 1966 by Good et al. [8]. These buffers contain chemicals with zwitterionic properties, along with one or more detergents and are known to be highly compatible to biological analyses. They are highly water soluble, have low salt-effects and minimal interference with biological functions [10]. The main drawback of these protocols is that they involve several expensive chemicals and some of the chemicals are suspected to be health damaging. In this study, a mixture of HEPES (zwitterionic agent) and CHAPS (detergent) was chosen as an example of this principle.

Proteins are often divided into four main classes, based on their solubility properties. The four classes are albumins that are water-soluble, globulins that are soluble in weak ionic solutions, glutelins that are soluble in weak acidic or basic solutions and prolamines that are soluble in 70% ethanol. Most foods are complex matrices, probably holding several of these protein classes and combining several solutes would probably optimize the extraction yield. In Mæhre et al. [9] it was shown that combining H_2O, 0.1 M NaOH and 3.5% NaCl at elevated temperature increased extraction yield compared to traditional extraction methods and this method was thus chosen as an example of a simpler protein extraction method in this study.

In Table 2, the extraction yields of the two different extraction methods are presented as protein content relative to the respective raw materials, calculated after amino acid analysis. As seen, extracting proteins using the HEPES/CHAPS protocol was quite ineffective, resulting in very low yields for all raw materials. Salt/alkaline extraction was shown to give significantly higher yields for all raw materials. Further, it was shown that the salt/alkaline protocol was very efficient for extracting proteins of animal origin, managing to extract more or less 100% of the protein. It was also quite efficient for extracting proteins from highly processed plant raw material (white flour), giving a yield of around 80%. The extraction yield was, however, lower in less processed and more complex plant materials, such as whole flour and dulse, the latter being comparable to a previous study [9].

Table 2. Protein extraction yields calculated from amino acid analysis of cod, salmon, shrimp, white and whole wheat flour and dulse.

	Raw Materials	
Species	**Extraction Yield Salt/Alkaline**	**Extraction Yield HEPES/CHAPS**
Cod	106.8 ± 11.2 [b]	13.8 ± 3.0 [a]
Salmon	99.9 ± 13.3 [b]	31.9 ± 2.1 [a]
Shrimp	113.3 ± 17.0 [b]	14.4 ± 2.2 [a]
White wheat flour	78.5 ± 11.1 [b]	15.0 ± 3.7 [a]
Whole wheat flour	66.1 ± 20.9 [b]	20.4 ± 3.2 [a]
Dulse (red seaweed)	34.9 ± 6.7 [b]	12.0 ± 1.6 [a]

Values are reported as mean value ± SD (n = 5) and in % of amino acid content in the respective raw materials (as shown in Table 1). Different letters within the same row indicate significant differences between the extraction yields of the different methods within each species. HEPES: 4-(2-hydroxyethyl)piperazine-1-ethanesulfonic acid; CHAPS: 3-((3-holamidopropyl)dimethylammonio)-1-propanesulfonate.

One of the differences between the two chosen extraction methods was that salt/alkaline extraction was performed at 60 °C while HEPES/CHAPS was performed on ice and this could contribute to the difference in extraction yield. Extraction on ice probably protects the proteins against degradation, while heat treatment could possibly accelerate this. Thus, the extraction method should be chosen based on the purpose of further use. However, in this study the goal was merely to examine the differences in protein extraction yield between methods, and thus, protein degradation was not analyzed.

Following extraction, protein determination using two different spectrophotometric methods, the Bradford method [18] and a modified Lowry method [17], was tested and compared with amino acid analysis. In the Bradford method, Coomassie G-250 dye reacts with ionizable groups on the protein disrupting the proteins tertiary structure and exposing the hydrophobic pockets. This is followed by the dye binding to the hydrophobic amino acids forming stable complexes that can be read at 595 nm [3]. The modified Lowry method is a combination of the Biuret method, where copper ions react with the peptide bonds within the protein, and a reaction between Folin-Ciocalteu reagent and the ring structure on aromatic amino acids. The total reaction forms a stable, dark blue complex that can be read at 650–750 nm [3].

3.2.4. Comparison between Amino Acid Analysis and the Spectrophotometric Methods

In Table 3, the results of protein determination by amino acid analysis, Bradford method and modified Lowry method are shown for the salt/alkaline extracts, while in Table 4 the same analyses are shown for the HEPES/CHAPS extracts. In the salt/alkaline extracts, both the Bradford method and the modified Lowry method gave higher protein estimates than the amino acid analysis for proteins of animal origin. The same was observed when using the modified Lowry method for plant proteins, while the Bradford method resulted in equal or lower protein estimates than the amino acid analysis. In the HEPES/CHAPS extracts, it was not possible to obtain any results using the modified Lowry method. The Bradford method gave higher protein estimates than the amino acid analysis for all raw materials except for the dulse.

Table 3. Protein content calculated from amino acid analysis, Bradford analysis and modified Lowry analysis in salt/alkaline protein extracts of cod, salmon, shrimp, white and whole wheat flour and dulse.

	Salt/Alkaline Protein Extracts		
Species	**Amino Acid Analysis**	**Bradford**	**Modified Lowry**
Cod	118.8 ± 9.8 [a]	198.7 ± 24.1 [b]	194.5 ± 11.4 [b]
Salmon	120.9 ± 13.9 [a]	234.6 ± 10.6 [c]	211.9 ± 9.2 [b]
Shrimp	93.9 ± 7.8 [a]	165.5 ± 10.3 [b]	151.2 ± 8.2 [b]
White flour (wheat)	60.9 ± 8.1 [b]	45.0 ± 6.6 [a]	85.7 ± 12.3 [c]
Whole flour (wheat)	58.2 ± 18.6 [a]	46.2 ± 18.5 [a]	88.8 ± 27.3 [b]
Dulse (red seaweed)	36.9 ± 9.1 [a]	48.4 ± 3.9 [a]	89.5 ± 15.9 [b]

Values are reported as mean value ± SD (n = 5) and in g protein kg^{-1} raw material. Different letters within the same row indicate significant differences between methods within each species.

Table 4. Protein content calculated from amino acid analysis, Bradford analysis and modified Lowry analysis in HEPES/CHAPS protein extracts of cod, salmon, shrimp, white and whole wheat flour and dulse.

	HEPES/CHAPS Protein Extracts		
Species	**Amino Acid Analysis**	**Bradford**	**Modified Lowry**
Cod	15.3 ± 2.7 [a]	39.4 ± 7.5 [b]	n.a.
Salmon	38.7 ± 3.3 [a]	98.4 ± 3.1 [b]	n.a.
Shrimp	11.9 ± 0.7 [a]	25.7 ± 1.6 [b]	n.a.
White flour (wheat)	11.6 ± 2.7 [a]	18.5 ± 3.9 [b]	n.a.
Whole flour (wheat)	18.0 ± 2.8 [a]	30.0 ± 5.0 [b]	n.a.
Dulse (red seaweed)	12.5 ± 1.4 [b]	7.8 ± 1.3 [a]	n.a.

Values are reported as mean value ± SD (n = 5) and in g protein kg^{-1} raw material. Different letters within the same row indicate significant differences between methods within each species. n.a.: not applicable.

The Lowry method has been widely used for protein determination for many decades, due to its simplicity and availability. However, besides aromatic amino acids, a wide range of other compounds react with the Folin–Ciocalteu reagent [30]. In complex food matrices, containing several interfering compounds, this normally leads to an overestimation of the protein content, which is also what the results from the salt/alkaline extraction show. One of the compounds that has been shown to interfere with the Lowry protocol is HEPES [31]. In this study, the reaction between the HEPES/CHAPS buffer and the Lowry reagent produced a very strong background color in both standards and extracts, making it impossible to obtain accurate protein results. Normally, the Bradford method has been recognized to be less prone to such interference. However, the results from this study show that also for several of the raw materials the protein estimates are very high compared to the amino acid analysis, indicating that there could be some kind of interference. Differences in extraction efficiency of different amino acids could also be a possible explanation. In the Bradford analysis, the basic amino acids contribute more to the final color than do other amino acids [3], while aromatic amino acid contribute more to the color development in the modified Lowry method. As seen in Table 5, there were some significant differences in the extraction efficiency between hydrophobic (proline, glycine, alanine, valine, isoleucine, leucine and phenylalanine), aromatic (tyrosine, phenylalanine, tryptophan and histidine) and basic (lysine, histidine and arginine) amino acids. However, this effect varied between species and the only significant effect when looking at all materials was that the HEPES/CHAPS method seemed to extract hydrophobic amino acids more efficiently than the other methods.

Table 5. Relative amount of hydrophobic, aromatic and basic amino acids (AA) in raw materials and protein extracts after extraction using HEPES/CHAPS and salt/alkaline in cod, salmon, shrimp, white wheat flour, whole wheat flour and dulse, as well as across all species.

		Raw Material	HEPES/CHAPS	Salt/Alkaline
Cod	Hydrophobic AA	37.8 ± 0.9 ab	42.8 ± 3.2 b	35.7 ± 1.3 a
	Aromatic AA	9.3 ± 1.7	9.4 ± 2.4	8.0 ± 2.7
	Basic AA	19.6 ± 0.3	19.2 ± 1.2	18.7 ± 0.6
Salmon	Hydrophobic AA	39.4 ± 1.0 a	44.3 ± 1.0 b	39.5 ± 1.6 a
	Aromatic AA	10.1 ± 1.4	9.0 ± 3.3	8.7 ± 3.0
	Basic AA	20.0 ± 0.2	18.2 ± 1.4	20.0 ± 3.1
Shrimp	Hydrophobic AA	38.4 ± 0.4 a	47.3 ± 1.2 b	38.0 ± 2.5 a
	Aromatic AA	10.2 ± 0.9 b	4.2 ± 0.9 a	8.7 ± 2.1 b
	Basic AA	19.8 ± 0.2 b	17.6 ± 1.1 a	19.8 ± 0.7 b
White flour (wheat)	Hydrophobic AA	41.2 ± 0.9 b	40.5 ± 1.4 b	37.4 ± 0.6 a
	Aromatic AA	8.9 ± 1.1 a	9.4 ± 2.7 ab	11.5 ± 0.5 b
	Basic AA	8.6 ± 0.3 a	11.0 ± 0.5 b	7.9 ± 0.7 a
Whole flour (wheat)	Hydrophobic AA	40.9 ± 0.5 b	38.6 ± 0.6 a	40.0 ± 1.7 ab
	Aromatic AA	9.4 ± 1.1	10.6 ± 0.3	8.1 ± 2.2
	Basic AA	10.2 ± 0.2 b	12.4 ± 0.5 c	8.4 ± 0.4 a
Dulse (red seaweed)	Hydrophobic AA	43.5 ± 1.4 b	34.6 ± 2.7 a	46.3 ± 2.3 b
	Aromatic AA	10.3 ± 0.5 b	1.6 ± 0.2 a	9.3 ± 1.8 b
	Basic AA	14.0 ± 1.7 c	4.4 ± 0.8 a	6.9 ± 0.7 b
All species	Hydrophobic AA	40.2 ± 2.1 ab	41.4 ± 4.5 b	39.5 ± 3.8 a
	Aromatic AA	9.7 ± 1.2	7.4 ± 3.8	9.1 ± 2.3
	Basic AA	15.4 ± 4.8	13.8 ± 5.4	13.6 ± 6.2

Values are given as mean ± SD (n = 5) and in % of total amino acids. Different letters within the same row indicate significant difference ($p < 0.05$) between amino acid compositions in raw materials and in the two extracts.

4. Conclusions

As shown, there are many methods available for protein determination and all of them have their advantages and disadvantages. This great assortment of methods makes direct comparison between studies difficult and the choice of analytical method should thus be justified depending on the purpose of the study.

The results from this study show that protein determination based on nitrogen analysis for most food matrices overestimates the protein content compared to amino acid analysis, whether or not the species-specific conversion factors are used. Spectrophotometric protein determination methods are often affected by interfering substances and could thus overestimate the protein content. Protein extraction often involves chemicals affecting both extraction yield and subsequent determination. This makes such protocols very dependent on the choice of buffers used and methods involving extraction should thus not be the primary choice for food purposes. However, if the purpose is further analytical use, such methods could be a suitable alternative. Amino acid analysis is the only protein analysis method where interfering substances do not affect the results. Although there is potential for improvement in regards to the hydrolysis method, we recommend that this method should be the preferred for food protein determination.

Acknowledgments: This work was supported by the Publication Fund of UIT The Arctic University of Norway.

Author Contributions: All authors conceived and designed the experiments; H.K.M., L.D., G.K.E. and I.-J.J. performed the experiments and analyzed the data; H.K.M. and I.-J.J. wrote the paper; all authors have discussed results, proof read and approved the manuscript.

Conflicts of Interest: The authors declare no conflict of interest.

References

1. Wu, G.Y.; Bazer, F.W.; Dai, Z.L.; Li, D.F.; Wang, J.J.; Wu, Z.L. Amino acid nutrition in animals: Protein synthesis and beyond. *Annu. Rev. Anim. Biosci.* **2014**, *2*, 387–417. [CrossRef] [PubMed]
2. Wolfe, R.R. The underappreciated role of muscle in health and disease. *Am. J. Clin. Nutr.* **2006**, *84*, 475–482. [PubMed]
3. Wilson, K.; Walker, J. *Principles and Techniques of Practical Biochemistry*; Cambridge University Press: Cambridge, UK, 2010.
4. Angell, A.R.; Mata, L.; de Nys, R.; Paul, N.A. The protein content of seaweeds: A universal nitrogen-to-protein conversion factor of five. *J. Appl. Phycol.* **2016**, *28*, 511–524. [CrossRef]
5. Imafidon, G.I.; Sosulski, F.W. Non-protein nitrogen contents of animal and plant foods. *J. Agric. Food Chem.* **1990**, *38*, 114–118. [CrossRef]
6. Jones, D.B. *Factors for Converting Percentages of Nitrogen in Foods and Feeds into Percentages of Protein*; US Department of Agriculture: Washington, DC, USA, 1941.
7. Mariotti, F.; Tome, D.; Mirand, P.P. Converting nitrogen into protein—Beyond 6.25 and Jones' factors. *Crit. Rev. Food Sci.* **2008**, *48*, 177–184. [CrossRef] [PubMed]
8. Good, N.E.; Winget, G.D.; Winter, W.; Connolly, T.N.; Izawa, S.; Singh, R.M.M. Hydrogen ion buffers for biological research. *Biochemistry* **1966**, *5*, 467–477. [CrossRef] [PubMed]
9. Maehre, H.K.; Jensen, I.J.; Eilertsen, K.E. Enzymatic pre-treatment increases the protein bioaccessibility and extractability in dulse (*Palmaria palmata*). *Mar. Drugs* **2016**, *14*, 1–10. [CrossRef] [PubMed]
10. Alhamdani, M.S.S.; Schröder, C.; Werner, J.; Giese, N.; Bauer, A.; Hoheisel, J.D. Single-step procedure for the isolation of proteins at near-native conditions from mammalian tissue for proteomic analysis on antibody microarrays. *J. Proteome Res.* **2010**, *9*, 963–971. [CrossRef] [PubMed]
11. Maehre, H.K.; Edvinsen, G.K.; Eilertsen, K.E.; Elvevoll, E.O. Heat treatment increases the protein bioaccessibility in the red seaweed dulse (*Palmaria palmata*), but not in the brown seaweed winged kelp (*Alaria esculenta*). *J. Appl. Phycol.* **2016**, *28*, 581–590.
12. Moore, S.; Stein, W.H. Chromatographic determination of amino acids by the use of automatic recording equipment. *Method Enzymol.* **1963**, *6*, 819–831.
13. Maehre, H.K.; Hamre, K.; Elvevoll, E.O. Nutrient evaluation of rotifers and zooplankton: Feed for marine fish larvae. *Aquacult. Nutr.* **2013**, *19*, 301–311. [CrossRef]
14. Latimer, G.W. *Official Methods of Analysis of AOAC International*; AOAC International: Gaithersburg, MD, USA, 2016.
15. Kjeldahl, J. Neue Methode zur Bestimmung des Stickstoffs in organischen Körpern. *Fresenius' J. Anal. Chem.* **1883**, *22*, 366–382. [CrossRef]
16. Lourenço, S.O.; Barbarino, E.; De-Paula, J.C.; Pereira, L.O.d.S.; Lanfer Marquez, U.M. Amino acid composition, protein content and calculation of nitrogen-to-protein conversion factors for 19 tropical seaweeds. *Phycol. Res.* **2002**, *50*, 233–241. [CrossRef]
17. Hartree, E.F. Determination of protein—Modification of Lowry method that gives a linear photometric response. *Anal. Biochem.* **1972**, *48*, 422–427. [CrossRef]
18. Bradford, M.M. Rapid and sensitive method for quantitation of microgram quantities of protein utilizing principle of protein-dye binding. *Anal. Biochem.* **1976**, *72*, 248–254. [CrossRef]
19. Pickering, M.V.; Newton, P. Amino acid hydrolysis—Old problems, new solutions. *LC GC Mag. Sep. Sci.* **1990**, *8*, 778–780.
20. FAO. *Food Energy—Methods of Analysis and Conversion Factors*; Food and Agriculture Organization of the United Nations: Rome, Italy, 2003.
21. Dumas, J.B.A. Procedes de l'analyse organique. *Ann. Chim. Phys.* **1831**, *T47*, 198–213.
22. Sosulski, F.W.; Imafidon, G.I. Amino acid composition and nitrogen-to-protein conversion factors for animal and plant foods. *J. Agric. Food Chem.* **1990**, *38*, 1351–1356. [CrossRef]
23. Moore, J.C.; DeVries, J.W.; Lipp, M.; Griffiths, J.C.; Abernethy, D.R. Total protein methods and their potential utility to reduce the risk of food protein adulteration. *Compr. Rev. Food Sci. F* **2010**, *9*, 330–357. [CrossRef]
24. Ingelfinger, J.R. Melamine and the global implications of food contamination. *N. Engl. J. Med.* **2008**, *359*, 2745–2748. [CrossRef] [PubMed]

25. Fleurence, J. Seaweed proteins: Biochemical, nutritional aspects and potential uses. *Trends Food Sci. Technol.* **1999**, *10*, 25–28. [CrossRef]
26. Taboada, C.; Millan, R.; Miguez, I. Evaluation of marine algae Undaria pinnatifida and Porphyra purpurea as a food supplement: Composition, nutritional value and effect of intake on intestinal, hepatic and renal enzyme activities in rats. *J. Sci. Food Agric.* **2013**, *93*, 1863–1868. [CrossRef] [PubMed]
27. Biancarosa, I.; Espe, M.; Bruckner, C.G.; Heesch, S.; Liland, N.; Waagbo, R.; Torstensen, B.; Lock, E.J. Amino acid composition, protein content, and nitrogen-to-protein conversion factors of 21 seaweed species from Norwegian waters. *J. Appl. Phycol.* **2017**, *29*, 1001–1009. [CrossRef]
28. Maehre, H.K.; Malde, M.K.; Eilertsen, K.E.; Elvevoll, E.O. Characterization of protein, lipid and mineral contents in common Norwegian seaweeds and evaluation of their potential as food and feed. *J. Sci. Food Agric.* **2014**, *94*, 3281–3290. [CrossRef] [PubMed]
29. Karsten, U. Seaweed acclimation to salinity and desiccation stress. In *Seaweed Biology: Novel Insights into Ecophysiology, Ecology and Utilization*; Wiencke, C., Bischof, K., Eds.; Springer: Heidelberg, Germany, 2012; pp. 87–107.
30. Everette, J.D.; Bryant, Q.M.; Green, A.M.; Abbey, Y.A.; Wangila, G.W.; Walker, R.B. Thorough study of reactivity of various compound classes toward the Folin-Ciocalteu reagent. *J. Agric. Food Chem.* **2010**, *58*, 8139–8144. [CrossRef] [PubMed]
31. Lleu, P.L.; Rebel, G. Interference of Good buffers and other biological buffers with protein determination. *Anal. Biochem.* **1991**, *192*, 215–218. [CrossRef]

foods

MDPI

Article

Comparison of Conventional and Microwave Treatment on Soymilk for Inactivation of Trypsin Inhibitors and In Vitro Protein Digestibility

Brinda Harish Vagadia [1], Sai Kranthi Vanga [1,*], Ashutosh Singh [2], Yvan Gariepy [1] and Vijaya Raghavan [1]

[1] Department of Bioresource Engineering, Faculty of Agriculture and Environmental Studies, McGill University, Sainte-Anne-de-Bellevue, QC H9X 3V9, Canada; brinda.vagadia@mail.mcgill.ca (B.H.V.); yvan.gariepy@mcgill.ca (Y.G.); vijaya.raghavan@mcgill.ca (V.R.)

[2] School of Engineering, University of Guelph, Guelph, ON N1G 2W1, Canada; asingh47@uoguelph.ca

* Correspondence: sai.vanga@mail.mcgill.ca

Received: 12 December 2017; Accepted: 27 December 2017; Published: 8 January 2018

Abstract: Soymilk is lower in calories compared to cow's milk, since it is derived from a plant source (no cholesterol) and is an excellent source of protein. Despite the beneficial factors, soymilk is considered as one of the most controversial foods in the world. It contains serine protease inhibitors which lower its nutritional value and digestibility. Processing techniques for the elimination of trypsin inhibitors and lipoxygenase, which have shorter processing time and lower production costs are required for the large-scale manufacturing of soymilk. In this study, the suitable conditions of time and temperature are optimized during microwave processing to obtain soymilk with maximum digestibility with inactivation of trypsin inhibitors, in comparison to the conventional thermal treatment. The microwave processing conditions at a frequency of 2.45 GHz and temperatures of 70 °C, 85 °C and 100 °C for 2, 5 and 8 min were investigated and were compared to conventional thermal treatments at the same temperature for 10, 20 and 30 min. Response surface methodology is used to design and optimize the experimental conditions. Thermal processing was able to increase digestibility by 7% (microwave) and 11% (conventional) compared to control, while trypsin inhibitor activity reduced to 1% in microwave processing and 3% in conventional thermal treatment when compared to 10% in raw soybean.

Keywords: soymilk; microwave processing; thermal processing; trypsin inhibitors; response surface methodology

1. Introduction

Soymilk is a high protein liquid with considerable amounts of carbohydrates, fats, essential vitamins and mineral, generally produced by grinding soaked soybeans in excess water, which is then filtered to separate out the milk from solids and fiber. It is a stable oil in water emulsion, where the continuous phase is formed by dispersed soybean protein. Soymilk is composed of 94% water, 3% protein, 1.5% fat and 1.5% of carbohydrates. It also contains 7.36 and 0.33 mg/100 mL of riboflavin and thiamin, respectively, a composition similar to cow's milk but with little-saturated fat and no cholesterol [1–3]. The nutritional profile of soymilk and cow's milk (3.25% milkfat) are summarized in Table 1 [4].

Table 1. Nutritional profile of soymilk and cow's milk (unfortified).

Nutrients	Soymilk	Cow Milk
Water	85.61	88.13
Protein	2.26	3.15
Dietary fiber	0.4	0.0
Calcium	0.025	0.113
Carbohydrates	9.95	4.80
Sugars	7.86	5.05
Potassium	0.143	0.132
Cholesterol	0	10
Trans fatty acids	0	-

All the values reported are per 100 g. Report: 01077 and 16166, United States Department of Agriculture (USDA) Database.

In recent years, the consumption of soymilk has increased, especially among consumers who are lactose intolerant, vegetarian, vegan and/or seeking healthy diets. It is also considered safe for children with galactosemia [5], as galactose is absent in soymilk. In developing countries, soymilk is used as a low-cost substitute for cow's milk in many food preparations [6,7]. This increase in consumption of soymilk can also be attributed to the presence of high-quality protein and low-fat content [2]. In 1999, the U.S. Food and Drug Administration approved the health claim for soy protein, which states that its consumption may reduce the risk of heart diseases by lowering the levels of low-density lipoproteins adding to its acceptance by a wide variety of consumers [8]. Several researchers have also associated the consumption of soy products to reduced risks of coronary heart diseases, atherosclerosis, type 2 diabetes, colorectal cancer, breast cancer and prostate cancer [9–11].

Despite all the beneficial factors, the nutritional value of soy milk is reduced by the presence of a variety of anti-nutritional factors such as Kunitz trypsin inhibitors (KTI), Bowman-Birk inhibitors (BBI) and lipoxygenase (LOX). Soybean contains the highest amount of protease inhibitors that accounts for two to six percent of whole soybean protein [12]. These proteases (KTI and BBI) inhibit the enzymatic activity of trypsin and chymotrypsin, the primary digestive enzymes responsible for reducing the proteins into dipeptides and tripeptides. The KTI has a molecular weight of 20 kDa with two disulfide bridges and exhibits specificity to inhibit trypsin. BBI has a molecular weight of eight kDa with seven disulfide bonds and exhibits specificity to inhibit chymotrypsin and trypsin [13,14]. Rouhana et al. reported 60% of soymilk trypsin inhibition activity (TIA) was from KTI [15]. High levels of active KTI have been shown to reduce protein digestibility and cause pancreatic carcinogenesis upon consumption [16]. In animals, protease inhibitors have been associated with growth suppression and pancreatic hypertrophy, emphasizing the need for identification and development of effective techniques to reduce their presence in soy products [17–19]. Soybean trypsin inhibitors are heat stable and require a long processing time. According to Yuan et al., TIA values decreased to 13% of the original raw soymilk TIA values when processed by the traditional thermal treatment (heating at 100 °C for 20 min) [20]. However, the long processing time may affect the other nutritional properties of soy products and hence should be avoided [20,21]. At the same time, 100% inactivation of trypsin inhibitors (TI) causes overheating and damages the proteins by destroying lysine, tryptophan and cysteine in soymilk. Thus, extended periods of thermal treatment inactivate TI effectively, but it denatures essential soybean proteins resulting in amino acid degradation, browning reaction and other deteriorative reactions [20,21]. The flavor, color, and vitamin content are also affected depending on the type of heat treatment used [21,22]. Hence, processing plays an essential role in the sensory appeal and nutritive value of soybean and soy products including soymilk. The various factors to be considered for a good quality soymilk during processing are yield, nutritional quality, anti-nutritional profile, color attributes, particle size, texture profile and organoleptic quality [14,23].

Moreover, there are still questions concerning the ideal processing conditions to produce commercially sterile soymilk with minimum nutrient degradation. Manufacturing techniques are required that have shorter processing time, are energy efficient (environment-friendly), have lower production costs and maintain the quality of soymilk [24]. Autoclaving, batch boiling and steam injection [25,26], Ultra-High Temperature (UHT) [27], High Temperature and Pressure combination [28], Ohmic heating [29], and High-Pressure Processing (HPP) [17] are processing methods that have been explored for inactivation of TI in soymilk.

Industrial scale dielectric heat treatment technology at 42 MHz (Radio frequency) and 2450 MHz (Microwave) were found to be effective against TI in soybean and these methods also improved the overall quality of the protein. The processing time required to reach safe levels of TI inactivation is less in microwave treatment when compared to conventional methods for soybeans [30]. Studies by Barac and his team showed that the TI levels were reduced to 13% of the initial value in soybean during microwave roasting at 2.45 GHz for two min [31]. In a study conducted by Yoshida et al., the inactivation of the anti-nutritional factors to safe limits of soaked soybean at 2.45 GHz requires only four min [32]. In comparison, the conventional batch boiling process takes 15 min at 100 °C to inactivate the levels of TI to 20% [15]. To the best of our knowledge, no studies have been done on the inactivation of soybean trypsin inhibitor in soymilk using microwave processing despite the upsides of using this dielectric processing technique. This can be regarded as a potential alternative to existing conventional processing methods in the food industry for inactivation of anti-nutritional factors.

This study reports the effect of microwave processing on the reduction of TIA in comparison to the conventional thermal processing of soymilk. In vitro Protein digestibility (IVPD) studies were also performed to assess the effects of microwave processing on its digestibility at different time and temperature conditions. Optimization of these processing techniques was performed using Response Surface Analysis.

2. Materials and Methods

Soybeans (*Glycine max*) was procured from Goliath, QC, Canada. Initial moisture content was found to be 10.1% on a wet basis. The moisture content was determined by AOAC official method for moisture content in soybean flour by hot air oven method. Soybean flour (5 g) was dried in an oven at 130 °C ± 3 °C for two hours, after which the weight became constant [33]. Fresh soymilk was prepared from fresh soybean before performing thermal and microwave processing.

2.1. Soymilk Preparation

Soybeans were washed, cleaned and soaked in distilled water in the ratio 1:10 (*w:v*) (bean:water) for 18 h at room temperature (25 °C) for complete hydration. The soymilk was prepared by wet grinding the hydrated soybeans along with water for three mins at high speed in a stainless-steel blender (Nutri Bullet, NutriBullet LLC, Pacoima, CA, USA). The slurry was filtered through a double layer of cheesecloth to separate out the solids from soymilk. Raw soymilk obtained had a pH of 6.5 [8,34].

2.2. Solvents and Reagents

All reagents and solvent used were of High-Performance Liquid Chromatography (HPLC) grade and were purchased from Fisher Scientific (Ottawa, ON, Canada). The enzymes used for in vitro Protein Digestibility (IVPD %) determination and trypsin inhibitor assay were purchased from Sigma Aldrich (Oakville, ON, Canada).

2.3. Conventional Thermal Treatment

For conventional thermal treatment, 30 mL of soymilk was placed in a water bath which was pre-set and maintained at the processing temperatures of 70 °C, 85 °C and 100 °C. The samples were treated for 10, 20 and 30 min in the water bath. All the experiments were conducted in triplicate. After cooling at room temperature, the samples were collected, stored overnight at 40 °C and later freeze-dried in a laboratory freeze-dryer (Gamma 1-16 LSC Freeze dryer, Martin Christ

Gefriertrocknungsanlagen GmbH, Osterode am Harz, Germany) and stored in opaque air-tight containers at −20 °C until further analysis was conducted.

2.4. Microwave Processing

The microwave processing was conducted using the MiniWAVE digestion system (SCP Science, Baie-D'Urfe, QC, Canada) that operates at a frequency of 2.45 GHz at 1000 watts. The soymilk samples were heated in cylindrical quartz reactor vessels. The experiments were conducted at processing temperatures of 70 °C, 85 °C and 100 °C for 2, 5 and 8 min. The sample temperature was monitored using Infra-red (IR) sensors located on the sidewalls and displayed in real time on the controller screen during the run. The MiniWAVE system uses a single magnetron located below the floor of the chamber. After the treatment, the reactor vessels were cooled to room temperature gradually by the cooling unit of the microwave system. The samples were stored until further analysis in the same manner as that of conventionally treated samples.

2.5. Chemical Analysis

2.5.1. In Vitro Protein Digestibility (Multi Enzyme Method)

The In-vitro Protein Digestibility (IVPD) of soybean protein was evaluated using the multi-enzyme method. The working protein suspension was prepared by dissolving samples to yield 312.5 mg of protein in 50 mL of distilled water, whose pH was adjusted to 8.0 using 0.1 N NaOH and 0.1 N HCl. A multi-enzyme mixture was prepared, containing 1.6 mg/mL trypsin, 3.6 mg/mL chymotrypsin, and 1.3 mg/mL peptidase and its pH was adjusted to 8.0. The mixture was placed in an ice-bath and continuously stirred [35–38]. Five milliliters of the multi-enzyme solution were added to the samples, which were maintained at 37 °C in a water bath for the digestion with continuous stirring. The pH was measured after 10 min of the digestion and IVPD was calculated using Equation (1) [37].

$$\text{IVPD}\ \% = 210.46 - (18.10 \times \text{pH}_{10\text{min}}) \quad (1)$$

2.5.2. Trypsin Inhibitor Assay

In this study, the total Trypsin inhibitor assay was assessed using the procedure followed by Hamerstrand et al. [39,40] with some modifications. Freeze dried soy milk (0.5 g) was extracted with 50 mL of 0.01 M NaOH for three hours, with constant stirring at room temperature. The suspension was then allowed to stand for two hours at 4 °C. The supernatant from each sample was collected and diluted, such that 2 mL of the extract could produce 40–60% trypsin inhibitor activity.

Trypsin (type 1× from bovine pancreas, Sigma Chemical Co.) was used as a standard. Diluted soymilk supernatant (1 mL) was pipetted into test tubes in triplicates containing 2 mL of trypsin solution (20 mg in 0.001 M HCl). The control sample (blank) consisted of diluted sample extract and distilled water. The tubes were preheated at 37 °C for 10 min and then, 5 mL of benzyl-DL-arginine-para-nitroanilide (BAPNA), pre-warmed to 37 °C, and was added to each of the tubes and vortexed. After incubating this mixture at 37 °C for 10 min, the reaction was stopped by adding 1 mL of acetic acid (30%). The samples were centrifuged at 3000 g for 10 min. The absorbance of the clear supernatant was measured using spectrophotometer at 410 nm [41,42]. TIA is calculated in terms pure trypsin/g sample as weighed (mg/g).

$$\text{TIA} = (2.632 \times D \times A_I)/S \quad (2)$$

where D is the dilution factor (factor by which the original soymilk sample was diluted to obtain an inhibition between 40% and 60% by 1 mL of the diluted extract), S is the sample weight and A_I is the change in absorbance due to trypsin inhibitor/mL diluted sample extracted.

2.6. Statistical Design and Analysis

In this study, a response surface methodology including the design of experiments, fitting of mathematical model and optimization of processing condition for soymilk samples was employed. The central composite design (CCD) with uniform precision was applied for two independent factors, namely temperature (X_1) and time (X_2), each at three levels (−1, 0, and +1) as shown in Table 2. The design used to plan experiments consisted of a total of 14 combinations with six central, four factorial and four axial points combinations as shown in Table 3. The responses: TIA and IVPD were recorded. JMP software version 11 (SAS Institute Inc., Cary, NC, USA) was used for the experimental design and analysis. The functional relationship between the factors (X_i, X_j, X_k, etc.) and responses (Y) was unknown, hence a regression model (Equation (3)) was used to analyze the actual response surfaces [43–45].

$$Y = \beta_0 + \sum_{i=1}^{i=n} \beta_i X_i + \sum_{i=1}^{i=n} \beta_{ii} X_i^2 + \sum_{i=1}^{i=n} \sum_{j=1}^{j=n} \beta_{ij} X_i X_j \quad (3)$$

where β_0 is the constant coefficient, β_i is the linear coefficient, β_{ii} is the quadratic coefficient for main process parameters and β_{ij} is the second order interaction coefficient of variables i and j, respectively. The statistical design was prepared taking the temperature in °C and time in min. Separate CCD was prepared for both conventional thermal processing and microwave processing method. F value and its significance, Lack of Fit (LOF), and the coefficient of determination (R^2) were assessed and the ANOVA analysis of the predictive model with the corresponding significant terms were reported in Tables 5 and 6, Tables 8 and 9. The differences among the treatments were also detected using Duncan multiple-range test using the probability level 0.05 [46].

Table 2. Central composite design for processing of soymilk with independent variables and their coded and actual values.

Process Parameters	Units	Coded Levels		
		−1	0	+1
Temperature	°C	70	85	100
Time (Microwave processing)	Min	2	5	8
Time (Conventional water bath)	Min	10	20	30

Table 3. Central composite design showing different combinations of temperature and time for processing of soymilk.

Experimental Run	Temperature (°C)	Time (min) Microwave Processing	Time (min) Conventional Processing
1	70	2	10
2	70	5	20
3	70	8	30
4	85	2	10
5	85	8	30
6	100	2	10
7	100	5	20
8	100	8	30
9–14	85	5	20

3. Results and Discussion

3.1. Optimization of Conditions for IVPD during Microwave and Conventional Processing

Legumes are known to have a lower protein digestibility, which is attributed to the presence of anti-nutritional factors [14,47]. On average, IVPD of microwave processed and conventionally treated soymilk samples was 85 ± 1.5% and 88 ± 2.0%, respectively (shown in Table 4). The nutritional quality of soybean protein cannot be determined by its amino acid composition alone, but its digestibility in the small intestine and determining the bioavailability should also be considered. Our investigations of the results for microwave and conventional processing of soymilk indicated that both the independent factors; temperature (temp) and time (t) significantly ($p \leq 0.05$) affected the IVPD of soy proteins (Tables 5 and 6). Overall, the regression model developed after ANOVA analysis was significant ($p < 0.05$) for both the treatments with insignificant lack of fit ($p > 0.05$). For both treatments, it can be interpreted that, by increasing the treatment temperature and time, an increase in IVPD was observed. In the case of microwave processing, from the values of parameter estimates or regression coefficients it was concluded that the most influential factor affecting the IVPD is the temperature with the highest regression coefficient of 1.694, followed by a square term of time, 1.250, a linear term of time, 0.3519, and lastly the interactive term of temperature and time with −0.620 regression coefficient. The negative regression coefficient for the interactive term (temp × time) suggested that as the microwave processing time was increased at any processing temperature a slight decrease in IVPD was observed till the time reached approximately 5 min and later the IVPD increased (Figure 1). This observation led to a conclusion that can be related to the changes in the conformation of proteins under the influence of oscillating electric field of microwave (2.4 GHz). As the sample was subjected to microwave processing a change in the confirmation of soymilk protein would reduce its susceptibility to the digestive enzymes but as the processing time increased the protein would denature and digestion would proceed as desired. This conclusion is based on previous observations made by the researchers though molecular modelling studies conducted to evaluate the effect of oscillating and static electric fields on various food proteins including peanut and soybean hydrophobic proteins [48–50]. Similar studies were conducted to understand the structure and digestibility in other legumes such as dry beans (*Phaseolus vulgaris*) and green peas (*Pisum sativum*) [51], sorghum (*Sorghum bicolor*) and maize (*Zea mays*) [52,53].

Table 4. Summarized statistics for in vitro protein digestibility of soymilk processing using process parameters according to central composite design.

	Microwave Processing	**Conventional Processing**
Range (% digestibility)	82–89	84–92
Average	85.2183	88
Standard deviation	±1.4586	±2.018

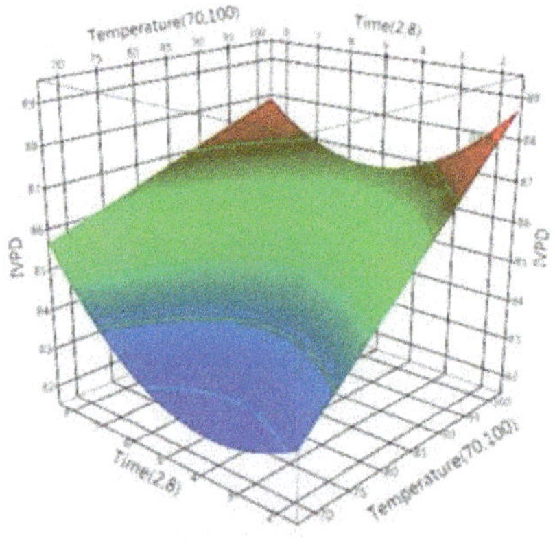

Figure 1. Effect of time (min) and temperature (°C) on in vitro protein digestibility (IVPD) of soymilk during microwave processing.

Table 5. ANOVA for effect of time (t) and temperature (temp) for in vitro protein digestibility for microwave processing.

Source	DF *	Sum of Squares	Mean Square	F Ratio	p-Value
Model	4	74.5361	18.6340	46.5061	<0.0001
Temperature (temp)	1	51.6754		128.9695	<0.0001
Time (t)	1	2.2296		5.5645	0.0237
Temp × t	1	4.5892		11.4537	0.0017
t^2	1	16.0418		40.0366	<0.0001
Lack of fit	4	3.4025	0.8506	2.4575	0.0648
Error	37	14.8251	0.4007		
C. Total	41	89.3613			
Pure Error	33	11.4225	0.3461		
Total Error	37	14.8251			

* DF: degrees of freedom.

The predictive quadratic model (Equation (4)) generated for microwave processing of soymilk was significant ($p < 0.0001$) with R^2 value of 0.83 and insignificant lack of fit.

$$\begin{aligned}\text{IVPD}_{\text{microwave}} &= 84.683 + 1.694\left[\frac{(\text{Temperature}-85)}{15}\right] + 0.351\left[\frac{(\text{Time}-5)}{3}\right] + \\ &\left[\frac{(\text{Temperature}-85)}{15} \times \left[\frac{(\text{Time}-5)}{3}\right] \times -0.620\right] + \left[\left[\frac{(\text{Time}-5)}{3}\right] \times \left[\frac{(\text{Time}-5)}{3}\right] \times 1.250\right]\end{aligned} \quad (4)$$

As in the case of conventional processing, it was observed that the most influential factor was the temperature with a regression coefficient of 2.127 followed by time with 1.765 regression coefficient. None of the other model parameters including the cross terms and square terms were significant leading to a linear regression model (Equation (5)) with R^2 of 0.80. Several researchers have suggested that the treatment temperature is the key determinant of food protein digestibility [54,55]. A similar linear relationship between treatment time and temperature were observed by Wallace et al. (1971) in their study on the effect of different heat processing conditions on the TIA and the IVPD of various soymilk preparation (Figure 2). They reported that digestibility of proteins increased with an increase of the heat treatment and it also coincided with a decrease in the TIA [26]. Our study showed similar results, maximum digestibility of soymilk proteins occurred at 100 °C for 30 min of conventional processing. Other studies establishing the relationship of increase in digestibility due to a decrease in anti-nutritional factors were seen in rice [56], cowpea [57], chickpeas [58], moth beans [59], and common beans [60].

$$\text{IVPD}_{\text{connventional}} = 88.731 + 2.127\left[\frac{(\text{Temperature} - 85)}{15}\right] + 1.765\left[\frac{(\text{Time} - 20)}{10}\right] \quad (5)$$

Table 6. ANOVA for effect of time (t) and temperature (temp) for in vitro protein digestibility for conventional processing.

Source	DF	Sum of Squares	Mean Square	F Ratio	p-Value
Model	2	137.5613	68.7807	79.9635	<0.0001
Temperature (temp)	1	81.4675		94.6810	<0.0001
Time (t)	1	56.0940		65.1919	<0.0001
Lack of fit	6	4.2183	0.7030	0.7908	0.5836
Error	39	33.5572	0.8604		
C. Total	41	171.1185			
Pure Error	33	29.3388	0.8891		
Total Error	39	33.5572			

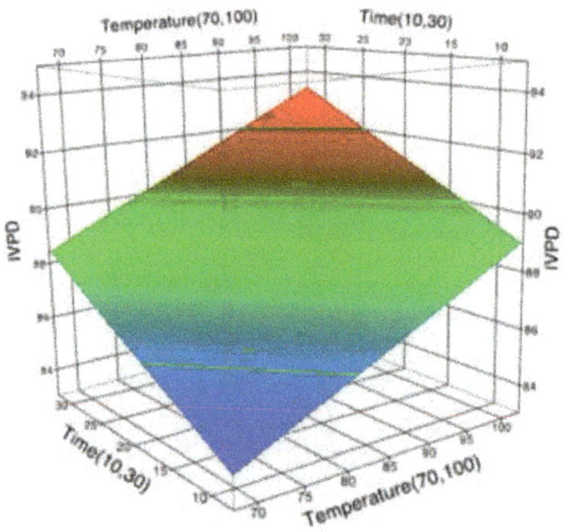

Figure 2. Effect of time (min) and temperature (°C) on in vitro protein digestibility (IVPD) of soymilk during conventional thermal treatment.

3.2. *Optimization of Conditions for TIA during Microwave and Conventional Processing*

Trypsin inhibitor activity (TIA) governs the nutritional value of soymilk protein [61,62]. The average values for TIA during microwave and conventional processing is mentioned in Table 7. It has been reported by several researchers that overheating for complete removal of TIA reduces the overall nutritive value of soybeans [63]. Hence, a precise, controlled thermal process or a novel process is required for preparation of soymilk with maximum nutritive value. In this study analysis of the effect of microwave and conventional processing of soymilk on TIA revealed that both temperature and time play a significant ($p < 0.05$) role in determining it. Table 8 presents the ANOVA analysis for the effect of temperature and time on TIA for microwave processing. From the table, we can observe that regression model developed was significant with ($p < 0.0001$), insignificant lack of fit ($p > 0.05$) and R^2 of 0.91 and that the independent factors temperature, its square term and time played a significant role. The regression coefficient analysis also supported the aforementioned observation, where temperature with regression coefficient estimate of −0.558, followed by its cross term with regression coefficient estimate of −0.192 and lastly time with a regression coefficient of −0.108 suggested that TIA decreases with increase in temperature and time (Figure 3), but overall presence of the significant temperature square term suggested temperatures significant influence on TIA. According to Rajko et al., inactivation of trypsin inhibitors requires more absorbed heat energy for longer processing time [64]. Similar experimental results were obtained by Alajaji and El-Adawy, during the microwave oven cooking of chickpea on high temperature for 15 min [58]. According to the studies by Oliveria and Haghighi, reduction in TIA was more pronounced in samples at higher temperatures as was expected because soybean TI loses activity irreversibly in the temperature range 80–110 °C [65]. In addition, studies by Esaka et al., TIA was not detectable after microwave heating at 120 °C for 5 min in case of winged bean seeds [66]. The predictive quadratic equation for effect of microwave processing on TIA was obtained as (Equation (6)).

$$\begin{aligned}\mathrm{TIA}_{\mathrm{Microwave}} = 1.245 + \left(-0.558\left[\tfrac{(\mathrm{Temperature}-85)}{15}\right]\right) + \left(-0.108\left[\tfrac{(\mathrm{Time}-5)}{3}\right]\right) + \\ \left(\left[\tfrac{(\mathrm{Temperature}-85)}{15}\right] \times \left(\left[\tfrac{(\mathrm{Temperature}-85)}{15}\right] \times -0.191\right)\right)\end{aligned} \quad (6)$$

For the conventional process of soymilk and its effect on trypsin inhibitor activity it was observed that all the independent factors and its quadratic or cross terms significantly affected it ($p < 0.05$). As seen in Table 9, the predictive model generated was significant ($p < 0.0001$) with insignificant lack of fit and R^2 of 0.77. The response surface graph presented in Figure 4 shows that highest TIA was obtained at lowest temperature for shortest processing time. As the processing time was increased, the TIA decreased at any given temperature. This observation can also be confirmed with analysis of regression coefficients of process parameters, which suggested that the most influential parameter was a temperature with regression coefficient estimate of −0.173, followed by time with −0.138, and their cross term with 0.099 and lastly with a square term of time, 0.102, and temperature, 0.099. Similar results were observed in lentil, chickpea and pea flours that were thermally processed by boiling in a water bath at 90 °C for 20 min, which significantly reduced the levels of trypsin inhibitors [67]. According to the results of Osman et al., hydrothermal treatment of Tepary bean extract at 100 °C for 60 min completely inactivated the TI [68]. Similar results were observed by Andrade et al., recommending a dry heat of 200 °C for 20 min for soy flour samples, showing temperature plays an essential role in inactivation of TI [69]. The predictive quadratic regression model generated for conventional processing is presented by Equation (7).

$$\begin{aligned} TIA_{conventional} = 0.228 + \left(-0.173\left[\frac{(Temperature-85)}{15}\right]\right) + \left(-0.138\left[\frac{(Time-20)}{10}\right]\right) + \\ \left(\left[\frac{(Temperature-85)}{15}\right] \times \left[\frac{(Time-20)}{10}\right] \times 0.099\right) + \left(\left[\frac{(Temperature-85)}{15}\right] \times \left[\frac{(Temperature-85)}{15}\right] \times 0.099\right) \\ +\left(\left[\frac{(Time-20)}{10}\right] \times \left[\frac{(Time-20)}{10}\right] \times 0.102\right) \end{aligned} \quad (7)$$

Thus, to obtain soymilk with maximum digestibility with inactivation of TI, microwave processing (2.45 GHz) at 100 °C for 8 min is recommended, in comparison to conventional processing at 100 °C for 30 min.

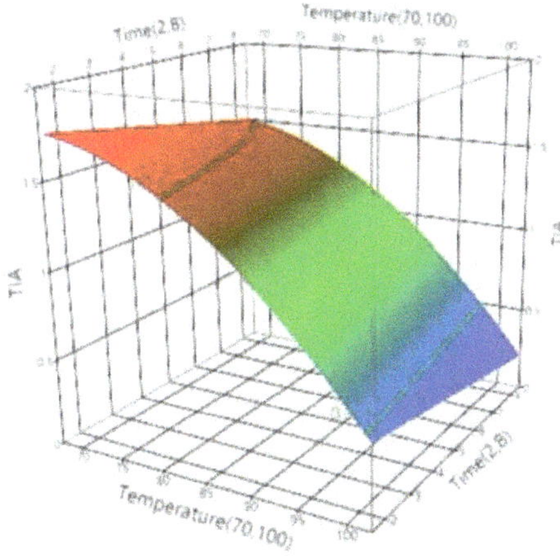

Figure 3. Effect of time (min) and temperature (°C) on trypsin inhibitor activity (TIA) of soymilk during microwave processing.

Table 7. Summarized statistics for trypsin inhibition activity of soymilk processing using process parameters according to central composite design.

	Microwave Processing	Conventional Processing
Range (% inhibition)	3–18.8	1–9
Average	11.6	3.1
Standard deviation	±4.0333	±2.002

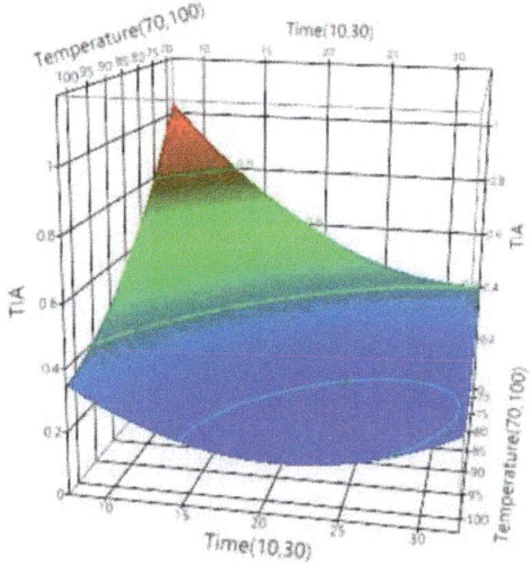

Figure 4. Effect of time (min) and temperature (°C) on trypsin inhibitor activity (TIA) of soymilk during conventional thermal treatment.

Table 8. ANOVA for effect of time (*t*) and temperature (temp) on trypsin inhibitor activity for microwave processing.

Source	DF	Sum of Squares	Mean Square	F Ratio	*p*-Value
Model	3	6.2115	2.0705	126.6718	<0.0001
Temperature (temp)	1	5.6224		343.9744	<0.0001
Time (*t*)	1	0.2112		12.9241	0.0009
Temp^2	1	0.3778		23.1169	<0.0001
Lack of fit	5	0.0814	0.0163	0.9956	0.4356
Error	38	0.5871	0.0163		
C. Total	41	6.8326			
Pure Error	33	0.5397	0.0163		
Total Error	38	0.6211			

Table 9. ANOVA for effect of time (*t*) and temperature (temp) for trypsin inhibitor activity for conventional processing.

Source	DF	Sum of Squares	Mean Square	F Ratio	*p*-Value
Model	5	1.3021	0.2604	24.5188	<0.0001
Temperature (temp)	1	0.5408		50.9176	<0.0001
Time (*t*)	1	0.3472		32.6918	<0.0001
Temp^2	1	0.1180		11.1108	0.0020
Temp $\times$ t	1	0.0835		7.8624	0.0081
t^2	1	0.0892		8.4001	0.0064
Lack of fit	3	0.2551	0.0850	22.0541	<0.0001
Error	36	0.3823	0.0106		
C. Total	41	1.6844			
Pure Error	33	0.1272	0.0040		
Total Error	36	0.3823			

4. Conclusions

In this study, microwave processing (2450 MHz, 1000 Watts) at different conditions of temperature (70 °C, 85 °C and 100 °C) and time (2, 5 and 8 min) were applied to soymilk samples, in comparison to conventional thermal treatments at the same temperatures and time (10, 20 and 30 min). The IVPD increased with increase in time and temperature during microwave processing (100 °C for 8 min) and conventional processing (100 °C for 30 min) to 87% and 92%, respectively. This is higher compared to an initial digestibility of raw soymilk, which was estimated to be 80.5%. Similarly, TIA for conventional treatment (100 °C for 30 min) is 1% and for microwave processing (100 °C for 8 min) is 3% from an initial TIA of 10% of raw soymilk. Hence, microwave processing can be used as a potential alternative method of processing soymilk for increased digestibility and elimination of anti-nutritional factors.

Acknowledgments: The authors would like to acknowledge the financial support of NSERC (Natural Sciences and Engineering Research Council of Canada) and MAPAQ (Ministère de l'Agriculture, des Pêcheries et de l'Alimentation du Québec) for this study. The authors are grateful to Lawrence Goodridge and Mark Lefsurd for their technical support.

Author Contributions: B.H.V., S.K.V. and Y.G. conceived and designed the experiments; B.H.V. and S.K.V. performed the experiments; B.H.V. and A.S. analyzed the data; Y.G. and A.S. contributed reagents/materials/analysis tools; B.H.V., S.K.V. and A.S. wrote the paper; V.R. is the supervisor under whom the project was conceived and executed

Conflicts of Interest: The authors declare no conflict of interest.

References

1. Buzzell, R. Composition and properties of soymilk and tofu made from Ontario light hilum soybeans. *Can. Inst. Food Sci. Technol. J.* **1987**, *20*, 363–367.
2. Lakshmanan, R.; De Lamballerie, M.; Jung, S. Effect of soybean-to-water ratio and pH on pressurized soymilk properties. *J. Food Sci.* **2006**, *71*, E384–E391. [CrossRef]
3. Jasti, L.; Lavanya, K.; Fadnavis, N. Adsorption induced denaturation: Application to denaturation of soybean trypsin inhibitor (SBTI) and lipoxygenase (LOX) in soymilk. *Biotechnol. Lett.* **2015**, *37*, 147–151. [CrossRef] [PubMed]
4. USDA. *USDA National Nutrient Database for Standard Reference*; US Department of Agriculture, Agricultural Research Service, Nutrient Data Laboratory: Beltsville, MD, USA, 2015.
5. Friedman, M.; Brandon, D.L. Nutritional and health benefits of soy proteins. *J. Agric. Food Chem.* **2001**, *49*, 1069–1086. [CrossRef] [PubMed]
6. Reilly, J.K.; Lanou, A.J.; Barnard, N.D.; Seidl, K.; Green, A.A. Acceptability of soymilk as a calcium-rich beverage in elementary school children. *J. Am. Diet. Assoc.* **2006**, *106*, 590–593. [CrossRef] [PubMed]
7. Cruz, N.; Capellas, M.; Hernández, M.; Trujillo, A.J.; Guamis, B.; Ferragut, V. Ultra high pressure homogenization of soymilk: Microbiological, physicochemical and microstructural characteristics. *Food Res. Int.* **2007**, *40*, 725–732. [CrossRef]
8. Min, S.; Yu, Y.; Martin, S.S. Effect of soybean varieties and growing locations on the physical and chemical properties of soymilk and tofu. *J. Food Sci.* **2005**, *70*, C8–C21. [CrossRef]
9. Anderson, R.L.; Wolf, W.J. Compositional changes in trypsin inhibitors, phytic acid, saponins and isoflavones related to soybean processing. *J. Nutr.* **1995**, *125*, 581S–588S. [PubMed]
10. Kerwin, S. Soy saponins and the anticancer effects of soybeans and soy-based foods. *Curr. Med. Chem. Anti-Cancer Agents* **2004**, *4*, 263–272. [CrossRef] [PubMed]
11. Hwang, Y.W.; Kim, S.Y.; Jee, S.H.; Kim, Y.N.; Nam, C.M. Soy food consumption and risk of prostate cancer: A meta-analysis of observational studies. *Nutr. Cancer* **2009**, *61*, 598–606. [CrossRef] [PubMed]
12. Vinh, L.T.; Dworschak, E. Trypsin and chymotrypsin inhibitor activities in plant foods from vietnam and hungary. *Mol. Nutr. Food Res.* **1986**, *30*, 53–58. [CrossRef]
13. Liener, I.E. Implications of antinutritional components in soybean foods. *Crit. Rev. Food Sci. Nutr.* **1994**, *34*, 31–67. [CrossRef] [PubMed]
14. Vagadia, B.H.; Vanga, S.K.; Raghavan, V. Inactivation methods of soybean trypsin inhibitor—A review. *Trends Food Sci. Technol.* **2017**, *64*, 115–125. [CrossRef]

15. Rouhana, A.; Adler-Nissen, J.; Cogan, U.R.I.; FrØKiÆR, H. Heat inactivation kinetics of trypsin inhibitors during high temperature-short time processing of soymilk. *J. Food Sci.* **1996**, *61*, 265–269. [CrossRef]
16. Xiao, C.W.; Wood, C.M.; Robertson, P.; Gilani, G.S. Protease inhibitor activities and isoflavone content in commercial soymilks and soy-based infant formulas sold in Ottawa, Canada. *J. Food Compos. Anal.* **2012**, *25*, 130–136. [CrossRef]
17. Van Der Ven, C.; Matser, A.M.; Van Den Berg, R.W. Inactivation of soybean trypsin inhibitors and lipoxygenase by high-pressure processing. *J. Agric. Food Chem.* **2005**, *53*, 1087–1092. [CrossRef] [PubMed]
18. Grant, G. Anti-nutritional effects of soyabean: A review. *Prog. Food Nutr. Sci.* **1988**, *13*, 317–348.
19. Friedman, M.; Brandon, D.L.; Bates, A.H.; Hymowitz, T. Comparison of a commercial soybean cultivar and an isoline lacking the kunitz trypsin inhibitor: Composition, nutritional value, and effects of heating. *J. Agric. Food Chem.* **1991**, *39*, 327–335. [CrossRef]
20. Yuan, S.; Chang, S.K.C.; Liu, Z.; Xu, B. Elimination of trypsin inhibitor activity and beany flavor in soy milk by consecutive blanching and ultrahigh-temperature (UHT) processing. *J. Agric. Food Chem.* **2008**, *56*, 7957–7963. [CrossRef] [PubMed]
21. Adams, J. Review: Enzyme inactivation during heat processing of food-stuffs. *Int. J. Food Sci. Technol.* **1991**, *26*, 1–20. [CrossRef]
22. Agrahar-Murugkar, D.; Jha, K. Effect of drying on nutritional and functional quality and electrophoretic pattern of soyflour from sprouted soybean (*Glycine max*). *J. Food Sci. Technol.* **2010**, *47*, 482–487. [CrossRef] [PubMed]
23. Sancho, F.; Lambert, Y.; Demazeau, G.; Largeteau, A.; Bouvier, J.-M.; Narbonne, J.-F. Effect of ultra-high hydrostatic pressure on hydrosoluble vitamins. *J. Food Eng.* **1999**, *39*, 247–253. [CrossRef]
24. Chen, Y.; Xu, Z.; Zhang, C.; Kong, X.; Hua, Y. Heat-induced inactivation mechanisms of kunitz trypsin inhibitor and Bowman-Birk inhibitor in soymilk processing. *Food Chem.* **2014**, *154*, 108–116. [CrossRef] [PubMed]
25. Murugkar, D. Effect of different process parameters on the quality of soymilk and tofu from sprouted soybean. *J. Food Sci. Technol.* **2015**, *52*, 2886–2893. [CrossRef] [PubMed]
26. Wallace, G.M.; Bannatyne, W.R.; Khaleque, A. Studies on the processing and properties of soymilk: II—Effect of processing conditions on the trypsin inhibitor activity and the digestibility in vitro of proteins in various soymilk preparations. *J. Sci. Food Agric.* **1971**, *22*, 526–531. [CrossRef] [PubMed]
27. Kwok, K.-C.; Liang, H.-H.; Niranjan, K. Mathematical modelling of the heat inactivation of trypsin inhibitors in soymilk at 121–154 °C. *J. Sci. Food Agric.* **2002**, *82*, 243–247. [CrossRef]
28. Guerrero-Beltrán, J.A.; Estrada-Girón, Y.; Swanson, B.G.; Barbosa-Cánovas, G.V. Pressure and temperature combination for inactivation of soymilk trypsin inhibitors. *Food Chem.* **2009**, *116*, 676–679. [CrossRef]
29. Lu, L.; Zhao, L.; Zhang, C.; Kong, X.; Hua, Y.; Chen, Y. Comparative effects of ohmic, induction cooker, and electric stove heating on soymilk trypsin inhibitor inactivation. *J. Food Sci.* **2015**, *80*, C495–C503. [CrossRef] [PubMed]
30. Kala, B.; Mohan, V. Effect of microwave treatment on the antinutritional factors of two accessions of velvet bean, Mucuna Pruriens (L.) DC. var. Utilis (Wall. ex Wight) Bak. ex Burck. *Int. Food Res. J.* **2012**, *19*, 961–969.
31. Barać, M.; Stanojević, S. The effect of microwave roasting on soybean protein composition and components with trypsin inhibitor activity. *Acta Alimentaria* **2005**, *34*, 23–31. [CrossRef]
32. Yoshida, H.; Kajimoto, G. Effects of microwave treatment on the trypsin inhibitor and molecular species of triglycerides in soybeans. *J. Food Sci.* **1988**, *53*, 1756–1760. [CrossRef]
33. Horwitz, W.; Latimer, G.W. *Official Methods of Analysis of AOAC International*; AOAC International: Gaithersburg, MD, USA, 2006.
34. Li, Y.-Q.; Chen, Q.; Liu, X.-H.; Chen, Z.-X. Inactivation of soybean lipoxygenase in soymilk by pulsed electric fields. *Food Chem.* **2008**, *109*, 408–414. [CrossRef] [PubMed]
35. Astwood, J.D.; Leach, J.N.; Fuchs, R.L. Stability of food allergens to digestion in vitro. *Nat. Biotechnol.* **1996**, *14*, 1269–1273. [CrossRef] [PubMed]
36. Bodwell, C.; Satterlee, L.; Hackler, L. Protein digestibility of the same protein preparations by human and rat assays and by in vitro enzymic digestion methods. *Am. J. Clin. Nutr.* **1980**, *33*, 677–686. [PubMed]
37. Hsu, H.; Vavak, D.; Satterlee, L.; Miller, G. A multienzyme technique for estimating protein digestibility. *J. Food Sci.* **1977**, *42*, 1269–1273. [CrossRef]

38. Vanga, S.K.; Singh, A.; Kalkan, F.; Gariepy, Y.; Orsat, V.; Raghavan, V. Effect of thermal and high electric fields on secondary structure of peanut protein. *Int. J. Food Prop.* **2015**, *19*, 1259–1271. [CrossRef]
39. Hamerstrand, G.; Black, L.; Glover, J. Trypsin inhibitors in soy products: Modification of the standard analytical procedure. *Cereal Chem.* **1981**, *58*, 42–45.
40. Embaby, H.E.-S. Effect of heat treatments on certain antinutrients and in vitro protein digestibility of peanut and sesame seeds. *Food Sci. Technol. Res.* **2010**, *17*, 31–38. [CrossRef]
41. Jiang, S.; Cai, W.; Xu, B. Food quality improvement of soy milk made from short-time germinated soybeans. *Foods* **2013**, *2*, 198–212. [CrossRef] [PubMed]
42. Kakade, M.; Rackis, J.; McGhee, J.; Puski, G. Determination of trypsin inhibitor activity of soy products: A collaborative analysis of an improved procedure. *Cereal Chem.* **1974**, *51*, 376–382.
43. Myers, R.H.; Montgomery, D.C.; Anderson-Cook, C.M. *Response Surface Methodology: Process and Product Optimization Using Designed Experiments*; John Wiley & Sons: Hoboken, NJ, USA, 2009; Volume 705.
44. Singh, A.; Vanga, S.K.; Nair, G.R.; Gariepy, Y.; Orsat, V.; Raghavan, V. Electrohydrodynamic drying (EHD) of wheat and its effect on wheat protein conformation. *LWT-Food Sci. Technol.* **2015**, *64*, 750–758. [CrossRef]
45. Singh, A.; Lahlali, R.; Vanga, S.K.; Karunakaran, C.; Orsat, V.; Raghavan, V. Effect of high electric field on secondary structure of wheat gluten. *Int. J. Food Prop.* **2015**, *19*, 1217–1226. [CrossRef]
46. Su, G.; Chang, K.C. Trypsin inhibitor activity in vitro digestibility and sensory quality of meat-like yuba products as affected by processing. *J. Food Sci.* **2002**, *67*, 1260–1266. [CrossRef]
47. Vanga, S.K.; Raghavan, V. How well do plant based alternatives fare nutritionally compared to cow's milk? *J. Food Sci. Technol.* **2017**. [CrossRef]
48. Vanga, S.K.; Singh, A.; Raghavan, V. Effect of thermal and electric field treatment on the conformation of Ara h 6 peanut protein allergen. *Innov. Food Sci. Emerg. Technol.* **2015**, *30*, 79–88. [CrossRef]
49. Singh, A.; Orsat, V.; Raghavan, V. Soybean hydrophobic protein response to external electric field: A molecular modeling approach. *Biomolecules* **2013**, *3*, 168–179. [CrossRef] [PubMed]
50. Vagadia, B.H.; Vanga, S.K.; Singh, A.; Raghavan, V. Effects of thermal and electric fields on soybean trypsin inhibitor protein: A molecular modelling study. *Innov. Food Sci. Emerg. Technol.* **2016**, *35*, 9–20. [CrossRef]
51. Deshpande, S.S.; Damodaran, S. Structure-digestibility relationship of legume 7S proteins. *J. Food Sci.* **1989**, *54*, 108–113. [CrossRef]
52. Duodu, K.; Taylor, J.; Belton, P.; Hamaker, B. Factors affecting sorghum protein digestibility. *J. Cereal Sci.* **2003**, *38*, 117–131. [CrossRef]
53. Duodu, K.; Tang, H.; Grant, A.; Wellner, N.; Belton, P.; Taylor, J. FTIR and solid state ^{13}C-NMR spectroscopy of proteins of wet cooked and popped sorghum and maize. *J. Cereal Sci.* **2001**, *33*, 261–269. [CrossRef]
54. Bax, M.L.; Aubry, L.; Ferreira, C.; Daudin, J.D.; Gatellier, P.; Rémond, D.; Santé-Lhoutellier, V. Cooking temperature is a key determinant of in vitro meat protein digestion rate: Investigation of underlying mechanisms. *J. Agric. Food Chem.* **2012**, *60*, 2569–2576. [CrossRef] [PubMed]
55. Li, H.; Zhu, K.; Zhou, H.; Peng, W.; Guo, X. Comparative study about some physical properties, in vitro digestibility and immunoreactivity of soybean protein isolate for infant formula. *Plant Foods Hum. Nutr.* **2013**, *68*, 124–130. [CrossRef] [PubMed]
56. Sagum, R.; Arcot, J. Effect of domestic processing methods on the starch, non-starch polysaccharides and in vitro starch and protein digestibility of three varieties of rice with varying levels of amylose. *Food Chem.* **2000**, *70*, 107–111. [CrossRef]
57. Laurena, A.; Garcia, V.; Mae, E.; Mendoza, T. Effects of heat on the removal of polyphenols and in vitro protein digestibility of cowpea (*Vigna Unguiculata* (L.) Walp.). *Plant Foods Hum. Nutr.* **1987**, *37*, 183–192. [CrossRef]
58. Alajaji, S.A.; El-Adawy, T.A. Nutritional composition of chickpea (*Cicer Arietinum* L.) as affected by microwave cooking and other traditional cooking methods. *J. Food Compos. Anal.* **2006**, *19*, 806–812. [CrossRef]
59. Negi, A.; Boora, P.; Khetarpaul, N. Effect of microwave cooking on the starch and protein digestibility of some newly released moth bean (*Phaseolus Aconitifolius* Jacq.) cultivars. *J. Food Compos. Anal.* **2001**, *14*, 541–546. [CrossRef]
60. Marconi, E.; Ruggeri, S.; Cappelloni, M.; Leonardi, D.; Carnovale, E. Physicochemical, nutritional, and microstructural characteristics of chickpeas (*Cicer Arietinum* L.) and common beans (Phaseolus Vulgaris L.) following microwave cooking. *J. Agric. Food Chem.* **2000**, *48*, 5986–5994. [CrossRef] [PubMed]

61. Hackler, L.R.; Van Buren, J.P.; Streinkraus, K.H.; El Rawi, I.; Hand, D.B. Effect of heat treatment on nutritive value of soymilk protein fed to weanling rats. *J. Food Sci.* **1965**, *30*, 723–728. [CrossRef]
62. Lei, M.G.; Bassette, R.; Reeck, G.R. Effect of cysteine on heat inactivation of soybean trypsin inhibitors. *J. Agric. Food Chem.* **1981**, *29*, 1196–1199. [CrossRef] [PubMed]
63. Skrede, A.; Krogdahl, A. Heat affects nutritional characteristics of soybean meal and excretion of proteinases in mink and chicks. *Nutr. Rep. Int.* **1985**, *32*, 479–489.
64. Rajkó, R.; Szabó, G.; Vidal-Valverde, C.; Kovács, E. Designed experiments for reducing antinutritive agents in soybean by microwave energy. *J. Agric. Food Chem.* **1997**, *45*, 3565–3569. [CrossRef]
65. Oliveira, L.S.; Haghighi, K. Conjugate heat and mass transfer in convective drying of multiparticle systems part II: Soybean drying. *Dry. Technol.* **1998**, *16*, 463–483. [CrossRef]
66. Esaka, M.; Suzuki, K.; Kubota, K. Effects of microwave heating on lipoxygenase and trypsin inhibitor activities, and water absorption of winged bean seeds. *J. Food Sci.* **1987**, *52*, 1738–1739. [CrossRef]
67. Ma, Z.; Boye, J.I.; Simpson, B.K.; Prasher, S.O.; Monpetit, D.; Malcolmson, L. Thermal processing effects on the functional properties and microstructure of lentil, chickpea, and pea flours. *Food Res. Int.* **2011**, *44*, 2534–2544. [CrossRef]
68. Osman, M.A.; Reid, P.M.; Weber, C.W. Thermal inactivation of tepary bean (*Phaseolus Acutifolius*), soybean and lima bean protease inhibitors: Effect of acidic and basic pH. *Food Chem.* **2002**, *78*, 419–423. [CrossRef]
69. Andrade, J.C.; Mandarino, J.M.G.; Kurozawa, L.E.; Ida, E.I. The effect of thermal treatment of whole soybean flour on the conversion of isoflavones and inactivation of trypsin inhibitors. *Food Chem.* **2016**, *194*, 1095–1101. [CrossRef] [PubMed]

foods

MDPI

Review

Algal Proteins: Extraction, Application, and Challenges Concerning Production

Stephen Bleakley [1,2] and Maria Hayes [1,*]

1 Food Biosciences Department, Teagasc Ashtown Food Research Centre, Ashtown, Dublin D15 KN3K, Ireland; Stephen.Bleakley@teagasc.ie
2 School of Biological Sciences, College of Sciences and Health and Environment, Sustainability and Health Institute, Dublin Institute of Technology, Kevin Street, Dublin D08 NF82, Ireland
* Correspondence: Maria.Hayes@teagasc.ie; Tel.: +353-1-805-9957

Academic Editor: Christopher J. Smith
Received: 21 March 2017; Accepted: 20 April 2017; Published: 26 April 2017

Abstract: Population growth combined with increasingly limited resources of arable land and fresh water has resulted in a need for alternative protein sources. Macroalgae (seaweed) and microalgae are examples of under-exploited "crops". Algae do not compete with traditional food crops for space and resources. This review details the characteristics of commonly consumed algae, as well as their potential for use as a protein source based on their protein quality, amino acid composition, and digestibility. Protein extraction methods applied to algae to date, including enzymatic hydrolysis, physical processes, and chemical extraction and novel methods such as ultrasound-assisted extraction, pulsed electric field, and microwave-assisted extraction are discussed. Moreover, existing protein enrichment methods used in the dairy industry and the potential of these methods to generate high value ingredients from algae, such as bioactive peptides and functional ingredients are discussed. Applications of algae in human nutrition, animal feed, and aquaculture are examined.

Keywords: seaweed; microalgae; peptides; phycobiliproteins; biorefinery; bioavailability; extraction methods; legislation

---

1. Introduction

The global population is expected to increase by over a third (2.3 billion people) by 2050, requiring an estimated 70% increase in food production [1]. A combination of improved agricultural food production methods and an increase of average per capita income have led to a decrease in global hunger over the last half-century, despite a doubling of the world's population [1]. However, worldwide food production is now facing a greater challenge than ever before. Previously utilised methods of intensifying agriculture will soon no longer be an option due to the high impact trade-offs they have on the environment, including fragmenting natural habitats and threatening biodiversity, production of greenhouse gases from land clearing, fertilisers and animal livestock production, and nutrient run-off from fertiliser damaging marine, freshwater and terrestrial ecosystems [2]. In particular, protein is one of the main nutrients that will be in short supply in the future. Alternative protein sources and production methods are required to fulfil the demand of consumers and to meet predicted global protein requirements.

Seaweed and microalgae are considered a viable source of protein. Some species of seaweed and microalgae are known to contain protein levels similar to those of traditional protein sources, such as meat, egg, soybean, and milk [3,4]. Algae use for protein production has several benefits over traditional high-protein crop use in terms of productivity and nutritional value. Seaweed and microalgae have higher protein yield per unit area (2.5–7.5 tons/Ha/year and 4–15 tons/Ha/year, respectively) compared to terrestrial crops, such as soybean, pulse legumes, and wheat (0.6–1.2 tons/Ha/year, 1–2 tons/Ha/year, and 1.1 tons/Ha/year, respectively) [5]. Terrestrial agriculture already requires approximately 75% of the total global freshwater with animal protein in particular requiring 100 times more water than if the equivalent amount of protein was produced from plant sources [6,7]. Marine algae do not require freshwater or arable land to grow, maximising resources that can be used for additional food production or other cash crops [5]. Furthermore, due to their harsh environment and phototropic life, algae are often exposed to high oxidative and free-radical stresses [8]. This has led to the evolution of natural protective systems, such as the production of pigments (e.g., carotenes, chlorophylls, and phycobiliproteins) and polyphenols (e.g., catechins, flavonols, and phlorotannins), which can impart health benefits to the consumer when eaten [9,10].

However, widespread use of seaweed and microalgae is limited by a number of factors including; harvesting access and rights, seasonality and geographical location of algae, as well as the availability of scalable production methods for protein isolation from algae. Current processes of algal protein isolation are time-consuming and economically unviable [11]. The objective of this paper is therefore to discuss the value of algal proteins as a source of human nutrition, functional foods and animal feed, as well as describe current extraction methods and novel processing technologies that are used in dairy processing which may be employed to make algae a viable source of protein ingredients.

1.1. Characteristics of Seaweed

Algae are a diverse group of species which can be broadly described as oxygen-producing, photosynthetic, unicellular or multicellular organisms excluding embryophyte terrestrial plants and lichens [12]. Macroalgae can be divided into three main taxonomic groups based on their pigmentation; Phaeophyta (brown algae), Chlorophyta (green algae), and Rhodophyta (red algae) [13].

According the Food Balance sheets published by the Food and Agriculture Organisation of the United Nations (FAO), the Republic of Korea is the greatest consumer of seaweed (22.41 kg/capita/year in 2013), followed by China and Japan [14]. Production of farmed seaweed has more than doubled worldwide since 2000, with particular expansion seen in Indonesia due to their vast areas of shallow sunlight coasts suitable for culture sites [15]. The culture of Japanese kelp, *Laminaria japonica*, has traditionally been the most extensively farmed cold-water species. This was surpassed by the tropical *Eucheuma* seaweeds (*Kappaphycus alvarezii* and *Eucheuma* spp.) in 2010. The other most commonly farmed seaweed species are *Gracilaria* spp., *Undaria* sp., and *Porphyra* spp. [15].

Brown algae are distinguished by the presence of the pigment fucoxanthin, which is responsible for the distinctive olive-brown colour that lends this group its name [13]. Brown algae are also unique among algae as they are only found in multicellular form [16]. There are approximately 1500–2000 species of brown algae worldwide [17]. Some species, such as *Macrocystis pyrifera* (giant kelp), play an important role in the ecosystem, growing up to 20 m and forming underwater kelp forests [18]. There are also many other species that have been exploited for human consumption, including *Undaria pinnatifida* (wakame), *Hizikia fusiformis* (hijiki), and *Laminaria japonica* (kombu) [19]. Several types of brown algae are used for animal feed, including *Laminaria digitata* (oarweed), *Ascophyllum nodosum* (rockweed) and *Fucus vesiculosus* (bladder wrack) [3].

Green algae are a diverse group of approximately 8000 species, consisting of the divisions Chlorophyta and Charophyta [20]. Chlorophyta are a large group of unicellular and multicellular algae, while charophyta are a smaller group of exclusively freshwater, multicellular green algae from which it is believed that Embryophyta (terrestrial plants) evolved from [21]. Green algae obtain their pigmentation from chlorophyll a and b, as well as other pigments including β-carotene and xanthophylls [13]. The most commonly consumed species of green algae are *Ulva* spp. including *U. lactuca* (sea lettuce), *U. intestinalis*, and *U. compressa*.

Red algae are a large group of mostly multicellular macroalgae with approximately 6000 species [13]. Red algae are characterised by the presence of phycobilins, which are responsible for their red colour [13]. Species of Irish moss, such as *Chondrus crispus* and *Mastocarpus stellatus*, are exploited for their production of carrageenan [22]. Other species of red algae, including *Porphyra tenera* (nori) and *Palmaria palmata* (dulse), are among the highest consumed species of seaweed in Asia, as well as Western countries, due to their high protein content and their delectable flavour [23]. In particular, *Porphyra tenera* is used in the production of sushi.

1.2. Characteristics of Microalgae

Microalgae are unicellular, microscopic organisms that are also considered as a viable alternative protein source. The most abundant microalgal divisions are Bacillariophyta (diatoms), Chlorophyta (green algae), Chrysophyta (golden algae), and Cyanophyta (blue-green algae). Microalgae are a hugely diverse group containing approximately 200,000 species [24]. Several of these species are currently exploited for a variety of biotechnological purposes, including cosmeceuticals, animal feed, fatty acids, alginates, wastewater treatment, and biofuel [9,25,26]. Furthermore, *Arthrospira platensis* (*Spirulina*), and *Chlorella vulgaris* (*Chlorella*) are also sold as functional foods due to their high vitamin and mineral content and as they are generally regarded as safe (GRAS) by the European Food Safety Authority (EFSA) [25]. Despite the relatively low quantity of microalgae produced annually compared to that of seaweed (5000 tonnes dry matter per year versus 7.5×10^6 tonnes dry matter per year, respectively), whole microalgal biomass and added-value compounds are economically valuable, representing a global turnover of about US $\$1.25 \times 10^9$ per year, compared to annual seaweed turnover of US $\$6 \times 10^9$ [9].

Arthrospira platensis is a filamentous Cyanobacterium that has among the highest recorded protein content of any whole food [27]. *Arthrospira* sp. was originally referred to as *Spirulina* sp. until a re-examination showed that they are actually a distinct genus [9]. However, due to its widely publicised use as a food and dietary supplement, the term *Spirulina* is often used interchangeably with *Arthrospira*.

Chlorella spp. are spherical members of the phylum Chlorophyta (green algae) that have also seen increased popularity as a food supplement in recent times. *Chlorella vulgaris* is the most commonly exploited industry species due to its high protein content (51%–58% dry weight; dw) and favourable essential amino acid composition [28]. *Chlorella* also contains many other beneficial nutrients, including β-1,3-glucan, vitamins (B-complex and ascorbic acid), minerals (potassium, sodium, magnesium, iron, and calcium), β-carotene, chlorophyll, and *Chlorella* growth factor (CGF) [29].

2. Protein Quality

2.1. Amino Acid Composition

The quality of proteins can vary dramatically, depending on digestibility and the availability of essential amino acids [30]. Animal sources of protein are generally considered as complete proteins, as they are a rich source of essential amino acids (EAAs) which the human body is unable to biosynthesise. Alternatively, plant proteins are often considered an incomplete protein source as they commonly lack one or more of the essential amino acids, including histidine, isoleucine, leucine, lysine, methionine, phenylalanine, threonine, tryptophan, and valine [31]. However, the lacking essential amino acid(s) in plant-based proteins can differ, meaning that an individual should be able to obtain a sufficient quantity of all essential amino acids if they consume a varied diet of plant proteins from fruit, vegetables, grains, and legumes [32]. Plant-based proteins are also typically harder to digest than animal proteins, due to their high concentration of insoluble polysaccharides. Despite this, there are increasing concerns about the high levels of saturated fats and cholesterol found in foods of animal origin, which are linked to the development of cardiovascular disease and diabetes. This has led to nutritionists and organisations, such as the FAO, recommending a more varied diet rich in plant-based proteins [33].

Algae are generally regarded as a viable protein source, with EAA composition meeting FAO requirements and they are often on par with other protein sources, such as soybean and egg [3,33]. The lack of widespread consumption of marine algae has led to a shortage of in vivo research regarding the ileal digestion of algae, thus limiting the comparison of algal protein quality between different algae species, as well as with other protein sources [34]. Nevertheless, tryptophan and lysine are often limiting amino acids in most algae species [35–37]. Furthermore, leucine and isoleucine are commonly found in low concentrations in red species of algae, while methionine, cysteine, and lysine are often limiting in brown algae species [35,38]. Cysteine typically occurs at low levels in many seaweed species, and is often not detectable [39]. Aspartic acid and glutamic acid constitute a relatively large proportion of the total amino acids in many seaweed species, largely contributing to the distinctive 'umami' taste associated with seaweed [40]. For example, these two amino acids have been reported to represent 22%–44% of total amino acids in *Fucus* sp. and 26%–32% in *Ulva* sp. [3].

2.2. Algal Protein Digestibility

Bioavailability can be described as the fraction of ingested food components that is available at the target site of action for utilisation in various physiological functions [41]. Bioavailability entails the entire process following food element consumption, including digestibility and solubility of the food element in the gastrointestinal tract, absorption/assimilation of the food element across the intestinal epithelial cells and into the circulatory system, and finally, incorporation into the target site of utilisation (Figure 1) [42]. Studies examining the bioavailability of food elements are therefore required to incorporate in vivo experiments. One in vivo study evaluated the bioavailability of *P. tenera* and *U. pinnatifida* in Wistar rats, which reported that the fibre in seaweed had a negative impact on protein intake digestibility and food efficiency [43]. Similarly, *L. japonica* was also reported to decrease protein digestibility in rats, although interestingly, digestibility ended up being comparable with the control diet after 3 weeks as the rats appeared to adapt to the high fibre diet [44]. It is thought that phlorotannins and high polysaccharide content are the main factors which negatively impact the digestibility of algal proteins [45,46].

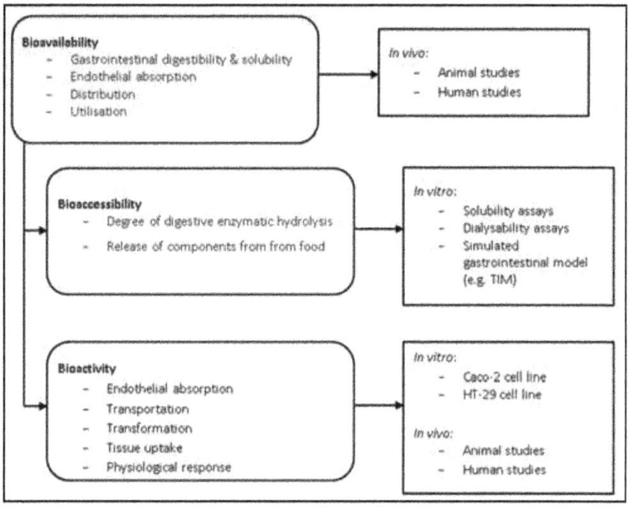

Figure 1. Schematic representation of digestion and methods that may be used to determine bioavailability, bio-accessibility and bioactivity of proteins and other foods. Adapted from Carbonell-Capella et al. [47]. TIM: TNOASR's intestinal model; TNOASR: The Netherlands Organization for Applied Scientific Research.

Bioavailability can be further divided into two different stages; bioaccessibility and bioactivity. Bioaccessibility involves examining the fraction of particular components that are released from the whole food matrix within the gastrointestinal tract in order to identify elements that are accessible for further absorption [48]. Bioactivity refers to the assimilation of a food elemen14 t across intestinal cells, transport of the element to the target site, interaction of the element with the target site, any necessary biotransformation of the food element, and the physiological response created as a result of incorporation of the element with the target site (Figure 1) [49]. There are many factors affecting digestibility which can make in vitro studies difficult, including the macronutrient composition, enzyme specificity, anti-nutritional factors, fibre, and varying absorptive capacities at different stages within the gastrointestinal tract [30]. Nevertheless, in vitro studies serve as a useful preliminary screening tool to identify promising food matrices, growing conditions, and processing methods [48,50,51].

Methods for assessing in vitro digestibility are typically divided into four categories, including solubility, dialysability, gastrointestinal chambers, and cell models [48]. Only small, soluble molecules can be absorbed in the small intestine, which can be evaluated by methods such as atomic absorption spectrophotometry, mass spectrometry, or high-performance liquid chromatography [48]. Dialysability is a direct measure of a food components ability to cross a membrane, although dialysability assays have typically been carried out on micronutrients, including iron, zinc, magnesium and calcium [52].

Bioaccessibility in vitro studies are typically accomplished using either static or dynamic measuring systems. Static systems are the more basic of the two, measuring the release of free amino acids from dietary proteins following hydrolysis from gastrointestinal enzymes under discrete pH and temperature [49]. Static systems have the advantage of being easily implemented, low cost, and high throughput, but have the disadvantage of being unrealistic for normal gastrointestinal physiological processes. Alternatively, dynamic systems use computer systems to tightly regulate pH, temperature, enzyme addition, mixing and residence times within chambers in order to more closely mimic gastrointestinal digestion [48]. These systems model gastric physiology more accurately, but have the drawbacks of being costly and low throughput, limiting their routine use.

The Netherlands Organization for Applied Scientific Research (TNOASR) has developed two similar dynamic gastrointestinal models, called TNOASR's intestinal model (TIM-1 and TIM-2) [48]. The TIM-1 system contains several compartments used to mimic the effect of the stomach and small intestine (including duodenum, jejunum, and ileum) [53]. TIM-2 focuses on the large intestine, serving as a tool to study the effect of microbial fermentation and nutrient absorption in the colon [54]. In both models, aliquots can be taken from any chamber at any given time [48]. Additional in vitro gastrointestinal models have also been described to study digestion and microbiota colonisation [55–58]. However, one such issue that plagues digestibility studies is the lack of consistency occurring in differing methodologies, making the resulting data difficult to compare [59]. INFOGEST is an action granted by the European Cooperation in Science and Technology (COST) and was developed to help overcome this hurdle. INFOGEST is a static in vitro digestion model which aims to harmonise the methods used to assess digestibility, allowing for better comparisons between studies [60,61].

Various in vitro cell culture methods have been utilised to simulate a food component's ability to be assimilated within the intestine, the first component within bioactivity (Figure 1). Caco-2 cells, a cell line derived from human colonic adenocarcinoma, are by far the most commonly used [52,62]. HT-29 is another human colon carcinoma cell line that has been used to study epithelial transport, although it is rarely used [63]. The co-culture of Caco-2 cells with a human mucous-producing cell line, such as HT29-MTX, has been suggested to more closely resemble in vivo conditions [64].

Similar to the previously described in vivo protein quality studies [43,44], in vitro bioaccessibility studies also appear to suggest that unprocessed seaweed proteins have reduced digestibility compared to that of other protein sources. For example, the seaweed species *P. tenera*, *U. pinnatifida*, and *Ulva pertusa* have reported in vitro bioaccessibility of 78%, 87%, and 95%, respectively, expressed as a percentage of casein bioaccessibility (100%) [3]. *U. lactuca* has been shown to have an in vitro digestibility of 85.7% ± 1.9%, while the red seaweeds *Hypnea charoides* and *H. japonica* have high digestibility of 88.7% ± 0.7% and 88.9% ± 1.4%, respectively [65]. Comparable results for *U. lactuca* were found with in vitro simulated ileal digestibility of 82.3% [66]. Tibbetts and colleagues (2016) reported significantly greater in vitro digestibility in red seaweeds (83%–87%) compared to brown seaweeds (78.7%–82%) [67]. These results demonstrate that seaweed proteins have comparable in vitro digestibility compared to that of other commonly consumed plants, including grains (69%–84%), legumes (72%–92%), fruits (72%–92%), and vegetables (68%–80%) [67].

The digestibility of microalgae is poorly examined within the literature, including in vitro bioaccessibility studies. However, microalgae appear to have similar digestibility to that of seaweed, with *Scenedesmus obliquus*, *Spirulina* sp., *Chlorella* sp. having digestibility coefficient values of 88.0%, 77.6%, and 76.6%, respectively [28]. This is in comparison to protein sources such as casein and egg with a digestibility coefficient of 95.1% and 94.2%.

3. Protein Extraction Methods

3.1. Conventional Protein Extraction Methods

Seaweed and microalgae have poor protein digestibility in their raw, unprocessed form and it is for this reason that great emphasis has been placed on developing improved methods for algal protein extraction in order to improve their bioavailability. Algal proteins and their extraction is a relatively poorly studied topic compared to proteins from other crops [68]. Algal proteins are conventionally extracted by means of aqueous, acidic, and alkaline methods, followed by several rounds of centrifugation and recovery using techniques such as ultrafiltration, precipitation, or chromatography [69]. Chemical extraction methods, such as two-phase acid and alkali treatments, have been especially efficient for extracting proteins from *A. nodosum*, *Ulva* spp. and *L. digitata* (Table 1) [69–71].

However, the successful extraction of algal proteins can be greatly influenced by the availability of the protein molecules, which can be substantially hindered by high viscosity and anionic cell-wall polysaccharides, such as alginates in brown seaweed and carrageenans in red seaweed [72]. Cell disruption methods and the inclusion of selected chemical reagents are therefore used in order to improve the efficiency of algal protein extraction. Some examples of conventional methods that are commonly utilised include mechanical grinding, osmotic shock, ultrasonic treatment, and polysaccharidases-aided hydrolysis (Table 1) [73].

Table 1. Conventional pre-treatment cell disruption methods and extraction methods for precipitating proteins from seaweed. Dry weight; dw.

Extraction Method	Species	Extraction Name	Reagents	Protein Yield	Reference
Enzymatic hydrolysis	*Palmaria palmata*	Polysaccharidase degradation	Cellulase (Cellucast®) and xylanase (Shearzyme®)	Factor 3.3 compared to control	[46]
	Chondrus crispus, *Gracilaria verrucosa*, and *Palmaria palmata*	Polysaccharidase degradation	κ-carrageenase, β-agarase, xylanase, cellulase	-	[74]
	Palmaria palmata	Polysaccharidase degradation	Cellulase (Cellucast®), xylanase (Shearzyme®) and Ultraflo® (β-glucanase)	11.57 ± 0.08 g/100 g dw (67% yield)	[73]
Physical Process	*Porphyra acanthophora* var. *acanthophora*, *Sargassum vulgare* and *Ulva fasciata*	Aqueous treatment and Potter homogenisation	Ultra-pure water	8.9 g/100 g dw, 6.9 g /100 g dw, 7.3 g /100 g dw	[68]
	Palmaria palmata	Osmotic stress	-	6.77 ± 0.22 g/100 g dw (39% yield)	[73]
		High shear force	-	6.92 ± 0.12 g/100 g dw (40% yield)	
Chemical extraction	*Ascophyylum nodosum*	Acid-alkaline treatment	0.4 M HCl and 0.4 M NaOH	59.76% yield	[69]
	Ulva rigida *Ulva rotunda*	Two-phase system	NaOH and 2-mercaptoethanol	-	[70]
	Laminaria digitata	Two-phase system	Polyethylene glycol (PEG) and potassium carbonate	-	[71]
	Palmaria palmata	Alkaline and aqueous	NaOH and N-acetyl-L-cysteine (NAC)	4.16 g/100 g dw (24% yield)	[73]

3.1.1. Physical Processes

Barbarino and Lourenço (2005) reported that physical grinding with the use of a Potter homogeniser significantly increased protein extraction yield from *Porphyra acanthophora* var. *acanthophora*, *Sargassum vulgare*, and *Ulva fasciata* following immersion in ultra-pure water (Table 1) [68]. Alternatively, osmotic stress has also been reported to improve extraction of algal proteins efficiency [45,75]. Osmotic shock was reported to yield a significantly higher concentration of water soluble proteins from *P. palmata* (1.02 ± 0.07 g/100 g) compared to high shear force with an Ultra-turrax® T25 Basic tool (IKA®, Staufen, Germany) (0.74 ± 0.02 g/100 g) [73]. However, there was no significant difference in the amount of total protein extracted between the two methods (6.77 versus 6.92 g/100 g). Alternatively, the use of polysaccharidases was reported to be a more promising method of protein extraction, with a concentration of 11.57 ± 0.08 g/100 g *P. palmata*, equating to a yield of 67% (Table 1) [73].

3.1.2. Enzymatic Hydrolysis

Seaweed is rich in several types of polysaccharides, including cellulose, galactans, xylans, fucoidan, laminarin, alginates, carrageenans, and floridean starch [22]. These polysaccharides can reduce the availability of algal proteins and decrease protein extraction efficiency [68]. Enzymes such as polysaccharidases can therefore be applied as a cell disruption treatment prior to protein extraction in order to increase protein yield (Table 1). Several polysaccharidases (κ-carrageenase, β-agarase, xylanase, cellulase) were used in protein extractions from the red seaweed species *C. crispus*, *Gracilaria verrucosa*, and *P. palmata* as a method of combating the tough cell wall [74]. Hydrolysis of *C. crispus* with carrageenase and cellulase increased protein yield ten-fold compared to the enzyme-free procedure, while the highest yield from *P. palmata* was obtained with xylanase. Similarly, hydrolysis of *P. palmata* with xylanase and cellulase was demonstrated to yield a ten-fold increase in the phycoerythrin pigment protein compared to mechanical extraction [46]. Harnedy and Fitzgerald (2013) also increased protein yield from *P. palmata* with xylanase, although they reported that the high enzyme:substrate concentration required (48.0×10^3 units/100 g) may not be commercially feasible at an industrial scale. Combining multiple extraction methods may also help to improve algal protein extraction. Combining enzymatic hydrolysis with alkaline extraction increased protein yield 1.63-fold compared to alkaline extraction alone in *P. palmata* [76].

3.2. *Current Protein Extraction Methods*

Protein extraction methods used on algae to date are limited for commercial use due to concerns with up-scaling. Conventional mechanical and enzymatic methods for protein extraction may also affect the integrity of extracted algal proteins due to the release of proteases from cytosolic vacuoles [77]. Furthermore, these methods are also laborious and time consuming [69]. Improved extraction methods of cell disruption and extraction are therefore required. Pre-treatment with cell-disruption techniques aid the breakdown of the tough algal cell wall, increasing the availability of proteins and other high-value components for later protein extraction. Some examples of novel protein extraction methods include ultrasound-assisted extraction, pulsed electric field, and microwave-assisted extraction [13].

3.2.1. Ultrasound-Assisted Extraction

Ultrasound-assisted extraction (UAE) can be applied to food sources for a number of applications, including modification of plant micronutrients to improve bioavailability, simultaneous extraction and encapsulation, quenching radical sonochemistry to avoid degradation of bioactives, and increasing bioactivity of phenolics and carotenoids by targeted hydroxylation [78]. The degradative effect of radical sonochemistry, which is the most relevant aspect in terms of improving bioavailability of algal proteins, is not produced by the ultrasound waves, but rather by the formation, growth, and implosion of bubbles formed by what is known as acoustic cavitation [79]. The violent implosion of these bubbles creates microscopic regions of extreme pressure and temperature, resulting in the chemical excitation of the sonicated liquid and its contents, facilitating the particle breakdown and degradation of the target compound [80]. The major advantages of UAE are its fast processing time, non-thermal properties, and low solvent consumption, resulting in a higher purity final product with reduced downstream processing required [81].

Ultrasound pre-treatment was reported to increase protein extraction of *Ascophyllum nodosum* with acid and alkaline treatment alone by 540% and 27%, respectively, as well as reduce processing time from 60 min to 10 min [69]. This dramatic increase in protein yield was suggested to be due to acid hydrolysis alone being insufficient to erode the tough cell wall. Ultrasound-aided extraction was also evaluated in microalgae for a number of value-added components, although there have been relatively few studies that have focused on ultrasound for improved protein extraction [82,83]. Keris-Sen and colleagues (2014) reported that ultrasound at a power intensity of 0.4 $kWh \cdot L^{-1}$ yielded the optimum concentrations of proteins from wastewater treatment microalgae from the Chlorococcales order of the Chlorophyceae class (e.g., *Scenedesmus* sp.) [84]. Ultrasound treatment of *C. vulgaris* significantly increased crude protein digestibility in rats compared to electroporated and untreated spray-dried *C. vulgaris* (56.7% $\pm$ 13.7%, 44.3% $\pm$ 7.5%, and 46.9% $\pm$ 12.7%, respectively), as well as significantly improving protein efficiency ratio and nitrogen balance [85]. Additionally, there were no adverse effects to the histology of major organs upon prolonged consumption of ultrasound-treated microalgae, and thus it has been suggested as a viable pre-treatment method for the food industry [86]. Alternating two counter-current frequencies has also been suggested as a viable method for further improving protein extraction, as demonstrated by a 50% increase in yield and 18% reduction in extraction time using 15 and 20 kHz in *Porphyra yezoensis* compared to mono-frequency ultrasound-assisted extraction [87].

3.2.2. Pulsed Electric Field

Pulsed electric field (PEF) has been used as a cell disruption technique in microalgae, although its primary use has thus far been for the extraction of lipids for conversion to biofuel [88]. PEF involves applying high electric currents in order to perforate a cell wall or cell membrane, causing reversible or irreversible electroporation [88]. Electroporation enables the introduction of various foreign components to cells, including DNA, proteins, and drugs [89]. PEF is a fast and green technology for inactivating microorganisms by irreversible electroporation and aiding the release of intracellular contents of plant cells [90,91]. However, conductivity and electrode gap are factors that could possibly limit this technology for up-scaling [92].

Goettel and colleagues (2013) were among the first to report the use of PEF as a means of extracting multiple intracellular components from algae [93]. Since then, PEF has been demonstrated to increase the yield of several high-value microalgae components, including lipids, carbohydrates, carotenoids and chlorophyll [94–98]. Protein yield from *Chlorella* sp. and *Spirulina* sp. was reported to increase by 27% and 13%, respectively, following PEF-treatment at 15 kV/cm and 100 kJ/kg [99]. Coustets and colleagues (2013) also reported significantly increased protein extraction in *C. vulgaris* and *Nannochloropsis salina* following PEF-assisted extraction, allowing for the extraction of intact cytosolic proteins [100].

3.2.3. Other

Microwave-assisted extraction (MAE) involves heating a material, causing moisture to evaporate, thus creating bubbles under high pressure which can then rupture to disrupt cell contents [82]. Increased levels of soluble proteins were extracted from a microalgae biomass containing green microalgae (*Stigeoclonium* sp. and *Monoraphidium* sp.) and diatoms (*Nitzschia* sp. and *Navicula* sp.) using microwave pre-treatment compared to ultrasound [101]. MAE has attracted attention for extraction of compounds due low energy efficiency, although its use in algae may be limited by impaired function with dried samples [81].

Alternatively, sub- and supercritical fluid extraction techniques have gained popularity in recent decades as extraction methods. Subcritical water extraction (SWE) involves using hot water (100–374 °C) under high pressure (~10 bar) to maintain water in a liquid state [102]. Alternatively, supercritical fluid extraction (SFE) is a technique that heats a fluid above its critical point, making it supercritical. Under supercritical conditions, the properties of the fluid become indistinguishable from its gaseous state, with a density similar to a fluid, but viscosity matching that of gas [102]. SFE typically utilises carbon dioxide (CO_2), making it a relatively 'green' technology with low solvent consumption [81]. However, SWE and SFE both require high investment costs for equipment and have typically only been used in algae to date for the extraction of lipids [102].

3.3. Enrichment Methods—Membrane Filtration

Membrane technologies are widely used in the dairy industry to recover whey proteins from milk released as a result of the cheese-making process [103]. Membrane technology refers to the use of a semi-permeable membrane to separate a liquid into two different fractions by selectively allowing some compounds to pass through while impeding other compounds, typically based on molecular weight. Membrane technologies are promising alternative methods of enriching algal proteins, as well as developing novel techno-functional and bioactive ingredients. They have the advantage of being non-thermal and environmentally-friendly [103]. The most commonly used membrane technologies include microfiltration, ultrafiltration, nanofiltration, and reverse osmosis.

Membrane technologies could act as an alternative method for enriching algal proteins when used in conjunction with a cell disruption technique, such as polysaccharidase hydrolysis, UAE, or PEF. Disruption of the tough cell wall is a critical step required in order to increase the availability of algal proteins for extraction [82]. Membrane technologies are well suited for use with seaweed as part of a cascading biorefinery process to maximise valorisation of all components within algae, while avoiding the presence of heavy metals in the final product [104]. A combination of membrane technologies could be used to isolate algal proteins using the same principles of molecular weight cut-offs used in the dairy industry. In the dairy industry, microfiltration (MF) is used to extend the shelf-life of milk without any thermal treatment by removing microorganisms, while preserving overall taste and sensory attributes [105]. MF could be used to remove algae cell wall components and bacteria with a molecular weight greater than 200 kDa. Ultrafiltration (UF) could then be used to isolate proteins and other macromolecules between 1 and 200 kDa, similar to the way it is used in the dairy industry to generate enriched fractions less than 10, 5, 3 and 1 kDa. Nanofiltration (NF) could then be used to remove monovalent salts to minimise osmotic pressure, followed by reverse osmosis (RO) to reduce fluid volume [103].

Indeed, membrane technologies have already been used to isolate whole microalgae cells and several seaweed components. Tangential flow microfiltration was reported as an efficient method for recovering 70%–89% of algal biomass from wastewater treatments [106]. UF was previously used in conjunction with supercritical CO_2 extraction and ultrasound to isolate *Sargassum pallidum* polysaccharides [107]. Polysaccharides with antioxidant activities were isolated from *U. fasciata* utilising hot water extraction followed by several stages of ultrafiltration with increasingly smaller pore sizes [108]. Furthermore, UF was used to isolate phycoerythrin protein from *Grateloupia turuturu* following cell homogenisation, which was reported to retain 100% of the protein without denaturation [109]. Alternatively, a two-stage ultrafiltration could be applied for algal protein enrichment, as demonstrated by the separation of polysaccharide components in *Tetraselmis suecica* using different sized pore membranes following high-pressure homogenisation [110].

4. Applications

4.1. Human Nutrition

Protein is an essential nutritional component in the diet of athletes, required to repair and build muscle tissue broken down during exercise, with the American College of Sports and Medicine recommending between 1.2 and 1.7 g protein per kg body weight [111]. Seaweed and microalgae are rich sources of protein and contain all of the essential amino acids at various concentrations [112]. Algae could therefore represent a valuable resource for athletes requiring high levels of protein, especially for vegan athletes for whom eggs and dairy whey protein may not be suitable [113].

Seaweed and microalgae have been used as a source of human nutrition for thousands of years in some indigenous populations [114]. One of the main reasons for the high consumption of seaweed and microalgae is due to their significant protein content, which is comparable to, or even greater than, some plant sources [28]. Some species of red seaweeds (Rhodophyta), such as *P. palmata* and *P. tenera*, have been reported to contain as much as 33% and 47% dw, respectively [3]. Similarly, some species of microalgae have been reported to contain even higher levels, as high as 63% dw in *Spirulina* sp. [115]. Microalgae are typically consumed as a dietary supplement in the form of powder, pills, or tablets [9]. However, they have also been incorporated into a number of functional foods, including noodles, bread, biscuits, drinks, sweets, and beer [116]. Several businesses have been set up for the sale of algal products, such as AlgaVia® (www.algavia.com), which produces protein- and lipid-rich algal flour from *Chlorella protothecoides*.

There are several species of seaweed that have traditionally been consumed, largely due to their high protein content. For example, *P. tenera* (nori) is commonly used as a sushi wrap in several Asian cultures [22]. Many species of seaweed are particularly high in the amino acids aspartic acid and glutamic acid, which exhibit a unique and interesting flavour that led to the discovery of the taste sensation referred to as 'umami' [40]. The flavour enhancer monosodium glutamate was first discovered in the brown seaweed *L. japonica* (kombu), which has been found to particularly appeal to the umami taste sensation [117].

Although the consumption of seaweed in humans is currently underdeveloped, especially in Western countries, the high protein content and favourable essential amino acid profile makes seaweed a promising source of protein that is ripe for future expansion [40]. Seaweed has been successfully incorporated as a functional ingredient into several foods at the laboratory scale. *U. pinnatifida* (wakame) integrated into pasta has antioxidant activity and acceptable sensory attributes at levels up to 10% [118]. Bread containing 4% *A. nodosum* can significantly reduce energy intake in overweight individuals in the meal following enriched bread consumption [119]. Bread incorporating similar concentrations of renin-inhibitory peptides from *P. palmata* hydrolysates also had acceptable sensory attributes, with the bioactive properties reported as having survived the baking process [120].

Spirulina is the most highly consumed microalgae due to its high protein content and added nutritional benefits, including anti-hypertension, renal protective, anti-hyperlipidaemia, and anti-hyperglycaemic [121]. As well as being a rich source of proteins, *Spirulina* contains high levels of hypocholesterolemic γ-linoleic acid (GLA), B-vitamins, and free-radical scavenging phycobiliproteins [122]. It has therefore been given the label of a 'super food' by the World Health Organisation (WHO) and has even been sent to space by the National Aeronautics and Space Administration (NASA) due to its nutrient-dense properties [123]. As a demonstration of this, *Spirulina* has 180% more calcium than milk, 670% more protein than tofu, 3100% more β-carotene than carrots, and 5100% more iron than spinach [27]. The world's largest producer of *Spirulina* is Hainan Simai Enterprising Ltd., which is located in the Hainan province of China [121]. This cultivation farm produces an annual 200 tonnes dried biomass, accounting for 25% of the national output and a considerable 10% of the global production. Alternatively, the Earthrise Company has the largest *Spirulina* production plant, which is located in California, USA, and covers 440,000 m^2 [121].

Chlorella is another widely consumed microalga, with global sales exceeding US $38 billion [124]. The largest producer of *Chlorella* is Taiwan Chlorella Manufacturing and Co. (Taipei, Taiwan), which produces 400 tons dried biomass annually [121]. The main substance found in *Chlorella* that is beneficial for human health is β-1,3-glucan, which is an active immunostimulant, free-radical scavenger and reducer of blood lipids [125]. *Chlorella* is also rich in proteins (48% dw), polyunsaturated fatty acids (PUFAs) (39% of total lipids), and phosphorus (1761.5 mg/100g dw) [115].

4.2. Industrial Applications

Lectins and phycobiliproteins are two families of bioactive algal proteins which have been exploited for several industrial applications. Lectins are most commonly extracted from macroalgal sources, while phycobiliproteins are typically isolated from microalgae [126]. Phycobiliproteins, especially phycoerythrin, can constitute a significant proportion of the overall protein content in red algae, with levels of 1.2% (total dw) reported for *P. palmata* [127]. Lectins are found at similar levels and a yield of 1% lectins was obtained previously from *Eucheuma serra* (Rhodophyta) [128].

4.2.1. Lectins

Lectins are glycoproteins known for their aggregation and high specificity binding with carbohydrates without initiating modification through associated enzymatic activity [129]. Lectins are involved in several biological processes, including host-pathogen interactions, cell–cell communication, induction of apoptosis, cancer metastasis and antiviral activities [130]. Due to their carbohydrate binding capacity with high specificity, lectins are used in blood grouping, anti-viral (including human immunodeficiency virus type 1(HIV-1)), cancer biomarkers, and targets for drug delivery [131]. Lectins derived from algae have not received the same level of characterisation compared to other plant-derived lectins. Nevertheless, some of the bioactivities that have been observed in algal lectins include mitogenic, cytotoxic, antibacterial, anti-nociceptive, anti-inflammatory, anti-viral (HIV-1), platelet aggregation inhibition, and anti-adhesion [132].

4.2.2. Phycobiliproteins

Phycobiliproteins are water-soluble proteins with an important role in photosynthesis within cyanobacteria, Rhodophyta, and cryptomonads [133]. Phycobiliproteins are components of phycobilisomes, which are large light energy-capturing complexes anchored to thylakoid membranes [134]. There are four main divisions of phycobiliproteins which are grouped based on their colour and absorption characteristics, namely, phycoerythrin, phycocyanin, allophycocyanin, and phycoerythrocyanin [135]. The main commercial producers are *Spirulina* sp. (cyanobacterium) and *Porphydrium* sp. (Rhodophyta macroalgae) [121].

Phycobiliproteins are used in fluorescent labelling, flow cytometry, fluorescent microscopy, and fluorescent immunohistochemistry [136,137]. However, the primary commercial application of these phycobiliproteins appears to be as natural dyes, with phycocyanin in particular used as a blue pigment used in products such as chewing gum, popsicles, confectionary, soft drinks, dairy products, and wasabi, as well as cosmetic products, such as lipstick and eyeliner [121]. Several patents concerning health beneficial bioactivities of phycobiliproteins have also already been filed for nutraceutical applications such as anti-oxidative, anti-inflammatory, anti-viral, anti-tumour, neuroprotective, and hepatoprotective activities [135].

4.3. Animal Feed

The high protein content of algae can also be beneficial for use as animal feed, including aquaculture, farm animals, and pets. An estimated 30% of global algal production is estimated to be used for animal feed, with 50% of *Spirulina* biomass in particular used as feed supplement due to its excellent nutritional profile [124,138]. Several species of microalgae including *Spirulina*, *Chlorella*, and *Schizochytrium* sp., and seaweed including *Laminaria* sp. and *Ulva* sp. can be incorporated as protein sources into the diets of poultry, pigs, cattle, sheep, and rabbits [4,139]. Most of the research on the incorporation of algae as animal feed has been carried out with poultry, likely due to their promising prospects for improved commerciality [138].

Tasco® is an example of a proprietary seaweed meal derived from *A. nodosum*, produced by Acadian Seaplants in Nova Scotia, Canada (http://www.acadianseaplants.com), which has demonstrated beneficial properties when included in animal feed [140]. Tasco® has four main identified benefits for animal production, including resistance to stressors, improved immune system, increased productivity/quality, and a reduction in pathogenic microorganisms in the final meat product [141–144]. These benefits have been observed in several species, including monogastric and ruminant species, when at feed inclusion levels of 2% on a daily basis [145].

4.3.1. Poultry

Supplementing poultry feed with microalgae as a protein source can improve their health, productivity, and value. This has been demonstrated using a variety of species, including *Chlorella* sp., *Arthrospira* sp., *Porphyridium* sp., and *Haematococcus* sp. [86,139,146,147]. Chickens fed with supplemented *Spirulina* have been reported to have increased viability, improved overall health and reduced plasma concentrations of cholesterol, triglycerides, and fatty acids [148]. These birds also appeared to have an improved immune system as demonstrated by a significant increase in white blood cell count and enhanced macrophage phagocytic activity [148,149]. Ross and Dominy (1990) reported that feeding White Leghorn cockerel chicks, Hubbard by Hubbard male broiler chicks, and Japanese quail with varying concentrations of *Spirulina* in dietary feed slightly delayed growth rates, but did not affect final growth at concentrations less than 10% [150]. Furthermore, this study also reported that the inclusion of *Spirulina* also increased fertility rates, as well as increasing the intensity of the egg-yolk colour [150]. These results have been confirmed by several other studies, indicating that the inclusion of *Spirulina* at a concentration of 2%–2.5% in the feed intensifies the colour of egg yolks to make it more esthetically pleasing for consumers [151,152]. The intensified colouration of the yolk is thought to be due to β-carotene [153]. The inclusion of *Spirulina* can also further valorise egg products by decreasing their cholesterol and saturated fatty acid content, and replacing it with increased levels of beneficial omega-3 polyunsaturated fatty acids [146,154].

Incorporation of 3% *U. lactuca* in broiler chickens increased breast muscle yield compared to birds solely fed corn diet, as well as decreased serum lipids, cholesterol, and uric acid concentrations [155]. The red seaweed *Polysiphonia* spp. was also demonstrated to improve pellet binding in duck feed at concentrations of 3%, as well as improve its overall nutrient profile [156]. The incorporation of red seaweeds *C. crispus* and *Sarcodiotheca gaudichaudii* was also reported to effectively act as a prebiotic to improve chicken gut health, productivity, and egg quality [157].

4.3.2. Pigs

The replacement of up to 33% of soy proteins with proteins from *Arthrospira maxima*, *A. platensis*, and *C. vulgaris* in pig feed has been reported as being suitable without any adverse effects [158]. The effect of feed processing appears to play a role in the utility of *Spirulina* in pig's feed. The addition of *Spirulina* to pellets was reported to decrease average daily gain, whereas incorporation of *Spirulina* to meal diets actually increased average daily gain [159]. Addition of *Spirulina* to the diet has also been suggested to improve fertility in pigs, increasing sperm motility and storage viability [160].

Supplementation of the brown seaweed *L. digitata* increased pig body weight gain by 10%, as well as increase the concentration of iodine in fresh muscle by 45%, thus increasing its valorisation [161]. Similarly, *A. nodosum* has also been reported to increase iodine content in pig tissue, while also increasing the concentrations of beneficial bacteria within the gut [162]. However, these results are in contrast to the findings of Reilly and colleagues (2008), who reported that the brown seaweeds *Laminaria hyperborea* and *L. digitata* actually decreased the biodiversity of beneficial microbial populations within the pig's gut, although this did not significantly affect the pig's performance [163].

4.3.3. Ruminants

Of all the animals evaluated for algae supplementation, ruminants are the most promising in terms of digesting the high fibre content for the greatest extraction efficiency of algal proteins [139]. This is in contrast to mono-gastric animals, for which it has been suggested that some form of prior processing may be required in order for animals (and humans) to utilise algal proteins more efficiently [164]. Cattle will preferentially drink water containing 20% suspended *Spirulina*, increasing their daily water intake by 24.8 g/kg [165]. Furthermore, this study also reported that 20% of the consumed *Spirulina* bypasses degradation within the rumen, allowing for increased digestion and absorption of protein and nutrients within the abomasum [165]. Incorporation of 200 g/day *Spirulina* with cattle feed was reported to be an economically effective method of increasing animal body weight (8.5%–11%) and daily milk production (21%) [166]. As well as increasing milk quantity, *Spirulina* supplementation has also been demonstrated to increase milk quality by decreasing saturated fatty acids, while simultaneously increasing monounsaturated fatty acids and polyunsaturated fatty acids [167]. Similar results were observed in other studies, as well as with supplementation of *Schizochytrium* sp. [168,169].

Sheep have also been demonstrated to benefit from microalgae as a protein source, with lambs reported to have increased average daily gains upon consumption of 10 g of *Spirulina* per day [170]. Similarly, *Spirulina* diet supplementation increased the feed intake of rabbits, as well as improve the quality of rabbit meat with higher levels of GLA [171]. The green seaweed *U. lactuca* has been reported as a suitable low-energy, high-protein foodstuff for sheep and goats [172,173]. However, seaweed may be not be suitable for supplementation in pregnant ewes, having been reported to interfere with passive immunity in lambs and increasing mortality rate [174].

4.4. *Aquaculture*

Microalgae are vital for the artificial reproduction of several aquaculture species, especially molluscs [138]. Microalgae also play an important role in aquaculture, other than as a food source for zooplanktons by stabilising pH, reducing bacterial growth, and improving the quality of rearing medium [175]. This leads to improved survival and growth compared to that of clear water fed with artificial diets [176]. Microalgae are the natural base of the entire aquatic food chain. This has led to their widespread incorporation as an important food source and feed additive in the commercial rearing of many aquatic animals, including molluscs, shrimp, and rotifers [177]. Filtering molluscs are by far the greatest consumer of microalgae in aquaculture, with 10.1×10^6 tonnes produced in 1999, compared to shrimp (1.2×10^6 tonnes), and small larvae fish, such as sea breams and flatfish (177,400 tonnes) [178].

Replacements of live microalgae are already commercially available (such as *Chaetoceros* 1000 "Premium Fresh" Instant Algae™ paste, Liqualife™ liquid larval feed, Zeigler™ E-Z Larvae liquid feed, and Zeigler™ Z-Plus feed), but typically provide inferior growth and survival rates [179]. For example, the survival rate of brown larval shrimp (*Farfantepenaeus aztecus*) significantly decreased from 90.86 ± 3.19% when fed live microalgae, to 14.865 ± 14.35% in shrimp fed with 100% replacement E-Z larvae [179].

Microalgae are also often used as a dietary supplement to refine the products of aquaculture and increase their valorisation. Carotenoid pigments, such as astaxanthin derived from *Haematococcus pluvalis*, are incorporated into the diets of salmonoids, shrimp, lobsters, and crayfish, to give them their characteristic pink flesh [180]. Similarly, inclusion of astaxanthin at a concentration of 30 ppm significantly increases the colour pattern and intensity of ornamental fish, including tetras, cichlids, gouramis, damos, goldfish, and koi, increasing their market value several fold [147].

Red seaweed has been suggested as a promising protein source feed additive. Incorporating 10% *Gracilaria chilensis* in the diet of Atlantic salmon (*Salmo solar*) was reported to significantly increase specific growth rate by 1.51% ± 0.12% compared to the control diet [181]. Including 1.0% and 10% *G. chilensis* in the diets of *S. solar* was also suggested to increase antiviral activity against the infectious salmon anaemia (ISA) virus. Similarly, inclusion of 5% and 15% *P. palmata* was reported to improve hepatic function and have a positive effect on body lipid content in *S. salar* compared to the control diets [182]. Wild abalone are opportunistic feeders that consume a variety of macroalgae species and typically have increased growth rate in captivity when fed a diet with several species compared to single species diets [183]. *Haliotis tuberculata coccinea* fed with a mixed diet including *Ulva rigida*, *Hypnea spinella*, and *Gracilaria cornea* displayed significantly greater growth rates, length, and weight gain than diets consisting of single algal species [184]. Seaweed is also often used in the feed of sea cucumber culture systems, which are used for human consumption in many Asian countries. *L. japonica* and *U. lactuca* have been reported as an economical additive to the diets of sea cucumber *Apostichopus japonicas* with low ammonia–nitrogen production and suitable digestibility [185].

4.5. Bioactive Peptides

Bioactive peptides are particular amino acid sequences that can have additional physiological health benefits beyond their basic nutritional value [126]. These peptides are typically between 2 and 30 amino acids in length and have hormone-like properties. Bioactive peptides are inactive within the parent proteins, but can be released through fermentation or hydrolysis. Milk proteins remain the most common source for bioactive peptides [186,187]. However, bioactive peptides have also been identified in a number of food sources, including meat, egg, fish, and blood, as well as plant sources, including rice, soybean, wheat, pea, broccoli, garlic, and algae [132,188–197].

Bioactive peptides have been found to have a multitude of beneficial effects, including anti-hypertensive, anti-oxidative, antithrombotic, hypocholesterolemic, opioid, mineral binding, appetite suppression, anti-microbial, immunomodulatory, and cytomodulatory properties [187]. Bioactive peptides from algae have also been found to display bioactive properties, although they are not as well characterised compared to peptides from other sources (Table 2) [198]. Antimicrobial peptides have been identified from *Saccharina longicruris* protein hydrolysates, significantly decreasing the growth rate of *Staphylococcus aureus* [199]. The hexapeptide Glu-Asp-Arg-Leu-Lys-Pro isolated from *Ulva* sp. was demonstrated to have mitogenic activity in skin fibroblasts [200].

Table 2. Angiotensin-I-converting enzyme (ACE)-I inhibitory bioactive peptides derived from seaweed and microalgae. The potency of the peptides is indicated by their IC_{50} values, which refers to the concentration required to inhibit enzyme activity by 50%.

Source	Hydrolytic Method	Peptide Sequence	IC_{50}	Reference
Undaria pinnatifida (wakame)	Pepsin	Ala-Ile-Tyr-Lys Tyr-Lys-Tyr-Tyr Lys-Phe-Tyr-Gly Tyr-Asn-Lys-Leu	213 μM 64.2 μM 90.5 μM 90.5 μM	[201]
Undaria pinnatifida (wakame)	Hot water extraction	Tyr-His Lys-Trp Lys-Tyr Lys-Phe Phe-Tyr Val-Trp Val-Phe Ile-Tyr Ile-Trp Val-Tyr	5.1 μM 10.8 μM 7.7 μM 28.3 μM 3.7 μM 10.8 μM 43.7 μM 2.7 μM 12.4 μM 11.3 μM	[202]
Undaria pinnatifida (wakame)	Protease S "Amano"	Val-Tyr Ile-Tyr Ala-Trp Phe-Tyr Val-Trp Ile-Trp Leu-Trp	35.2 μM 6.1 μM 18.8 μM 42.3 μM 3.3 μM 1.5 μM 23.6 μM	[203]
Ecklonia cava	Alcalase Flavourzyme Kojizyme Neutrase Protamex	Enzymatic digest Enzymatic digest Enzymatic digest Enzymatic digest Enzymatic digest	2.79 μg/mL 3.56 μg/mL 2.33 μg/mL 3.10 μg/mL 3.28 μg/mL	[204]

Table 2. *Cont.*

Source	Hydrolytic Method	Peptide Sequence	IC_{50}	Reference
Porphyra yezoensis		Ile-Tyr Met-Lys-Tyr Ala-Lys-Tyr-Ser-Tyr Leu-Arg-Tyr	2.69 μM 7.26 μM 1.52 μM 5.06 μM	[205]
Hizikia fusiformis		Gly-Lys-Tyr Ser-Val-Tyr Ser-Lys-Thr-Tyr	3.92 μM 8.12 μM 11.07 μM	[206]
Palmaria palmata (dulse)	Thermolysin	Val-Tyr-Arg-Thr Leu-Asp-Tyr Leu-Arg-Tyr Phe-Glu-Gln-Trp-Ala-Ser	0.14 μM 6.1 μM 0.044 μM 2.8 μM	[207]
Chlorella vulgaris	Pepsin	Ile-Val-Val-Glu Ala-Phe-Leu Phe-Ala-Leu Ala-Glu-Leu Val-Val-Pro-Pro-Ala	315.3 μM 63.8 μM 26.3 μM 57.1 μM 79.5 μM	[208]
Arthrospira platensis	Pepsin	Ile-Ala-Glu Phe-Ala-Leu Ala-Glu-Leu Ile-Ala-Pro-Gly Val-Ala-Phe	34.7 μM 11.4 μM 11.4 μM 11.4 μM 35.8 μM	[208]
Nannochloropsis oculata	Alcalase	Leu-Val-Thr-Val-Met	18.0 μM	[209]
Nannochloropsis oculata	Pepsin	Gly-Met-Asn-Asn-Leu-Thr-Pro Leu-Glu-Gln	123 μM 173 μM	[210]
Chlorella ellipsoidea	Protamex, Kojizyme, Neutrase, Flavourzyme, Alcalase, trypsin, α-chymotrypsin, pepsin, and papain	Val-Glu-Gly-Tyr	128.4 μM	[211]
Chlorella vulgaris	Flavourzyme, alcalase, papain, and pepsin	Val-Glu-Cys-Tyr-Gly-Pro-Asn-Arg-Pro-Gln-Phe	29.6 μM	[212]

Owing to the high levels of oxidative stresses and free radicals in their environment, microalgae and seaweed have developed many defensive systems that can stimulate antioxidant activity when consumed [213]. Antioxidant peptides have therefore been isolated from several species of microalgae, including *C. vulgaris*, *Navicula incerta*, and *Chlorella ellipsoidea* [214–217]. Antioxidant peptides have also been isolated from various Irish and Korean brown seaweeds, which indicated that those with higher phenolic content, such as *Ecklonia cava* and *Sargassum coreanum*, correlated with increased antioxidant activity [218–220]. Peptides displaying antioxidant and anticancer bioactivity have also been reported in Sri Lankan red and green seaweed, of which *Caulerpa racemosa* demonstrated the most promising free radical scavenging and anticancer bioactivity [221].

Microalgae have also been reported to display anticancer peptides, such as *Chlorella pyrenoidosa* antitumor polypeptide (CPAP) derived from *Chlorella pyrenoidosa* and polypeptide Y2 derived from *A. platensis* [222,223]. Protein hydrolysates from *Porphyra columbina* phycocolloid extraction by-products were reported to have immunosuppressive, anti-hypertensive, and antioxidant capacities [224]. Anti-inflammatory peptides have been isolated from microalgae, such as *Chlorella* 11-peptide derived from *C. pyrenoidosa* [225], as well as peptides derived from *A. maxima* (Leu-Asp-Ala-Val-Asn-Arg and Met-Met-Leu-Asn-Phe), which were additionally reported to display anti-atherosclerosis bioactivity [226,227]. Similarly, anti-atherosclerosis peptides have also been isolated from *P. palmata* and were shown to be non-toxic in Zebrafish at a concentration of 1 mg/mL [228].

Algal peptides have been reported to display several other bioactivities, including hepatoprotective, immunomodulatory, ultraviolet (UV) radiation-protective, anti-osteoporosis, and anti-coagulant [229–234]. Finally, it is important to note that several studies have reported that short algae-derived peptides are capable of resisting gastrointestinal digestion from enzymes such as trypsin, pepsin, and chymotrypsin [203,207,235]. This is an essential trait for bioactive peptides in order to achieve their physiological effect at their site of action [236].

Anti-Hypertensive Peptides

Hypertension is the single largest risk factor attributed to deaths worldwide, making it an ideal target for bioactive peptides [237]. Angiotensin-I-converting enzyme (ACE-I) is a proteolytic enzyme that affects vasoconstriction in two major blood pressure regulatory systems, namely renin-angiotensin–aldosterone system (RAAS) and kinin–kallikrein system, leading on to the development of hypertension (Figure 2) [126]. ACE-I inhibitors have therefore become one of the most commonly studied targets, and with global annual sales exceeding US $6 billion, ACE-I inhibitory drugs can be considered as one of the major protease inhibitor success stories [238]. Synthetic ACE-I inhibitor drugs, such as captopril, enalapril, and alacepril, often come with several side effects, including hypotension, dry cough, and impaired renal function [19]. Function foods with anti-hypertensive bioactivities have therefore become a popular alternative to synthetic drugs, especially for individuals who are borderline hypertensive and do not warrant the prescription of pharmaceutical drugs [239].

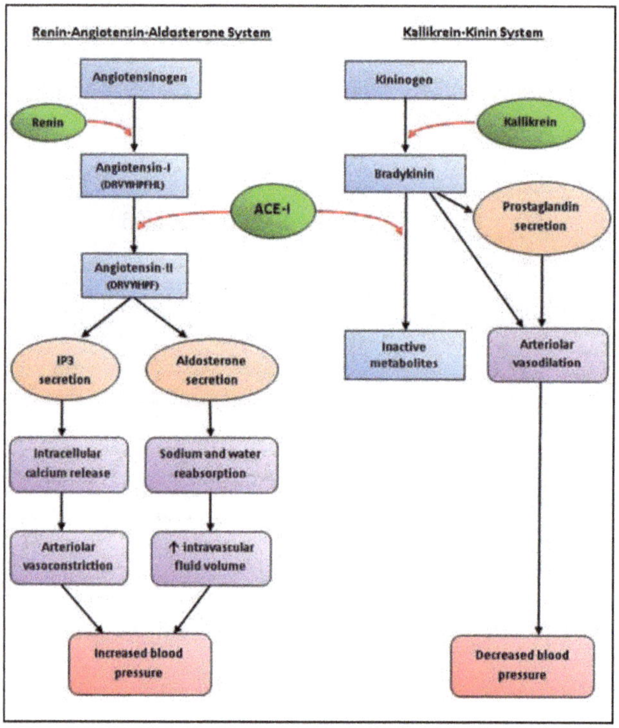

Figure 2. Schematic representation of the renin–angiotensin–aldosterone system (RAAS) and the hypertensive effect of angiotensin-I-converting enzyme (ACE-I). Angiotensinogen is converted to the decapeptide angiotensin-I by renin. ACE-I cleaves the C-terminal dipeptide His-Leu of angiotensin-I to form angiotensin-II. Binding of angiotensin-II to its receptor (AT1) stimulates the secretion of inositol 1,4,5-triphosphate (IP3) and aldosterone, which induce arteriolar vasoconstriction and increased intravascular fluid volume, respectively, resulting in increased blood pressure. Within the kallikrein–kinin system, kallikrein converts kininogen to bradykinin, which induces arteriolar vasodilation by prostaglandin secretion and binding of bradykinin with its receptor, resulting in decreased blood pressure. However, the hypotensive effect of bradykinin is largely dependent on the rate of degradation by ACE-I, which hydrolyzes bradykinin to form inactive metabolites.

ACE-I-inhibitory peptides have been isolated from a variety of seaweed and microalgae sources [132]. Four tetrapeptides were identified from peptic digests of *U. pinnatifida* which displayed in vitro ACE-I inhibitory and in vivo anti-hypertensive bioactivity in spontaneously hypertensive rats (SHRs) (Table 2) [201]. Suetsuna and colleagues also identified several dipeptides from *U. pinnatifida* using hot water extraction [202]. Both single administration and repeated dose administration of four of the dipeptides significantly reduced blood pressure in SHRs. For example, a single administration of 50 mg/kg body weight Tyr-His, Phe-Tyr, and Ile-Tyr lowered systolic blood pressure (SBP) by 50, 46, and 33 mm Hg, respectively, after 3 h [202]. The dipeptide Lys-Tyr took slightly longer to have the greatest effect, with a reduction in SBP of 45 mm Hg observed after 6 h. Similarly, one week of continuous oral administration of 10 mg/kg body weight/day Tyr-His, Lys-Tyr, Phe-Tyr, and Ile-Tyr reduced SBP by 34, 26, 34 and 25 mm Hg, respectively [202]. Similarly, Sato and colleagues (2002) also identified ACE-I inhibitory peptides from *U. pinnatifida* which displayed in vitro and in vivo bioactivity [203].

Dipeptides were isolated from wakame Protease S "Amano" hydrolysates by three-step high-performance liquid chromatography. The peptides showed strong ACE-I inhibitory potency and four of the dipeptides (Val-Tyr, Ile-Tyr, Phe-Tyr, and Ile-Trp) were reported to significantly reduce blood pressure in SHRs comparable to that of the positive control (captopril) when administered at a dose of 1 mg/kg body weight (Table 2) [203]. Cha and colleagues (2006) reported promising peptides derived from digestion of 70 °C aqueous extracts of *E. cava* using five proteases from Novo Co. (Novozyme Nordisk, Bagsvaerd, Denmark), including alcalase, flavourzyme, kojizyme, neutrase, and protamex [204]. ACE-I inhibitory peptides have also been isolated from *P. columbina*, *P. palmata*, *Bangia fusco-purpurea*, *P. yezoensis*, and *H. fusiformis* (Table 2) [205–207,235,240].

In addition to seaweed, ACE-I inhibitory bioactive peptides have also been isolated from a number of microalgae species. Suetsuna and Chen (2001) identified several peptides from *C. vulgaris* and *A. platensis* which displayed promising ACE-I inhibitory and anti-hypertensive activity in SHRs (Table 2) [208]. *Nannochloropsis oculata* has been shown to be a promising source of commercially exploitable biodiesel due to its high lipid content (up to 28.7% dw) [209]. In order to maximise its potential, while minimising waste disposal costs, biodiesel by-products were analysed for ACE-I inhibitory peptides. The commercial enzymes Alcalase, Neutrase, Flavourzyme, Pancreatic Trypsin Novo (PTN), and Protamex were used to generate hydrolysates, of which the Alcalase-generated peptide Leu-Val-Thr-Val-Met was identified as the most potent [209]. In a similar study, *N. oculata* was also digested with a variety of proteases, including pepsin, trypsin, α-chymotrypsin, papain, Alcalase, and Neutrase for the generation of ACE-I inhibitory hydrolysates [210]. The pepsin hydrolysates exhibited the highest ACE-I inhibitory bioactivity, of which two peptide sequences (Gly-Met-Asn-Asn-Leu-Thr-Pro and Leu-Glu-Gln) in particular were identified as having the greatest activity (Table 2). Hydrolysis of *C. ellipsoidea* using the proteases Protamax, Kojizyme, Neutrase, Flavourzyme, Alcalase, trypsin, α-chemotrypsin, pepsin, and papain generated several ACE-I inhibitory peptides, of which Val-Glu-Gly-Tyr was shown to be the most potent in vitro ACE-I inhibitor and in vivo anti-hypertensive with SHRs (Table 2) [211].

An anti-hypertensive peptide (Ile-Arg-Leu-Ile-Ile-Val-Leu-Met-Pro-Ile-Leu-Met-Ala) has also been isolated from *P. palmata*, which was reported to inhibit renin at concentrations similar to that that of the positive control [241]. Renin catalyzes the initial rate-limiting step within the RAAS, and is therefore an important target for the treatment of high blood pressure (Figure 2) [242].

5. Challenges

5.1. Access Rights

At present, all seaweed species are harvested by hand in Ireland. This includes the most economically valuable species brown seaweed *Ascophyllum nodosum* and the red calcified coralline seaweeds *Phymatolithon purpureum* and *Lithothamnion corallioides* (commonly called maërl) [243].

The legislation for harvesting seaweed in Ireland is based on the Foreshore Acts 1933–1998, which states that the Department of Communications Marine and Natural Resources is empowered to grant licences for seaweed harvesting on the seabed out to 12 nautical miles [244]. There are currently no restrictions on harvesting quantities in Ireland, nor are there any restrictions on harvesting times. The only exception to this is for maërl due to its extremely slow growth rate (0.6–1.5 mm per annum) [245]. However, the future implementation of mechanical harvesting methods will likely require a review of existing legislation to ensure appropriate access rights and sustainable harvesting yields are maintained. Mechanical harvesting of *Laminaria* spp. is already present in various European countries, such as France and Norway, providing a valuable resource for making any such revisions [243].

5.2. Variability

An additional challenge, which is particularly relevant in the production of bioactive peptides, are the high levels of variability in algal proteins. The protein content can vary by season, temperature, and location in which the seaweed is harvested [46]. The relative composition of particular proteins within the plant can also differ, changing the concentrations of amino acids and therefore altering the yield of desired peptides as a consequence [246]. For example, annual monitoring of *P. palmata* harvested on the French Atlantic coast showed that protein levels were highest in the winter and spring months, varying from 9 to 25%, and peaking in May [247]. Similarly, *Gracilaria cervicornis* and *S. vulgare* varied by season, with protein levels negatively correlating with temperature and salinity [248]. Different harvest locations of *U. pinnatifida* in New Zealand also significantly affected the protein content and amino acid composition [249].

5.3. Scalability

The scalability of protein extraction from algae is a further obstacle that needs to be overcome before seaweed and microalgae become a viable source. Algal protein extraction is still very much in its infancy, meaning that many of the methods that have been developed are still at small scale [132]. PEF and ultrasound have been suggested as being suitable for large scale algal protein extraction [250]. Membrane technologies may also be scalable for commercial applications, with ultrafiltration reported as being suitable for R-phycoerythrin extraction from the seaweed *G. turuturu* at the industrial scale [109].

5.4. Digestibility

One of the most important challenges for extracting proteins from algae is the high levels of cell wall anionic polysaccharides which can become bound to the proteins and create viscous medium, further increasing the difficulty of protein extraction [46]. The morphology of different seaweed species has been suggested to be an important factor in determining protein yield, with tougher thallus forms reported to require increased processing [68]. Several brown algae species (*Eisena bicyclis, H. fusiformis, U. pinnatifida*) were reported to have reduced in vitro digestibility (51.8%–57.1%) compared to red algae *P. palmata* and *P. tenera* (70.2% and 87.3%, respectively), which was suggested to have been due to increased levels of dietary fibre [34]. Furthermore, the type of polysaccharide has been shown to influence the level of protein digestion. *P. palmata* and *G. verrucosa*, both red seaweeds, were shown to have significantly different levels of in vitro corrected nitrogen digestibility (1.9% and 16.3% after 6 h, respectively) [251]. This was largely attributed to differences in polysaccharide fractions (pentoses and hexoses, respectively).

Treatments to disrupt the cellulosic cell wall could help to overcome this issue to make algal proteins and other cell components more accessible [28]. Heat treatment is one such option that can improve food taste, texture, safety in terms of allergenicity and microbial load, and preservation, while also increasing bioavailability and utilisation of proteins by partial denaturation and breakdown of proteins into peptides, allowing for easier access by proteolytic enzymes [252]. An example of this can be seen with boiling of *P. palmata* , resulting in a 64%–96% increase in liberated amino acids [253].

Alternatively, fermentation can also increase protein digestibility due to the degradation of insoluble fibres, such as xylan. This can be seen with the fermentation of *P. palmata* using the fungal mould *Trichoderma pseudokoningii* which was found to decrease the xylan content from 53% to 19% dw [75]. Washing in distilled water has also been shown to be a simple, yet effective method for improving in vitro digestibility by solubilising and removing mineral salts [75].

Alternatively, the method of drying seaweed has been shown to influence algal protein digestibility. Compared to freeze-drying, oven-drying was demonstrated to significantly improve ($p < 0.05$, two-way ANOVA) the protein extractability and in vitro bioaccessibility of brown seaweed species *Sargassum hemiphyllum*, *Sargassum henslowianum*, and *Sargassum patens* [45]. The increased extractability from oven-drying were suggested to have been due to the decomposition of phenolic compounds at the high temperatures, as well as increased disruption of anionic or neutral polysaccharides found within the cell wall of the seaweed [45].

Protein digestibility can also be largely influenced by the presence of phenolic compounds present in algae, especially phlorotannins in brown seaweeds [45]. Oxidised phenolic compounds can react with amino acids to form insoluble complexes, which may inhibit proteolytic enzymes and thus decrease their nutritive values [65]. The negative correlation between phenolic content and digestibility was demonstrated by *U. lactuca* which was reported to have higher phenolic content (38.8 ± 0.5%) compared to the red algae *H. charoides* (16.9% ± 1.0%) and *H. japonica* (16.3% ± 0.03%) and thus, lower bioaccessibility (85.7% ± 1.9%, 88.7% ± 0.7%, and 88.9% ± 1.4%, respectively) [65]. This is likely to be a greater issue in brown algae, as these are typically higher in phenolics, including catechins, flavanols, and phlorotannins [10,67].

5.5. Food Safety

While algae can be natural accumulators of vitamins and minerals, they can also gather toxic elements, such as heavy metals. There are therefore strict legal limits in Europe for the safe maximum exposure of heavy metals, such as mercury, arsenic, lead, and cadmium, in foods for human consumption. These limits are based upon the recommendations of the Joint FAO/WHO Expert Committee on Food Additives (JECFA) and the Panel on Contaminants in the Food Chain of the European Food Safety Authority (CONTAM Panel), and are outlined in the legislation under Commission Regulation (EC) No. 1881/2006 [254].

The tolerable weekly intake (TWI) of methylmercury, arsenic, lead, and cadmium are 1.6, 15, 25, and 25 μg/kg body weight, respectively [255–258]. The CONTAM panel has reported that the current lead TWI (15 μg/kg body weight) may no longer be appropriate due to low levels found in foodstuffs, while arsenic TWI (1.6 μg/kg body weight) should be lowered due to cancers and adverse effects observed at current legal limits [256,257]. Lead and mercury appear to occur in algae at levels that are safe for human consumption [259,260]. In contrast, arsenic and cadmium have been identified in algae at levels above the legal limits, with *H. fusiformis* in particular containing a high content of these toxic heavy metals [259,260]. Microalgae do not appear to exceed legal levels of heavy metals [261].

The lack of regulation for algae as a dietary supplement is of additional concern, leading to large variations between different batches from the same producer [262]. This has led to concerns being raised about quality and contamination at the production and processing stages [263]. Downstream applications for which the algal proteins are intended can play a role in determining processing methods. For example, food grade reagents and processes may be required for proteins intended for human consumption, whereas this may not be such an issue with proteins intended for commercial applications or animal feed [264]. Furthermore, any reagents used for the production or extraction of algal proteins must also be environmentally friendly with minimum impact on local ecosystems [265]. Following protein extraction, greater emphasis must additionally be placed on protein purification, with salt removal and buffer exchange being important steps [266]. Membrane technologies, such as ultrafiltration and nanofiltration, have already been implemented in this context within the dairy industry, although these technologies could also be used for other food applications, including algal proteins [267].

5.6. Price

A critical factor that will determine the commercial viability of microalgae and seaweed is their competitiveness compared to other sources on the market. For example, a major application for the production of microalgae is the extraction of lipids to be converted to biofuel. However, while they are a greener alternative for the environment, they are still not competitive compared to fossil-based petroleum fuels [266]. Algae competitiveness could be further increased by taking a holistic view and maximising the extraction of all available high-value components by cascading biorefinery. Conventional methods of algal protein extraction are often quite wasteful as they typically dispose of the algae by-products following processing [88]. Seaweed is ideally suited for cascading biorefinery because it contains many high-value components as well as bulky low-value components that are considered raw materials for the bio-based economy, such as xylose and glucose [268]. As several studies have shown, the by-products of algae can still have many applications that could be of economic value [209,215,224]. Further research is needed to develop new methods that will enable algal protein production, extraction and processing costs to be lowered.

6. Discussions

Seaweed and microalgae are generally considered to be a promising source of nutrition, rich in vitamins, minerals, and protein. Some species have even been reported to display protein at concentrations greater than traditional plant and animal protein sources [28]. Within these protein sequences, bioactive peptides have been identified with beneficial properties for human health, including antioxidant, anticancer, anti-hypertensive, immunosuppressive, anti-atherosclerosis, and hepatoprotective effects [198]. Algae have also been reported to contain high quality proteins in terms of EAA content [3], as well as displaying in vitro protein digestibility that is comparable to commonly consumed plants [67]. However, there are conflicting reports about the bioavailability of algal proteins when examined in vivo. Several studies have demonstrated that consumption of raw, unprocessed seaweed can result in reduced growth [43,44]. In contrast, other studies have reported opposite findings, with seaweed proteins suggested as being a beneficial dietary supplement for improved animal growth, meat quality, and valorisation [155,161,162]. Furthermore, microalgae have long been used as the basal diet in the rearing of animals in aquaculture, especially molluscs [138].

One of the major reasons for reduced protein digestibility was suggested to be due to the high fibre content that makes up the algal cell wall [46]. Algae typically contain high levels of polysaccharides, although this can vary significantly in seaweed (4%–76% dw) and microalgae (8%–64% dw) according to species and time of harvest [22,28]. Similarly, phenolic compounds can react with amino acids to form insoluble compounds [65]. Great emphasis has therefore been placed on researching methods for breaking down of the anionic cell wall in order to allow the release of the valuable intracellular contents [13].

Conventional protein extraction methods, such as enzymatic hydrolysis and physical processes, are laborious, time-consuming, and may require the use of solvents [13]. Novel protein extraction methods, such as UAE, PEF, and MAE, have thus been investigated as a means of overcoming these limitations. UAE is a cost-effective and low-solvent consumption method for extraction of high-value compounds. UAE is already widely used at commercial scale in food processing and could therefore also be implemented for the extraction of algal proteins [81]. PEF for the extraction of intracellular algal components is still in its infancy, having predominantly been used for lipid extraction [93]. Nevertheless, its lack of solvent consumption or heat gives it potential for the future extraction of algal proteins [100]. MAE has been poorly studied in algae, although the high temperature involved would likely make it unsuitable for extraction of proteins [81]. Alternatively, membrane filtration technologies, such as those used in the dairy industry, could also be considered a gentle method for protein enrichment when used in conjunction with cell disruption methods. MF has already been demonstrated for harvesting whole microalgal cells, while UF was reported for the extraction of algal polysaccharides [106,107]. However, there remains to be a distinct lack of studies investigating the use of membrane technologies in the enrichment of algal proteins.

In conclusion, algae have been reported as a valuable source of nutrition. However, there are inconclusive reports about the digestibility of algae and the bioavailability of the proteins within. More research is needed to investigate the bioavailability of algal proteins in vivo. Cell disruption techniques will likely play an essential role in the successful extraction and enrichment of algal protein ingredients at a commercial scale. The most promising methods that have been used in algae to date are UAE and PEF, both of which are non-thermal, require low solvent consumption, and have been implemented at commercial scale for the extraction of compounds in other food processes. Alternatively, membrane technologies are a method for isolating compounds which show great promise, but have yet to be investigated sufficiently in algae.

Acknowledgments: Stephen Bleakley is in receipt of a Teagasc Walsh Fellowship (Grant No. 2016073). This work forms part of the BioAlgae project funded by the Teagasc (Grant No. NFNY6889-142).

Conflicts of Interest: The authors declare no conflict of interest.

Abbreviations

EAA(s)	essential amino acid(s)
FAO	Food and Agriculture Organisation of the United Nations
TNOASR	The Netherland Organisation for applied scientific research
TIM	TNOASR's Intestinal Model
UAE	ultrasound-assisted extraction
PEF	pulsed electric field
MAE	microwave-assisted extraction
MF	microfiltration
UF	ultrafiltration
WHO	World Health Organisation
ACE-I	angiotensin-I-converting enzyme
RAAS	renin-angiotensin-aldosterone system
SHR(s)	spontaneously hypertensive rat(s)
SBP	systolic blood pressure

References

1. Godfray, H.C.J.; Beddington, J.R.; Crute, I.R.; Haddad, L.; Lawrence, D.; Muir, J.F.; Pretty, J.; Robinson, S.; Thomas, S.M.; Toulmin, C. Food security: The challenge of feeding 9 billion people. *Science* **2010**, *327*, 812–818. [CrossRef] [PubMed]
2. Tilman, D.; Balzer, C.; Hill, J.; Befort, B.L. Global food demand and the sustainable intensification of agriculture. *Proc. Natl. Acad. Sci. USA* **2011**, *108*, 20260–20264. [CrossRef] [PubMed]
3. Fleurence, J. Seaweed proteins: Biochemical, nutritional aspects and potential uses. *Trends Food Sci. Technol.* **1999**, *10*, 25–28. [CrossRef]
4. Gouveia, L.; Batista, A.P.; Sousa, I.; Raymundo, A.; Bandarra, N. Microalgae in Novel Food Products. In *Food Chemistry Research Development*; Konstantinos, N., Papadopoulos, P.P., Eds.; Nova Science Publishers: New York, NY, USA, 2008; pp. 75–112.
5. Van Krimpen, M.; Bikker, P.; Van der Meer, I.; Van der Peet-Schwering, C.; Vereijken, J. *Cultivation, Processing and Nutritional Aspects for Pigs and Poultry of European Protein Sources as Alternatives for Imported Soybean Products*; Wageningen UR Livestock Research: Lelystad, The Netherlands, 2013; p. 48.
6. Wallace, J. Increasing agricultural water use efficiency to meet future food production. *Agric. Ecosyst. Environ.* **2000**, *82*, 105–119. [CrossRef]
7. Pimentel, D.; Pimentel, M. Sustainability of meat-based and plant-based diets and the environment. *Am. J. Clin. Nutr.* **2003**, *78*, 660S–663S. [PubMed]

8. Sampath-Wiley, P.; Neefus, C.D.; Jahnke, L.S. Seasonal effects of sun exposure and emersion on intertidal seaweed physiology: Fluctuations in antioxidant contents, photosynthetic pigments and photosynthetic efficiency in the red alga *Porphyra umbilicalis* kützing (Rhodophyta, Bangiales). *J. Exp. Mar. Biol. Ecol.* **2008**, *361*, 83–91. [CrossRef]
9. Pulz, O.; Gross, W. Valuable products from biotechnology of microalgae. *Appl. Microbiol. Biotechnol.* **2004**, *65*, 635–648. [CrossRef] [PubMed]
10. Wang, T.; Jonsdottir, R.; Ólafsdóttir, G. Total phenolic compounds, radical scavenging and metal chelation of extracts from icelandic seaweeds. *Food Chem.* **2009**, *116*, 240–248. [CrossRef]
11. Wijffels, R.H.; Barbosa, M.J. An outlook on microalgal biofuels. *Science* **2010**, *329*, 796–799. [CrossRef] [PubMed]
12. Cavalier-Smith, T. Evolution and relationships. In *Unravelling the Algae: The Past, Present, and Future of Algal Systematics*; Brodie, J., Ed.; CRC Press: Boca Raton, FL, USA, 2007; p. 21.
13. Kadam, S.U.; Tiwari, B.K.; O'Donnell, C.P. Application of novel extraction technologies for bioactives from marine algae. *J. Agric. Food Chem.* **2013**, *61*, 4667–4675. [CrossRef] [PubMed]
14. Food and Agriculture Organization. *Food Balance Sheets*; Food and Agriculture Organisation of the United Nations: Rome, Italy, 2013.
15. Food and Agriculture Organization. *The State of World Fisheries and Aquaculture: Opportunties and Challenges*; Food and Agriculture Organisation of the United Nations: Rome, Italy, 2014.
16. Cock, J.M.; Sterck, L.; Rouzé, P.; Scornet, D.; Allen, A.E.; Amoutzias, G.; Anthouard, V.; Artiguenave, F.; Aury, J.-M.; Badger, J.H. The ectocarpus genome and the independent evolution of multicellularity in brown algae. *Nature* **2010**, *465*, 617–621. [CrossRef] [PubMed]
17. Hoek, C.; Mann, D.; Jahns, H.M. *Algae: An Introduction to Phycology*; Cambridge University Press: Cambridge, UK, 1995.
18. Foster, M.S.; Schiel, D.R. *Ecology of Giant Kelp Forests in California: A Community Profile*; San Jose State University, Moss Landing Marine Labs.: Moss Landing, CA, USA, 1985.
19. Fitzgerald, C.; Gallagher, E.; Tasdemir, D.; Hayes, M. Heart health peptides from macroalgae and their potential use in functional foods. *J. Agric. Food Chem.* **2011**, *59*, 6829–6836. [CrossRef] [PubMed]
20. Karol, K.G.; McCourt, R.M.; Cimino, M.T.; Delwiche, C.F. The closest living relatives of land plants. *Science* **2001**, *294*, 2351–2353. [CrossRef] [PubMed]
21. McCourt, R.M.; Delwiche, C.F.; Karol, K.G. Charophyte algae and land plant origins. *Trends Ecol. Evol.* **2004**, *19*, 661–666. [CrossRef] [PubMed]
22. Holdt, S.L.; Kraan, S. Bioactive compounds in seaweed: Functional food applications and legislation. *J. Appl. Phycol.* **2011**, *23*, 543–597. [CrossRef]
23. Mouritsen, O.G.; Dawczynski, C.; Duelund, L.; Jahreis, G.; Vetter, W.; Schröder, M. On the human consumption of the red seaweed dulse (*Palmaria palmata* (L.) weber & amp; mohr). *J. Appl. Phycol.* **2013**, *25*, 1777–1791.
24. Norton, T.A.; Melkonian, M.; Andersen, R.A. Algal biodiversity. *Phycologia* **1996**, *35*, 308–326. [CrossRef]
25. Chacón-Lee, T.; González-Mariño, G. Microalgae for "healthy" foods—Possibilities and challenges. *Compr. Rev. Food Sci. Food Saf.* **2010**, *9*, 655–675. [CrossRef]
26. Brennan, L.; Owende, P. Biofuels from microalgae—A review of technologies for production, processing, and extractions of biofuels and co-products. *Renew. Sustain. Energy Rev.* **2010**, *14*, 557–577. [CrossRef]
27. Capelli, B.; Cysewski, G.R. Potential health benefits of *Spirulina* microalgae. *Nutrafoods* **2010**, *9*, 19–26. [CrossRef]
28. Becker, E. Micro-algae as a source of protein. *Biotechnol. Adv.* **2007**, *25*, 207–210. [CrossRef] [PubMed]
29. Rodriguez-Garcia, I.; Guil-Guerrero, J.L. Evaluation of the antioxidant activity of three microalgal species for use as dietary supplements and in the preservation of foods. *Food Chem.* **2008**, *108*, 1023–1026. [CrossRef] [PubMed]
30. Boisen, S.; Eggum, B. Critical evaluation of in vitro methods for estimating digestibility in simple-stomach animals. *Nutr. Res. Rev.* **1991**, *4*, 141–162. [CrossRef] [PubMed]
31. Young, V.R.; Pellett, P.L. Plant proteins in relation to human protein and amino acid nutrition. *Am. J. Clin. Nutr.* **1994**, *59*, 1203S–1212S. [PubMed]
32. Hoffman, J.R.; Falvo, M.J. Protein-which is best. *J. Sports Sci. Med.* **2004**, *3*, 118–130. [PubMed]

33. Food and Agriculture Organization; World Health Organization. *Protein Quality Evaluation—Report of Joint FAO/WHO Expert Consultation*; FAO: Rome, Italy, 1991.
34. Mišurcová, L.; Kráčmar, S.; Klejdus, B.; Vacek, J. Nitrogen content, dietary fiber, and digestibility in algal food products. *Czech J. Food Sci.* **2010**, *28*, 27–35.
35. Dawczynski, C.; Schubert, R.; Jahreis, G. Amino acids, fatty acids, and dietary fibre in edible seaweed products. *Food Chem.* **2007**, *103*, 891–899. [CrossRef]
36. Kolb, N.; Vallorani, L.; Stocchi, V. Chemical composition and evaluation of protein quality by amino acid score method of edible brown marine algae Arame (*Eisenia bicyclis*) and Hijiki (*Hijikia fusiforme*). *Acta Aliment.* **1999**, *28*, 213–222. [CrossRef]
37. Volkmann, H.; Imianovsky, U.; Oliveira, J.L.; Sant'Anna, E.S. Cultivation of arthrospira (*Spirulina*) platensis in desalinator wastewater and salinated synthetic medium: Protein content and amino-acid profile. *Braz. J. Microbiol.* **2008**, *39*, 98–101. [CrossRef] [PubMed]
38. Mišurcová, L.; Buňka, F.; Ambrožová, J.V.; Machů, L.; Samek, D.; Kráčmar, S. Amino acid composition of algal products and its contribution to RDI. *Food Chem.* **2014**, *151*, 120–125. [CrossRef] [PubMed]
39. Kakinuma, M.; Park, C.S.; Amano, H. Distribution of free L-cysteine and glutathione in seaweeds. *Fish. Sci.* **2001**, *67*, 194–196. [CrossRef]
40. MacArtain, P.; Gill, C.I.; Brooks, M.; Campbell, R.; Rowland, I.R. Nutritional value of edible seaweeds. *Nutr. Rev.* **2007**, *65*, 535–543. [CrossRef] [PubMed]
41. Guerra, A.; Etienne-Mesmin, L.; Livrelli, V.; Denis, S.; Blanquet-Diot, S.; Alric, M. Relevance and challenges in modeling human gastric and small intestinal digestion. *Trends Biotechnol.* **2012**, *30*, 591–600. [CrossRef] [PubMed]
42. Ekmekcioglu, C. A physiological approach for preparing and conducting intestinal bioavailability studies using experimental systems. *Food Chem.* **2002**, *76*, 225–230. [CrossRef]
43. Urbano, M.G.; Goñi, I. Bioavailability of nutrients in rats fed on edible seaweeds, nori (*Porphyra tenera*) and wakame (*Undaria pinnatifida*), as a source of dietary fibre. *Food Chem.* **2002**, *76*, 281–286. [CrossRef]
44. Suzuki, T.; Nakai, K.; Yoshie, Y.; Shirai, T.; Hirano, T. Digestibility of dietary fiber in brown Alga, Kombu, by rats. *Bull. Jpn. Soc. Sci. Fish* **1993**, *59*, 879–884. [CrossRef]
45. Wong, K.; Cheung, P.C. Influence of drying treatment on three *Sargassum* species. *J. Appl. Phycol.* **2001**, *13*, 43–50. [CrossRef]
46. Joubert, Y.; Fleurence, J. Simultaneous extraction of proteins and DNA by an enzymatic treatment of the cell wall of *Palmaria palmata* (Rhodophyta). *J. Appl. Phycol.* **2008**, *20*, 55–61. [CrossRef]
47. Carbonell-Capella, J.M.; Buniowska, M.; Barba, F.J.; Esteve, M.J.; Frígola, A. Analytical methods for determining bioavailability and bioaccessibility of bioactive compounds from fruits and vegetables: A review. *Compr. Rev. Food Sci. Food Saf.* **2014**, *13*, 155–171. [CrossRef]
48. Etcheverry, P.; Grusak, M.A.; Fleige, L.E. Application of in vitro bioaccessibility and bioavailability methods for calcium, carotenoids, folate, iron, magnesium, polyphenols, zinc, and vitamins B6, B12, D, and E. *Front. Physiol.* **2012**, *3*, 317. [CrossRef] [PubMed]
49. Alegría, A.; Garcia-Llatas, G.; Cilla, A. Static digestion models: General introduction. In *The Impact of Food Bioactives on Health*; Springer: Heidelberg, Germany, 2015; pp. 3–12.
50. Fernández-García, E.; Carvajal-Lérida, I.; Pérez-Gálvez, A. In vitro bioaccessibility assessment as a prediction tool of nutritional efficiency. *Nutr. Res.* **2009**, *29*, 751–760. [CrossRef] [PubMed]
51. Butts, C.A.; Monro, J.A.; Moughan, P.J. In vitro determination of dietary protein and amino acid digestibility for humans. *Br. J. Nutr.* **2012**, *108*, S282. [CrossRef] [PubMed]
52. Hur, S.J.; Lim, B.O.; Decker, E.A.; McClements, D.J. In vitro human digestion models for food applications. *Food Chem.* **2011**, *125*, 1–12. [CrossRef]
53. Minekus, M.; Marteau, P.; Havenaar, R. Multicompartmental dynamic computer-controlled model simulating the stomach and small intestine. *Altern. Lab. Anim. ATLA* **1995**, *23*, 197–209.
54. Minekus, M.; Smeets-Peeters, M.; Bernalier, A.; Marol-Bonnin, S.; Havenaar, R.; Marteau, P.; Alric, M.; Fonty, G. A computer-controlled system to simulate conditions of the large intestine with peristaltic mixing, water absorption and absorption of fermentation products. *Appl. Microbiol. Biotechnol.* **1999**, *53*, 108–114. [CrossRef] [PubMed]

55. Barmpalia-Davis, I.M.; Geornaras, I.; Kendall, P.A.; Sofos, J.N. Differences in survival among 13 listeria monocytogenes strains in a dynamic model of the stomach and small intestine. *Appl. Environ. Microbiol.* **2008**, *74*, 5563–5567. [CrossRef] [PubMed]
56. Van den Abbeele, P.; Grootaert, C.; Marzorati, M.; Possemiers, S.; Verstraete, W.; Gérard, P.; Rabot, S.; Bruneau, A.; El Aidy, S.; Derrien, M. Microbial community development in a dynamic gut model is reproducible, colon region specific, and selective for Bacteroidetes and Clostridium cluster IX. *Appl. Environ. Microbiol.* **2010**, *76*, 5237–5246. [CrossRef] [PubMed]
57. Vardakou, M.; Mercuri, A.; Naylor, T.; Rizzo, D.; Butler, J.; Connolly, P.; Wickham, M.; Faulks, R. Predicting the human in vivo performance of different oral capsule shell types using a novel in vitro dynamic gastric model. *Int. J. Pharm.* **2011**, *419*, 192–199. [CrossRef] [PubMed]
58. Curto, A.L.; Pitino, I.; Mandalari, G.; Dainty, J.R.; Faulks, R.M.; Wickham, M.S.J. Survival of probiotic lactobacilli in the upper gastrointestinal tract using an in vitro gastric model of digestion. *Food Microbiol.* **2011**, *28*, 1359–1366. [CrossRef] [PubMed]
59. Dupont, D.; Bordoni, A.; Brodkorb, A.; Capozzi, F.; Cirkovic Velickovic, T.; Corredig, M.; Cotter, P.D.; de Noni, I.; Gaudichon, C.; Golding, M. An international network for improving health properties of food by sharing our knowledge on the digestive process. *Food Dig.* **2011**, *2*, 23–25. [CrossRef]
60. Minekus, M.; Alminger, M.; Alvito, P.; Ballance, S.; Bohn, T.; Bourlieu, C.; Carriere, F.; Boutrou, R.; Corredig, M.; Dupont, D. A standardised static in vitro digestion method suitable for food—An international consensus. *Food Funct.* **2014**, *5*, 1113–1124. [CrossRef] [PubMed]
61. Egger, L.; Menard, O.; Delgado-Andrade, C.; Alvito, P.; Assunção, R.; Balance, S.; Barberá, R.; Brodkorb, A.; Cattenoz, T.; Clemente, A.; et al. The harmonized infogest in vitro digestion method: From knowledge to action. *Food Res. Int.* **2016**, *88*, 217–225. [CrossRef]
62. Glahn, R.P.; Lee, O.A.; Yeung, A.; Goldman, M.I.; Miller, D.D. Caco-2 cell ferritin formation predicts nonradiolabeled food iron availability in an in vitro digestion/Caco-2 cell culture model. *J. Nutr.* **1998**, *128*, 1555–1561. [PubMed]
63. Rousset, M. The human colon carcinoma cell lines HT-29 and Caco-2: Two in vitro models for the study of intestinal differentiation. *Biochimie* **1986**, *68*, 1035–1040. [CrossRef]
64. Mahler, G.J.; Shuler, M.L.; Glahn, R.P. Characterization of Caco-2 and HT29-MTX cocultures in an in vitro digestion/cell culture model used to predict iron bioavailability. *J. Nutr. Biochem.* **2009**, *20*, 494–502. [CrossRef] [PubMed]
65. Wong, K.; Cheung, P.C. Nutritional evaluation of some subtropical red and green seaweeds part II. In vitro protein digestibility and amino acid profiles of protein concentrates. *Food Chem.* **2001**, *72*, 11–17. [CrossRef]
66. Bikker, P.; Krimpen, M.M.; Wikselaar, P.; Houweling-Tan, B.; Scaccia, N.; Hal, J.W.; Huijgen, W.J.; Cone, J.W.; López-Contreras, A.M. Biorefinery of the green seaweed *Ulva* lactuca to produce animal feed, chemicals and biofuels. *J. Appl. Phycol.* **2016**, *28*, 3511–3525. [CrossRef] [PubMed]
67. Tibbetts, S.M.; Milley, J.E.; Lall, S.P. Nutritional quality of some wild and cultivated seaweeds: Nutrient composition, total phenolic content and in vitro digestibility. *J. Appl. Phycol.* **2016**, *28*, 3575–3585. [CrossRef]
68. Barbarino, E.; Lourenço, S.O. An evaluation of methods for extraction and quantification of protein from marine macro-and microalgae. *J. Appl. Phycol.* **2005**, *17*, 447–460. [CrossRef]
69. Kadam, S.U.; Álvarez, C.; Tiwari, B.K.; O'Donnell, C.P. Extraction and characterization of protein from irish brown seaweed ascophyllum nodosum. *Food Res. Int.* **2016**. [CrossRef]
70. Fleurence, J.; Le Coeur, C.; Mabeau, S.; Maurice, M.; Landrein, A. Comparison of different extractive procedures for proteins from the edible seaweeds *Ulva* rigida and *Ulva* rotundata. *J. Appl. Phycol.* **1995**, *7*, 577–582. [CrossRef]
71. Jordan, P.; Vilter, H. Extraction of proteins from material rich in anionic mucilages: Partition and fractionation of vanadate-dependent bromoperoxidases from the brown algae *Laminaria* digitata and L. *Saccharina* in aqueous polymer two-phase systems. *Biochim. Biophys. Acta (BBA) Gen. Subj.* **1991**, *1073*, 98–106. [CrossRef]
72. Fleurence, J. The enzymatic degradation of algal cell walls: A useful approach for improving protein accessibility? *J. Appl. Phycol.* **1999**, *11*, 313–314. [CrossRef]
73. Harnedy, P.A.; FitzGerald, R.J. Extraction of protein from the macroalga *Palmaria palmata*. *LWT Food Sci. Technol.* **2013**, *51*, 375–382. [CrossRef]

74. Fleurence, J.; Massiani, L.; Guyader, O.; Mabeau, S. Use of enzymatic cell wall degradation for improvement of protein extraction from chondrus crispus, gracilaria verrucosa and *Palmaria palmata*. *J. Appl. Phycol.* **1995**, *7*, 393–397. [CrossRef]
75. Marrion, O.; Schwertz, A.; Fleurence, J.; Gueant, J.L.; Villaume, C. Improvement of the digestibility of the proteins of the red alga *Palmaria palmata* by physical processes and fermentation. *Mol. Nutr. Food Res.* **2003**, *47*, 339–344.
76. Maehre, H.K.; Jensen, I.-J.; Eilertsen, K.-E. Enzymatic pre-treatment increases the protein bioaccessibility and extractability in Dulse (*Palmaria palmata*). *Mar. Drugs* **2016**, *14*, 196. [CrossRef] [PubMed]
77. Ganeva, V.; Galutzov, B.; Teissié, J. High yield electroextraction of proteins from yeast by a flow process. *Anal. Biochem.* **2003**, *315*, 77–84. [CrossRef]
78. Vilkhu, K.; Mawson, R.; Simons, L.; Bates, D. Applications and opportunities for ultrasound assisted extraction in the food industry—A review. *Innov. Food Sci. Emerg. Technol.* **2008**, *9*, 161–169. [CrossRef]
79. Ashokkumar, M.; Sunartio, D.; Kentish, S.; Mawson, R.; Simons, L.; Vilkhu, K.; Versteeg, C.K. Modification of food ingredients by ultrasound to improve functionality: A preliminary study on a model system. *Innov. Food Sci. Emerg. Technol.* **2008**, *9*, 155–160. [CrossRef]
80. Mason, T.; Paniwnyk, L.; Lorimer, J. The uses of ultrasound in food technology. *Ultrason. Sonochem.* **1996**, *3*, S253–S260. [CrossRef]
81. Chemat, F.; Khan, M.K. Applications of ultrasound in food technology: Processing, preservation and extraction. *Ultrason. Sonochem.* **2011**, *18*, 813–835. [CrossRef] [PubMed]
82. Barba, F.J.; Grimi, N.; Vorobiev, E. New approaches for the use of non-conventional cell disruption technologies to extract potential food additives and nutraceuticals from microalgae. *Food Eng. Rev.* **2015**, *7*, 45–62. [CrossRef]
83. Parniakov, O.; Apicella, E.; Koubaa, M.; Barba, F.; Grimi, N.; Lebovka, N.; Pataro, G.; Ferrari, G.; Vorobiev, E. Ultrasound-assisted green solvent extraction of high-added value compounds from microalgae *Nannochloropsis* spp. *Bioresour. Technol.* **2015**, *198*, 262–267. [CrossRef] [PubMed]
84. Keris-Sen, U.D.; Sen, U.; Soydemir, G.; Gurol, M.D. An investigation of ultrasound effect on microalgal cell integrity and lipid extraction efficiency. *Bioresour. Technol.* **2014**, *152*, 407–413. [CrossRef] [PubMed]
85. Janczyk, P.; Wolf, C.; Souffrant, W.B. Evaluation of nutritional value and safety of the green microalgae *Chlorella* vulgaris treated with novel processing methods. *Arch Zootech* **2005**, *8*, 132–147.
86. Janczyk, P.; Halle, B.; Souffrant, W. Microbial community composition of the crop and ceca contents of laying hens fed diets supplemented with *Chlorella* vulgaris. *Poult. Sci.* **2009**, *88*, 2324–2332. [CrossRef] [PubMed]
87. Qu, W.; Ma, H.; Wang, T.; Zheng, H. Alternating two-frequency countercurrent ultrasonic-assisted extraction of protein and polysaccharide from *Porphyra* yezoensis. *Trans. Chin. Soc. Agric. Eng.* **2013**, *29*, 285–292.
88. Vanthoor-Koopmans, M.; Wijffels, R.H.; Barbosa, M.J.; Eppink, M.H. Biorefinery of microalgae for food and fuel. *Bioresour. Technol.* **2013**, *135*, 142–149. [CrossRef] [PubMed]
89. Fox, M.; Esveld, D.; Valero, A.; Luttge, R.; Mastwijk, H.; Bartels, P.; Van Den Berg, A.; Boom, R. Electroporation of cells in microfluidic devices: A review. *Anal. Bioanal. Chem.* **2006**, *385*, 474–485. [CrossRef] [PubMed]
90. Grahl, T.; Märkl, H. Killing of microorganisms by pulsed electric fields. *Appl. Microbiol. Biotechnol.* **1996**, *45*, 148–157. [CrossRef] [PubMed]
91. Corrales, M.; Toepfl, S.; Butz, P.; Knorr, D.; Tauscher, B. Extraction of anthocyanins from grape by-products assisted by ultrasonics, high hydrostatic pressure or pulsed electric fields: A comparison. *Innov. Food Sci. Emerg. Technol.* **2008**, *9*, 85–91. [CrossRef]
92. Joannes, C.; Sipaut, C.S.; Dayou, J.; Yasir, S.M.; Mansa, R.F. The potential of using pulsed electric field (pef) technology as the cell disruption method to extract lipid from microalgae for biodiesel production. *Int. J. Renew. Energy Res.* **2015**, *5*, 598–621.
93. Goettel, M.; Eing, C.; Gusbeth, C.; Straessner, R.; Frey, W. Pulsed electric field assisted extraction of intracellular valuables from microalgae. *Algal Res.* **2013**, *2*, 401–408. [CrossRef]
94. Zbinden, M.D.A.; Sturm, B.S.; Nord, R.D.; Carey, W.J.; Moore, D.; Shinogle, H.; Stagg-Williams, S.M. Pulsed electric field (PEF) as an intensification pretreatment for greener solvent lipid extraction from microalgae. *Biotechnol. Bioeng.* **2013**, *110*, 1605–1615. [CrossRef] [PubMed]
95. Lai, Y.S.; Parameswaran, P.; Li, A.; Baez, M.; Rittmann, B.E. Effects of pulsed electric field treatment on enhancing lipid recovery from the microalga, scenedesmus. *Bioresour. Technol.* **2014**, *173*, 457–461. [CrossRef] [PubMed]

96. Postma, P.; Pataro, G.; Capitoli, M.; Barbosa, M.; Wijffels, R.H.; Eppink, M.; Olivieri, G.; Ferrari, G. Selective extraction of intracellular components from the microalga *Chlorella* vulgaris by combined pulsed electric field–temperature treatment. *Bioresour. Technol.* **2016**, *203*, 80–88. [CrossRef] [PubMed]
97. Luengo, E.; Martínez, J.M.; Coustets, M.; Álvarez, I.; Teissié, J.; Rols, M.-P.; Raso, J. A comparative study on the effects of millisecond-and microsecond-pulsed electric field treatments on the permeabilization and extraction of pigments from *Chlorella* vulgaris. *J. Membr. Biol.* **2015**, *248*, 883–891. [CrossRef] [PubMed]
98. Parniakov, O.; Barba, F.J.; Grimi, N.; Marchal, L.; Jubeau, S.; Lebovka, N.; Vorobiev, E. Pulsed electric field assisted extraction of nutritionally valuable compounds from microalgae *Nannochloropsis* spp. Using the binary mixture of organic solvents and water. *Innov. Food Sci. Emerg. Technol.* **2015**, *27*, 79–85. [CrossRef]
99. Töpfl, S. *Pulsed Electric Fields (Pef) for Permeabilization of Cell Membranes in Food-and Bioprocessing–Applications, Process and Equipment Design and Cost Analysis*; Berlin University of Technology: Berlin, Germany, 2006.
100. Coustets, M.; Al-Karablieh, N.; Thomsen, C.; Teissié, J. Flow process for electroextraction of total proteins from microalgae. *J. Membr. Biol.* **2013**, *246*, 751–760. [CrossRef] [PubMed]
101. Passos, F.; Carretero, J.; Ferrer, I. Comparing pretreatment methods for improving microalgae anaerobic digestion: Thermal, hydrothermal, microwave and ultrasound. *Chem. Eng. J.* **2015**, *279*, 667–672. [CrossRef]
102. Herrero, M.; Cifuentes, A.; Ibanez, E. Sub-and supercritical fluid extraction of functional ingredients from different natural sources: Plants, food-by-products, algae and microalgae: A review. *Food Chem.* **2006**, *98*, 136–148. [CrossRef]
103. Kumar, P.; Sharma, N.; Ranjan, R.; Kumar, S.; Bhat, Z.; Jeong, D.K. Perspective of membrane technology in dairy industry: A review. *Asian-Australas. J. Anim. Sci.* **2013**, *26*, 1347. [CrossRef] [PubMed]
104. Yaich, H.; Garna, H.; Besbes, S.; Paquot, M.; Blecker, C.; Attia, H. Chemical composition and functional properties of *Ulva lactuca* seaweed collected in Tunisia. *Food Chem.* **2011**, *128*, 895–901. [CrossRef]
105. Pafylias, I.; Cheryan, M.; Mehaia, M.; Saglam, N. Microfiltration of milk with ceramic membranes. *Food Res. Int.* **1996**, *29*, 141–146. [CrossRef]
106. Petrusevski, B.; Bolier, G.; Van Breemen, A.; Alaerts, G. Tangential flow filtration: A method to concentrate freshwater algae. *Water Res.* **1995**, *29*, 1419–1424. [CrossRef]
107. Ye, H.; Wang, K.; Zhou, C.; Liu, J.; Zeng, X. Purification, antitumor and antioxidant activities in vitro of polysaccharides from the brown seaweed *Sargassum* pallidum. *Food Chem.* **2008**, *111*, 428–432. [CrossRef] [PubMed]
108. Shao, P.; Chen, M.; Pei, Y.; Sun, P. In intro antioxidant activities of different sulfated polysaccharides from chlorophytan seaweeds *Ulva* fasciata. *Int. J. Biol. Macromol.* **2013**, *59*, 295–300. [CrossRef] [PubMed]
109. Denis, C.; Massé, A.; Fleurence, J.; Jaouen, P. Concentration and pre-purification with ultrafiltration of a r-phycoerythrin solution extracted from macro-algae grateloupia turuturu: Process definition and up-scaling. *Sep. Purif. Technol.* **2009**, *69*, 37–42. [CrossRef]
110. Safi, C.; Liu, D.Z.; Yap, B.H.; Martin, G.J.; Vaca-Garcia, C.; Pontalier, P.-Y. A two-stage ultrafiltration process for separating multiple components of *Tetraselmis* suecica after cell disruption. *J. Appl. Phycol.* **2014**, *26*, 2379–2387. [CrossRef]
111. American College of Sports Medicine; American Dietetic Association; Dietitians of Canada. Joint Position Statement: Nutrition and athletic performance. American College of Sports Medicine, American Dietetic Association, and Dietitians of Canada. *Med. Sci. Sports Exerc.* **2000**, *32*, 2130–2145.
112. Fleurence, J.; Morançais, M.; Dumay, J.; Decottignies, P.; Turpin, V.; Munier, M.; Garcia-Bueno, N.; Jaouen, P. What are the prospects for using seaweed in human nutrition and for marine animals raised through aquaculture? *Trends Food Sci. Technol.* **2012**, *27*, 57–61. [CrossRef]
113. Du, J. Research on functionality sports nutrition and health food security issues based on *Circulation. Open Cybern. System. J.* **2015**, *9*, 1945–1949. [CrossRef]
114. Chakdar, H.; Jadhav, S.D.; Dhar, D.W.; Pabbi, S. Potential applications of blue green algae. *J. Sci. Ind. Res.* **2012**, *71*, 13–20.
115. Tokuşoglu, Ö.; Üunal, M. Biomass nutrient profiles of three microalgae: *Spirulina platensis*, *Chlorella vulgaris*, and *Isochrisis galbana*. *J. Food Sci.* **2003**, *68*, 1144–1148. [CrossRef]
116. Liang, S.; Liu, X.; Chen, F.; Chen, Z. Current Microalgal Health Food R & D Activities in China. In *Asian Pacific Phycology in the 21st Century: Prospects and Challenges*; Springer: Heidelberg, Germany, 2004; pp. 45–48.
117. Marcus, J.B. Enhancing umami taste in foods. In *Modifying Flavour in Food*; Taylor, A., Hort, J., Eds.; CRC Press: Cambridge, UK, 2007; pp. 202–220.

118. Prabhasankar, P.; Ganesan, P.; Bhaskar, N.; Hirose, A.; Stephen, N.; Gowda, L.R.; Hosokawa, M.; Miyashita, K. Edible *Japanese* seaweed, wakame (*Undaria pinnatifida*) as an ingredient in pasta: Chemical, functional and structural evaluation. *Food Chem.* **2009**, *115*, 501–508. [CrossRef]
119. Hall, A.; Fairclough, A.; Mahadevan, K.; Paxman, J. Ascophyllum nodosum enriched bread reduces subsequent energy intake with no effect on post-prandial glucose and cholesterol in healthy, overweight males. A pilot study. *Appetite* **2012**, *58*, 379–386. [CrossRef] [PubMed]
120. Fitzgerald, C.; Gallagher, E.; Doran, L.; Auty, M.; Prieto, J.; Hayes, M. Increasing the health benefits of bread: Assessment of the physical and sensory qualities of bread formulated using a renin inhibitory *Palmaria palmata* protein hydrolysate. *LWT-Food Sci. Technol.* **2014**, *56*, 398–405. [CrossRef]
121. Spolaore, P.; Joannis-Cassan, C.; Duran, E.; Isambert, A. Commercial applications of microalgae. *J. Biosci. Bioeng.* **2006**, *101*, 87–96. [CrossRef] [PubMed]
122. Sajilata, M.; Singhal, R.; Kamat, M. Fractionation of lipids and purification of γ-linolenic acid (GLA) from *Spirulina platensis*. *Food Chem.* **2008**, *109*, 580–586. [CrossRef]
123. Khan, Z.; Bhadouria, P.; Bisen, P. Nutritional and therapeutic potential of *Spirulina*. *Curr. Pharm. Biotechnol.* **2005**, *6*, 373–379. [CrossRef] [PubMed]
124. Yamaguchi, K. Recent advances in microalgal bioscience in Japan, with special reference to utilization of biomass and metabolites: A review. *J. Appl. Phycol.* **1996**, *8*, 487–502. [CrossRef]
125. Yaakob, Z.; Ali, E.; Zainal, A.; Mohamad, M.; Takriff, M.S. An overview: *Biomolecules* from microalgae for animal feed and aquaculture. *J. Biol. Res. Thessalon.* **2014**, *21*, 1. [CrossRef] [PubMed]
126. Hayes, M. Biological Activities of Proteins and Marine-Derived Peptides from Byproducts and Seaweeds. In *Marine Proteins and Peptides: Biological Activities and Applications*; Kim, S.-K., Ed.; John Wiley & Sons, Ltd.: Chichester, UK, 2013; pp. 139–165.
127. Wang, G.; Sun, H.; Fan, X.; Tseng, C. Large-scale isolation and purification of R-phycoerythrin from red alga *Palmaria palmata* using the expanded bed adsorption method. *Acta Bot. Sin.* **2001**, *44*, 541–546.
128. Kawakubo, A.; Makino, H.; Ohnishi, J.-I.; Hirohara, H.; Hori, K. The marine red alga *Eucheuma serra* J. Agardh, a high yielding source of two isolectins. *J. Appl. Phycol.* **1997**, *9*, 331–338. [CrossRef]
129. Weis, W.I.; Drickamer, K. Structural basis of lectin-carbohydrate recognition. *Annu. Rev. Biochem.* **1996**, *65*, 441–473. [CrossRef] [PubMed]
130. Ziółkowska, N.E.; Wlodawer, A. Structural studies of algal lectins with anti-HIV activity. *Acta Biochim. Pol.* **2006**, *53*, 617–626. [PubMed]
131. Naeem, A.; Saleemuddin, M.; Hasan Khan, R. Glycoprotein targeting and other applications of lectins in biotechnology. *Curr. Protein Pept. Sci.* **2007**, *8*, 261–271. [CrossRef] [PubMed]
132. Harnedy, P.A.; FitzGerald, R.J. Bioactive *Proteins, Peptides*, and *Amino Acids* from macroalgae. *J. Phycol.* **2011**, *47*, 218–232. [CrossRef] [PubMed]
133. Glazer, A.N. Phycobiliproteins—A family of valuable, widely used fluorophores. *J. Appl. Phycol.* **1994**, *6*, 105–112. [CrossRef]
134. De Marsac, N.T. Phycobiliproteins and phycobilisomes: The early observations. *Photosynth. Res.* **2003**, *76*, 193–205. [CrossRef] [PubMed]
135. Sekar, S.; Chandramohan, M. Phycobiliproteins as a commodity: Trends in applied research, patents and commercialization. *J. Appl. Phycol.* **2008**, *20*, 113–136. [CrossRef]
136. Kronick, M.; Grossman, P.D. Immunoassay techniques with fluorescent phycobiliprotein conjugates. *Clin. Chem.* **1983**, *29*, 1582–1586. [PubMed]
137. Aneiros, A.; Garateix, A. Bioactive peptides from marine sources: *Pharmacological* properties and isolation procedures. *J. Chromatogr. B* **2004**, *803*, 41–53. [CrossRef] [PubMed]
138. Becker, W. Microalgae in Human and *Animal Nutrition*. In *Handbook of Microalgal Culture: Biotechnology and Applied Phycology*; Richmond, A., Ed.; Wiley-Blackwell: Hoboken, NJ, USA, 2004; p. 312.
139. Holman, B.; Malau-Aduli, A. *Spirulina* as a livestock supplement and animal feed. *J. Anim. Physiol. Anim. Nutr.* **2013**, *97*, 615–623. [CrossRef] [PubMed]
140. Evans, F.; Critchley, A. Seaweeds for animal production use. *J. Appl. Phycol.* **2014**, *26*, 891–899. [CrossRef]
141. Allen, V.; Pond, K.; Saker, K.; Fontenot, J.; Bagley, C.; Ivy, R.; Evans, R.; Brown, C.; Miller, M.; Montgomery, J. Tasco-*Forage*: III. Influence of a seaweed extract on performance, monocyte immune cell response, and carcass characteristics in feedlot-finished steers. *J. Anim. Sci.* **2001**, *79*, 1032–1040. [CrossRef] [PubMed]

142. Saker, K.; Allen, V.; Fontenot, J.; Bagley, C.; Ivy, R.; Evans, R.; Wester, D. Tasco-*Forage*: II. Monocyte immune cell response and performance of beef steers grazing tall fescue treated with a seaweed extract. *J. Anim. Sci.* **2001**, *79*, 1022–1031. [CrossRef] [PubMed]
143. Montgomery, J.; Allen, V.; Pond, K.; Miller, M.; Wester, D.; Brown, C.; Evans, R.; Bagley, C.; Ivy, R.; Fontenot, J. Tasco-*Forage*: IV. Influence of a seaweed extract applied to tall fescue pastures on sensory characteristics, shelf-life, and vitamin e status in feedlot-finished steers. *J. Anim. Sci.* **2001**, *79*, 884–894. [CrossRef] [PubMed]
144. Braden, K.; Blanton, J.; Allen, V.; Pond, K.; Miller, M. Ascophyllum nodosum supplementation: A preharvest intervention for reducing *Escherichia* Coli O157: H7 and *Salmonella* spp. In feedlot steers. *J. Food Prot.* **2004**, *67*, 1824–1828. [CrossRef]
145. Allen, V.; Pond, K.; Saker, K.; Fontenot, J.; Bagley, C.; Ivy, R.; Evans, R.; Schmidt, R.; Fike, J.; Zhang, X. Tasco: Influence of a brown seaweed on antioxidants in forages and livestock—A review. *J. Anim. Sci.* **2001**, *79*, E21–E31. [CrossRef]
146. Ginzberg, A.; Cohen, M.; Sod-Moriah, U.A.; Shany, S.; Rosenshtrauch, A.; Arad, S.M. Chickens fed with biomass of the red microalga *Porphyridium* sp. Have reduced blood cholesterol level and modified fatty acid composition in egg yolk. *J. Appl. Phycol.* **2000**, *12*, 325–330. [CrossRef]
147. Lorenz, R.T.; Cysewski, G.R. Commercial potential for *Haematococcus* microalgae as a natural source of astaxanthin. *Trends Biotechnol.* **2000**, *18*, 160–167. [CrossRef]
148. Mariey, Y.; Samak, H.; Abou-Khashba, H.; Sayed, M.; Abou-Zeid, A. Effect of using *Spirulina* platensis algae as a feed additives for poultry diets: 2 productive performance of broiler. *Egypt. Poult. Sci.* **2014**, *34*, 245–258.
149. Al-Batshan, H.A.; Al-Mufarrej, S.I.; Al-Homaidan, A.A.; Qureshi, M.A. Enhancement of chicken macrophage phagocytic function and nitrite production by dietary *Spirulina* platensis. *Immunopharmacol. Immunotoxicol.* **2001**, *23*, 281–289. [CrossRef] [PubMed]
150. Ross, E.; Dominy, W. The nutritional value of dehydrated, blue-green algae (*Spirulina* plantensis) for poultry. *Poult. Sci.* **1990**, *69*, 794–800. [CrossRef] [PubMed]
151. Zahroojian, N.; Moravej, H.; Shivazad, M. Effects of dietary marine algae (*Spirulina* platensis) on egg quality and production performance of laying hens. *J. Agric. Sci. Technol.* **2013**, *15*, 1353–1360.
152. Mariey, Y.; Samak, H.; Ibrahem, M. Effect of using *Spirulina* platensis algae as a feed additive for poultry diets. 1. Productive and reproductive performances of local laying hens. *Egypt. Poult. Sci.* **2012**, *32*, 201–215.
153. Anderson, D.W.; Tang, C.S.; Ross, E. The xanthophylls of *Spirulina* and their effect on egg-yolk pigmentationthe xanthophylls of *Spirulina* and their effect on egg-yolk pigmentation. *Poult. Sci.* **1991**, *70*, 115–119. [CrossRef]
154. Sujatha, T.; Narahari, D. Effect of designer diets on egg yolk composition of 'white leghorn' hens. *J. Food Sci. Technol.* **2011**, *48*, 494–497. [CrossRef] [PubMed]
155. Abudabos, A.M.; Okab, A.B.; Aljumaah, R.S.; Samara, E.M.; Abdoun, K.A.; Al-Haidary, A.A. Nutritional value of green seaweed (*Ulva* lactuca) for broiler chickens. *Ital. J. Anim. Sci.* **2013**, *12*, e28. [CrossRef]
156. El-Deek, A.; Brikaa, M.A. Nutritional and biological evaluation of marine seaweed as a feedstuff and as a pellet binder in poultry diet. *Int. J. Poult. Sci.* **2009**, *8*, 875–881. [CrossRef]
157. Kulshreshtha, G.; Rathgeber, B.; Stratton, G.; Thomas, N.; Evans, F.; Critchley, A.; Hafting, J.; Prithiviraj, B. Feed supplementation with red seaweeds, *Chondrus* crispus and *Sarcodiotheca* gaudichaudii, affects performance, egg quality, and gut microbiota of layer hens. *Poult. Sci.* **2014**, *93*, 2991–3001. [CrossRef] [PubMed]
158. Gatrell, S.; Lum, K.; Kim, J.; Lei, X. Nonruminant *Nutrition Symposium*: Potential of defatted microalgae from the biofuel industry as an ingredient to replace corn and soybean meal in swine and poultry diets. *J. Anim. Sci.* **2014**, *92*, 1306–1314. [CrossRef] [PubMed]
159. Grinstead, G.; Tokach, M.; Dritz, S.; Goodband, R.; Nelssen, J. Effects of *Spirulina* platensis on growth performance of weanling pigs. *Anim. Feed Sci. Technol.* **2000**, *83*, 237–247. [CrossRef]
160. Granaci, V. *Achievements in the Artificial Insemination of Swine*; University of Agricultural Sciences and Veterinary Medicine: Cluj-Napoca, Romania, 2007.
161. He, M.; Hollwich, W.; Rambeck, W. Supplementation of algae to the diet of pigs: A new possibility to improve the iodine content in the meat. *J. Anim. Physiol. Anim. Nutr.* **2002**, *86*, 97–104. [CrossRef]
162. Dierick, N.; Ovyn, A.; De Smet, S. Effect of feeding intact brown seaweed *Ascophyllum* nodosum on some digestive parameters and on iodine content in edible tissues in pigs. *J. Sci. Food Agric.* **2009**, *89*, 584–594. [CrossRef]

163. Reilly, P.; O'doherty, J.; Pierce, K.; Callan, J.; O'sullivan, J.; Sweeney, T. The effects of seaweed extract inclusion on gut morphology, selected intestinal microbiota, nutrient digestibility, volatile fatty acid concentrations and the immune status of the weaned pig. *Animal* **2008**, *2*, 1465–1473. [CrossRef] [PubMed]
164. Angell, A.R.; Angell, S.F.; de Nys, R.; Paul, N.A. Seaweed as a protein source for mono-gastric livestock. *Trends Food Sci. Technol.* **2016**, *54*, 74–84. [CrossRef]
165. Panjaitan, T.; Quigley, S.; McLennan, S.; Poppi, D. Effect of the concentration of Spirulina (*Spirulina platensis*) algae in the drinking water on water intake by cattle and the proportion of algae bypassing the rumen. *Anim. Prod. Sci.* **2010**, *50*, 405–409. [CrossRef]
166. Kulpys, J.; Paulauskas, E.; Pilipavicius, V.; Stankevicius, R. Influence of cyanobacteria arthrospira (*Spirulina*) platensis biomass additive towards the body condition of lactation cows and biochemical milk indexes. *Agron. Res* **2009**, *7*, 823–835.
167. Christaki, E.; Karatzia, M.; Bonos, E.; Florou-Paneri, P.; Karatzias, C. Effect of dietary *Spirulina Platensis* on milk fatty acid profile of dairy cows. *Asian J. Anim. Vet. Adv.* **2012**, *7*, 597–604. [CrossRef]
168. Šimkus, A.; Oberauskas, V.; Laugalis, J.; Želvytė, R.; Monkevičienė, I.; Sederevičius, A.; Šimkienė, A.; Pauliukas, K. The effect of weed *Spirulina platensis* on the milk production in cows. *Vet. Zootech.* **2007**, *38*, 74–77.
169. Boeckaert, C.; Vlaeminck, B.; Dijkstra, J.; Issa-Zacharia, A.; Van Nespen, T.; Van Straalen, W.; Fievez, V. Effect of dietary starch or micro algae supplementation on rumen fermentation and milk fatty acid composition of dairy cows. *J. Dairy Sci.* **2008**, *91*, 4714–4727. [CrossRef] [PubMed]
170. Bezerra, L.; Silva, A.; Azevedo, S.; Mendes, R.; Mangueira, J.; Gomes, A. Performance of santa inês lambs submitted to the use of artificial milk enriched with *Spirulina Platensis*. *Ciênc. Anim. Bras.* **2010**, *11*, 258–263.
171. Peiretti, P.; Meineri, G. Effects of diets with increasing levels of *Spirulina Platensis* on the performance and apparent digestibility in growing rabbits. *Livest. Sci.* **2008**, *118*, 173–177. [CrossRef]
172. Arieli, A.; Sklan, D.; Kissil, G. A note on the nutritive value of *Ulva* lactuca for ruminants. *Anim. Sci.* **1993**, *57*, 329–331. [CrossRef]
173. Ventura, M.; Castañón, J. The nutritive value of seaweed (*Ulva* lactuca) for goats. *Small Rumin. Res.* **1998**, *29*, 325–327. [CrossRef]
174. Novoa-Garrido, M.; Aanensen, L.; Lind, V.; Larsen, H.J.S.; Jensen, S.K.; Govasmark, E.; Steinshamn, H. Immunological effects of feeding macroalgae and various vitamin e supplements in *Norwegian* white sheep-ewes and their offspring. *Livest. Sci.* **2014**, *167*, 126–136. [CrossRef]
175. Lio-Po, G.D.; Leaño, E.M.; Peñaranda, M.M.D.; Villa-Franco, A.U.; Sombito, C.D.; Guanzon, N.G. Anti-luminous vibrio factors associated with the 'green water'grow-out culture of the tiger shrimp *Penaeus* monodon. *Aquaculture* **2005**, *250*, 1–7. [CrossRef]
176. Chuntapa, B.; Powtongsook, S.; Menasveta, P. Water quality control using *Spirulina Platensis* in shrimp culture tanks. *Aquaculture* **2003**, *220*, 355–366. [CrossRef]
177. Borowitzka, M.A. Microalgae for aquaculture: Opportunities and constraints. *J. Appl. Phycol.* **1997**, *9*, 393–401. [CrossRef]
178. Müller-Fuega, A. Microalgae for *Aquaculture*: The current global situation and future trends. In *Handbook of Microalgal C Ulture: Applied Phycology and Biotechnology*, 2nd ed.; Richmond, A., Hu, Q., Eds.; John Wiley Sons, Ltd.: Oxford, UK, 2013; pp. 352–364.
179. Robinson, C.; Samocha, T.; Fox, J.; Gandy, R.; McKee, D. The use of inert artificial commercial food sources as replacements of traditional live food items in the culture of larval shrimp, *Farfantepenaeus* aztecus. *Aquaculture* **2005**, *245*, 135–147. [CrossRef]
180. Muller-Feuga, A. The role of microalgae in aquaculture: *Situation* and trends. *J. Appl. Phycol.* **2000**, *12*, 527–534. [CrossRef]
181. Lozano, I.; Wacyk, J.M.; Carrasco, J.; Cortez-San Martín, M.A. Red macroalgae *Pyropia* columbina and *Gracilaria* chilensis: Sustainable feed additive in the *Salmo* salar diet and the evaluation of potential antiviral activity against infectious salmon anemia virus. *J. Appl. Phycol.* **2016**, *28*, 1343–1351. [CrossRef]
182. Wan, A.H.; Soler-Vila, A.; O'Keeffe, D.; Casburn, P.; Fitzgerald, R.; Johnson, M.P. The inclusion of *Palmaria palmata* macroalgae in *Atlantic* salmon (*Ulva* lactuca) diets: Effects on growth, haematology, immunity and liver function. *J. Appl. Phycol.* **2016**, *28*, 3091–3100. [CrossRef]
183. Bansemer, M.S.; Qin, J.G.; Harris, J.O.; Howarth, G.S.; Stone, D.A. Nutritional requirements and use of macroalgae as ingredients in abalone feed. *Rev. Aquac.* **2016**, *8*, 121–135. [CrossRef]

184. Viera, M.; de Vicose, G.C.; Gómez-Pinchetti, J.; Bilbao, A.; Fernandez-Palacios, H.; Izquierdo, M. Comparative performances of juvenile abalone (*Haliotis* tuberculata coccinea *Reeve*) fed enriched vs non-enriched macroalgae: Effect on growth and body composition. *Aquaculture* **2011**, *319*, 423–429. [CrossRef]
185. Xia, S.; Yang, H.; Li, Y.; Liu, S.; Zhou, Y.; Zhang, L. Effects of different seaweed diets on growth, digestibility, and ammonia-nitrogen production of the sea cucumber *Apostichopus* japonicus (selenka). *Aquaculture* **2012**, *338*, 304–308. [CrossRef]
186. Saito, T. Antihypertensive peptides derived from *Bovine Casein* and *Whey Proteins*. In *Bioactive Components of Milk*; Bösze, Z., Ed.; Springer: New York, NY, USA, 2008; pp. 295–317.
187. Korhonen, H.; Pihlanto, A. Bioactive peptides: Production and functionality. *Int. Dairy J.* **2006**, *16*, 945–960. [CrossRef]
188. Vercruysse, L.; Van Camp, J.; Smagghe, G. *ACE* inhibitory peptides derived from enzymatic hydrolysates of animal muscle protein: A review. *J. Agric. Food Chem.* **2005**, *53*, 8106–8115. [CrossRef] [PubMed]
189. Miguel, M.; Aleixandre, A. Antihypertensive peptides derived from egg proteins. *J. Nutr.* **2006**, *136*, 1457–1460. [PubMed]
190. Jung, W.K.; Mendis, E.; Je, J.Y.; Park, P.J.; Son, B.W.; Kim, H.C.; Kim, S.K. Angiotensin I-converting enzyme inhibitory peptide from yellowfin sole (*Limanda aspera*) frame protein and its antihypertensive effect in spontaneously hypertensive ratsangiotensin I-converting enzyme inhibitory peptide from yellowfin sole (*Limanda aspera*) frame protein and its antihypertensive effect in spontaneously hypertensive rats. *Food Chem* **2006**, *94*, 26.
191. Yu, Y.; Hu, J.; Miyaguchi, Y.; Bai, X.; Du, Y.; Lin, B. Isolation and characterization of angiotensin I-converting enzyme inhibitory peptides derived from porcine hemoglobin. *Peptides* **2006**, *27*, 2950–2956. [CrossRef] [PubMed]
192. Li, G.H.; Qu, M.R.; Wan, J.Z.; You, J.M. Antihypertensive effect of rice protein hydrolysate with in vitro angiotensin I-converting enzyme inhibitory activity in spontaneously hypertensive rats. *Asia. Pac. J. Clin. Nutr.* **2007**, *16*, 275–280. [PubMed]
193. Rho, S.J.; Lee, J.S.; Chung, Y.I.; Kim, Y.W.; Lee, H.G. Purification and identification of an angiotensin I-converting enzyme inhibitory peptide from fermented soybean extract. *Process Biochem.* **2009**, *44*, 490. [CrossRef]
194. Motoi, H.; Kodama, T. Isolation and characterization of angiotensin I-converting enzyme inhibitory peptides from wheat gliadin hydrolysate. *Nahrung* **2003**, *47*, 354–358. [CrossRef] [PubMed]
195. Aluko, R.E. Determination of nutritional and bioactive properties of peptides in enzymatic pea, chickpea, and mung bean protein hydrolysates. *J. AOAC Int.* **2008**, *91*, 947–956. [PubMed]
196. Lee, J.-E.; Bae, I.Y.; Lee, H.G.; Yang, C.-B. Tyr-Pro-Lys, an angiotensin I-converting enzyme inhibitory peptide derived from broccoli (*Brassica* oleracea *Italica*). *Food Chem.* **2006**, *99*, 143–148. [CrossRef]
197. Suetsuna, K. Isolation and characterization of angiotensin I-converting enzyme inhibitor dipeptides derived from *Allium sativum* L. (garlic) *Isolation* and characterization of angiotensin I-converting enzyme inhibitor dipeptides derived from. *J. Nutr. Biochem.* **1998**, *9*, 415. [CrossRef]
198. Fan, X.; Bai, L.; Zhu, L.; Yang, L.; Zhang, X. Marine algae-derived bioactive peptides for human nutrition and health. *J. Agric. Food Chem.* **2014**, *62*, 9211–9222. [CrossRef] [PubMed]
199. Beaulieu, L.; Bondu, S.; Doiron, K.; Rioux, L.-E.; Turgeon, S.L. Characterization of antibacterial activity from protein hydrolysates of the macroalga *Saccharina* longicruris and identification of peptides implied in bioactivity. *J. Funct. Foods* **2015**, *17*, 685–697. [CrossRef]
200. Ennamany, R.; Saboureau, D.; Mekideche, N.; Creppy, E. Secma 1®, a mitogenic hexapeptide from *Ulva* algeae modulates the production of proteoglycans and glycosaminoglycans in human foreskin fibroblast. *Hum. Exp. Toxicol.* **1998**, *17*, 18–22. [CrossRef] [PubMed]
201. Suetsuna, K.; Nakano, T. Identification of an antihypertensive peptide from peptic digest of wakame (*Undaria pinnatifida*). *J. Nutr. Biochem.* **2000**, *11*, 450–454. [CrossRef]
202. Suetsuna, K.; Maekawa, K.; Chen, J.-R. Antihypertensive effects of *Undaria* pinnatifida (wakame) peptide on blood pressure in spontaneously hypertensive rats. *J. Nutr. Biochem.* **2004**, *15*, 267–272. [CrossRef] [PubMed]
203. Sato, M.; Hosokawa, T.; Yamaguchi, T.; Nakano, T.; Muramoto, K.; Kahara, T.; Funayama, K.; Kobayashi, A.; Nakano, T. Angiotensin I-converting enzyme inhibitory peptides derived from wakame (*Undaria pinnatifida*) and their antihypertensive effect in spontaneously hypertensive rats. *J. Agric. Food Chem.* **2002**, *50*, 6245–6252. [CrossRef] [PubMed]

204. Cha, S.-H.; Ahn, G.-N.; Heo, S.-J.; Kim, K.-N.; Lee, K.-W.; Song, C.-B.; Jeon, Y.-J. Screening of extracts from marine green and brown algae in *Jeju* for potential marine angiotensin-I converting enzyme (*ACE*) inhibitory activity. *J. Korean Soc. Food Sci. Nutr.* **2006**, *35*, 307–314.
205. Suetsuna, K. Purification and identification of angiotensin I-converting enzyme inhibitors from the red alga *Porphyra* yezoensis. *J. Mar. Biotechnol.* **1998**, *6*, 163–167. [PubMed]
206. Suetsuna, K. Separation and identification of angiotensin I-converting enzyme inhibitory peptides from peptic digest of *Hizikia* fusiformis protein. *Nippon. Suisan. Gakkaishi.* **1998**, *64*, 862–866. [CrossRef]
207. Furuta, T.; Miyabe, Y.; Yasui, H.; Kinoshita, Y.; Kishimura, H. Angiotensin *I Converting Enzyme Inhibitory Peptides Derived* from *Phycobiliproteins* of *Dulse Palmaria palmata*. *Mar. Drugs* **2016**, *14*, 32. [CrossRef] [PubMed]
208. Suetsuna, K.; Chen, J.-R. Identification of antihypertensive peptides from peptic digest of two microalgae, *Chlorella* vulgaris and *Spirulina Platensis*. *Mar. Biotechnol.* **2001**, *3*, 305–309. [CrossRef] [PubMed]
209. Qian, Z.-J.; Heo, S.-J.; Oh, C.H.; Kang, D.-H.; Jeong, S.H.; Park, W.S.; Choi, I.-W.; Jeon, Y.-J.; Jung, W.-K. Angiotensin I-Converting Enzyme (ACE) Inhibitory Peptide Isolated from Biodiesel Byproducts of Marine Microalgae, Nannochloropsis oculata. *J. Biobased Mater. Bioenergy* **2013**, *7*, 135–142. [CrossRef]
210. Samarakoon, K.W.; Kwon, O.-N.; Ko, J.-Y.; Lee, J.-H.; Kang, M.-C.; Kim, D.; Lee, J.B.; Lee, J.-S.; Jeon, Y.-J. Purification and identification of novel angiotensin-I converting enzyme (*ACE*) inhibitory peptides from cultured marine microalgae (*Nannochloropsis oculata*) protein hydrolysate. *J. Appl. Phycol.* **2013**, *25*, 1595–1606. [CrossRef]
211. Ko, S.-C.; Kang, N.; Kim, E.-A.; Kang, M.C.; Lee, S.-H.; Kang, S.-M.; Lee, J.-B.; Jeon, B.-T.; Kim, S.-K.; Park, S.-J. A novel angiotensin I-converting enzyme (ACE) inhibitory peptide from a marine *Chlorella* ellipsoidea and its antihypertensive effect in spontaneously hypertensive rats. *Process Biochem.* **2012**, *47*, 2005–2011. [CrossRef]
212. Sheih, I.-C.; Fang, T.J.; Wu, T.-K. Isolation and characterisation of a novel angiotensin I-converting enzyme (ACE) inhibitory peptide from the algae protein waste. *Food Chem.* **2009**, *115*, 279–284. [CrossRef]
213. Tierney, M.S.; Croft, A.K.; Hayes, M. A review of antihypertensive and antioxidant activities in macroalgae. *Bot. Mar.* **2010**, *53*, 387–408. [CrossRef]
214. Sheih, I.C.; Fang, T.J.; Wu, T.K.; Lin, P.H. Anticancer and antioxidant activities of the peptide fraction from algae protein waste. *J. Agric. Food Chem.* **2010**, *58*, 1202–1207. [CrossRef] [PubMed]
215. Sheih, I.C.; Wu, T.K.; Fang, T.J. Antioxidant properties of a new antioxidative peptide from algae protein waste hydrolysate in different oxidation systems. *Bioresour. Technol.* **2009**, *100*, 3419–3425. [CrossRef] [PubMed]
216. Kang, K.H.; Qian, Z.J.; Ryu, B.; Karadeniz, F.; Kim, D.; Kim, S.K. Antioxidant peptides from protein hydrolysate of microalgae *Navicula* incerta and their protective effects in HepG2/CYP2E1 cells induced by ethanol. *Phytother. Res.* **2012**, *26*, 1555–1563. [CrossRef] [PubMed]
217. Ko, S.-C.; Kim, D.; Jeon, Y.-J. Protective effect of a novel antioxidative peptide purified from a marine *Chlorella* ellipsoidea protein against free radical-induced oxidative stress. *Food Chem. Toxicol.* **2012**, *50*, 2294–2302. [CrossRef] [PubMed]
218. O'Sullivan, A.; O'Callaghan, Y.; O'Grady, M.; Queguineur, B.; Hanniffy, D.; Troy, D.; Kerry, J.; O'Brien, N. In vitro and cellular antioxidant activities of seaweed extracts prepared from five brown seaweeds harvested in spring from the west coast of Ireland. *Food Chem.* **2011**, *126*, 1064–1070. [CrossRef]
219. Heo, S.J.; Lee, G.W.; Song, C.B.; Jeon, Y.J. Antioxidant activity of enzymatic extracts from brown seaweeds. *Algae* **2003**, *18*, 71–81. [CrossRef]
220. Heo, S.-J.; Park, E.-J.; Lee, K.-W.; Jeon, Y.-J. Antioxidant activities of enzymatic extracts from brown seaweeds. *Bioresour. Technol.* **2005**, *96*, 1613–1623. [CrossRef] [PubMed]
221. Lakmal, H.C.; Samarakoon, K.W.; Lee, W.; Lee, J.-H.; Abeytunga, D.; Lee, H.-S.; Jeon, Y.-J. Anticancer and antioxidant effects of selected *Sri Lankan* marine algae. *J. Natl. Sci. Found. Sri Lanka* **2014**, *42*, 315–323. [CrossRef]
222. Wang, X.; Zhang, X. Separation, antitumor activities, and encapsulation of polypeptide from *Chlorella* pyrenoidosa. *Biotechnol. Progress* **2013**, *29*, 681–687. [CrossRef] [PubMed]
223. Zhang, B.; Zhang, X. Separation and nanoencapsulation of antitumor polypeptide from *Spirulina Platensis*. *Biotechnol. Progress* **2013**, *29*, 1230–1238. [CrossRef] [PubMed]
224. Cian, R.E.; Martínez-Augustin, O.; Drago, S.R. Bioactive properties of peptides obtained by enzymatic hydrolysis from protein byproducts of *Porphyra* columbina. *Food Res. Int.* **2012**, *49*, 364–372. [CrossRef]

225. Shih, M.F.; Chen, L.C.; Cherng, J.Y. *Chlorella* 11-peptide inhibits the production of macrophage-induced adhesion molecules and reduces endothelin-1 expression and endothelial permeability. *Mar. Drugs* **2013**, *11*, 3861–3874. [CrossRef] [PubMed]
226. Vo, T.-S.; Kim, S.-K. Down-regulation of histamine-induced endothelial cell activation as potential anti-atherosclerotic activity of peptides from *Spirulina maxima*. *Eur. J. Pharm. Sci.* **2013**, *50*, 198–207. [CrossRef] [PubMed]
227. Vo, T.-S.; Ryu, B.; Kim, S.-K. Purification of novel anti-inflammatory peptides from enzymatic hydrolysate of the edible microalgal *Spirulina maxima*. *J. Funct. Foods* **2013**, *5*, 1336–1346. [CrossRef]
228. Fitzgerald, C.; Gallagher, E.; O'Connor, P.; Prieto, J.; Mora-Soler, L.; Grealy, M.; Hayes, M. Development of a seaweed derived platelet activating factor acetylhydrolase (PAF-AH) inhibitory hydrolysate, synthesis of inhibitory peptides and assessment of their toxicity using the zebrafish larvae assay. *Peptides* **2013**, *50*, 119–124. [CrossRef] [PubMed]
229. Kang, K.-H.; Qian, Z.-J.; Ryu, B.; Karadeniz, F.; Kim, D.; Kim, S.-K. Hepatic fibrosis inhibitory effect of peptides isolated from navicula incerta on TGF-β1 Induced activation of LX-2 human hepatic stellate cells. *Prev. Nutr. Food Sci.* **2013**, *18*, 124–132. [CrossRef] [PubMed]
230. Ahn, G.; Hwang, I.; Park, E.; Kim, J.; Jeon, Y.-J.; Lee, J.; Park, J.W.; Jee, Y. Immunomodulatory effects of an enzymatic extract from *Ecklonia* cava on murine splenocytes. *Mar. Biotechnol.* **2008**, *10*, 278–289. [CrossRef] [PubMed]
231. Chen, C.-L.; Liou, S.-F.; Chen, S.-J.; Shih, M.-F. Protective effects of *Chlorella*-derived peptide on uvb-induced production of MMP-1 and degradation of procollagen genes in human skin fibroblasts. *Regul. Toxicol. Pharm.* **2011**, *60*, 112–119. [CrossRef] [PubMed]
232. Shih, M.-F.; Cherng, J.-Y. Potential protective effect of fresh grown unicellular green algae component (resilient factor) against PMA-and UVB-induced MMP1 expression in skin fibroblasts. *Eur. J. Dermatol.* **2008**, *18*, 303–307. [PubMed]
233. Nguyen, M.H.T.; Qian, Z.-J.; Nguyen, V.-T.; Choi, I.-W.; Heo, S.-J.; Oh, C.H.; Kang, D.-H.; Kim, G.H.; Jung, W.-K. Tetrameric peptide purified from hydrolysates of biodiesel byproducts of *Nannochloropsis* oculata induces osteoblastic differentiation through MAPK and SMAD pathway on MG-63 and D1 cells. *Process Biochem.* **2013**, *48*, 1387–1394. [CrossRef]
234. Athukorala, Y.; Lee, K.-W.; Kim, S.-K.; Jeon, Y.-J. Anticoagulant activity of marine green and brown algae collected from Jeju Island in Korea. *Bioresour. Technol.* **2007**, *98*, 1711–1716. [CrossRef] [PubMed]
235. Cian, R.E.; Garzón, A.G.; Ancona, D.B.; Guerrero, L.C.; Drago, S.R. Hydrolyzates from *Pyropia* columbina seaweed have antiplatelet aggregation, antioxidant and ACE I inhibitory peptides which maintain bioactivity after simulated gastrointestinal digestion. *LWT-Food Sci. Technol.* **2015**, *64*, 881–888. [CrossRef]
236. Vermeirssen, V.; Van Camp, J.; Verstraete, W. Bioavailability of angiotensin I converting enzyme inhibitory peptides. *Br. J. Nutr.* **2004**, *92*, 357–366. [CrossRef] [PubMed]
237. World Heath Organization. *Global Health Risks: Mortality and Burden of Disease Attributable to Selected Major Risks*; World Health Organization: Geneva, Switzerland, 2009.
238. Turk, B. Targeting proteases: Successes, failures and future prospects. *Nature Rev. Drug Discov.* **2006**, *5*, 785–799. [CrossRef] [PubMed]
239. Chen, Z.-Y.; Peng, C.; Jiao, R.; Wong, Y.M.; Yang, N.; Huang, Y. Anti-hypertensive nutraceuticals and functional foods. *J. Agric. Food Chem.* **2009**, *57*, 4485–4499. [CrossRef] [PubMed]
240. Wu, Q.; Cai, Q.-F.; Yoshida, A.; Sun, L.-C.; Liu, Y.-X.; Liu, G.-M.; Su, W.-J.; Cao, M.-J. Purification and characterization of two novel angiotensin I-converting enzyme inhibitory peptides derived from r-phycoerythrin of red algae (*Bangia fusco-purpurea*). *Eur. Food Res. Technol.* **2017**, *243*, 779–789. [CrossRef]
241. Fitzgerald, C.N.; Mora-Soler, L.; Gallagher, E.; O'Connor, P.; Prieto, J.; Soler-Vila, A.; Hayes, M. Isolation and characterization of bioactive pro-peptides with in vitro renin inhibitory activities from the macroalga *Palmaria palmata*. *J. Agric. Food Chem.* **2012**, *60*, 7421–7427. [CrossRef] [PubMed]
242. Verdecchia, P.; Angeli, F.; Mazzotta, G.; Gentile, G.; Reboldi, G. The renin angiotensin system in the development of cardiovascular disease: *Role* of aliskiren in risk reduction. *Vasc. Health Risk Manag.* **2008**, *4*, 971–981. [CrossRef] [PubMed]
243. Werner, A.; Kraan, S. *Review of the Potential Mechanisation of Kelp Harvesting in Ireland*; Marine Institute: Dublin, Ireland, 2004; p. 52.

244. Werner, A.; Clarke, D.; Kraan, S. *Strategic Review and the Feasibility of Seaweed Aquaculture in Ireland*; Irish Seaweed Centre, Martin Ryan Institute, National University of Ireland: Galway, Ireland, 2004; p. 120.
245. De Grave, S.; Fazakerley, H.; Kelly, L.; Guiry, M.; Ryan, M.; Walshe, J. *A Study of Selected MaërI Beds in Irish Waters and Their Potential for Sustainable Extraction*; Marine Institute: Dublin, Ireland, 2000.
246. Stengel, D.B.; Connan, S.; Popper, Z.A. Algal chemodiversity and bioactivity: Sources of natural variability and implications for commercial application. *Biotechnol. Adv.* **2011**, *29*, 483–501. [CrossRef] [PubMed]
247. Galland-Irmouli, A.-V.; Fleurence, J.; Lamghari, R.; Luçon, M.; Rouxel, C.; Barbaroux, O.; Bronowicki, J.-P.; Villaume, C.; Guéant, J.-L. Nutritional value of proteins from edible seaweed *Palmaria palmata* (dulse). *J. Nutr. Biochem.* **1999**, *10*, 353–359. [CrossRef]
248. Marinho-Soriano, E.; Fonseca, P.; Carneiro, M.; Moreira, W. Seasonal variation in the chemical composition of two tropical seaweeds. *Bioresour. Technol.* **2006**, *97*, 2402–2406. [CrossRef] [PubMed]
249. Zhou, A.Y.; Robertson, J.; Hamid, N.; Ma, Q.; Lu, J. Changes in total nitrogen and amino acid composition of New Zealand *Undaria pinnatifida* with growth, location and plant parts. *Food Chem.* **2015**, *186*, 319–325. [CrossRef] [PubMed]
250. Sheng, J.; Vannela, R.; Rittmann, B. Disruption of *Synechocystis* PCC 6803 for lipid extraction. *Water Sci. Technol.* **2012**, *65*, 567–573. [CrossRef] [PubMed]
251. Marrion, O.; Fleurence, J.; Schwertz, A.; Guéant, J.-L.; Mamelouk, L.; Ksouri, J.; Villaume, C. Evaluation of protein in vitro digestibility of *Palmaria palmata* and *Gracilaria* verrucosa. *J. Appl. Phycol.* **2005**, *17*, 99–102. [CrossRef]
252. Meade, S.J.; Reid, E.A.; Gerrard, J.A. The impact of processing on the nutritional quality of food proteins. *J. AOAC Int.* **2005**, *88*, 904–922. [PubMed]
253. Maehre, H.K.; Edvinsen, G.K.; Eilertsen, K.-E.; Elvevoll, E.O. Heat treatment increases the protein bioaccessibility in the red seaweed dulse (*Palmaria palmata*), but not in the brown seaweed winged kelp (*Alaria* esculenta). *J. Appl. Phycol.* **2016**, *28*, 581–590. [CrossRef]
254. Setting Maximum Levels for Certain Contaminants in Foodstuffs. 2006. Available online: https://www.fsai.ie/uploadedFiles/Consol_Reg1881_2006.pdf (accessed on 20 April 2017).
255. Opinion of the Scientific Panel on Contaminants in the Food Chain on a request from the Commission Related to Mercury and Methylmercury in Food. Available online: http://www.efsa.europa.eu/sites/default/files/scientific_output/files/main_documents/34.pdf (accessed on 20 April 2017).
256. European Food Safety Authority. Scientific Opinion on Arsenic in Food. EFSA Panel on Contaminants in the Food Chain. Available online: http://www.iss.it/binary/meta/cont/AsSummary2009en.pdf (accessed on 20 April 2017).
257. European Food Safety Authority. Scientific Opinion on Lead in Food. EFSA Panel on Contaminants in the Food Chain. Available online: http://www.iss.it/binary/meta/cont/Pb_Opinion2010.pdf (accessed on 20 April 2017).
258. European Food Safety Authority. Statement on Tolerable Weekly Intake for Cadmium. EFSA Panel on Contaminants in the Food Chain. Available online: http://www.megapesca.com/megashop/FH201102_tgf/EFSA_Scientific_Opinion_Cadmium.pdf (accessed on 20 April 2017).
259. Almela, C.; Algora, S.; Benito, V.; Clemente, M.; Devesa, V.; Suner, M.; Velez, D.; Montoro, R. Heavy metal, total arsenic, and inorganic arsenic contents of algae food products. *J. Agric. Food Chem.* **2002**, *50*, 918–923. [CrossRef] [PubMed]
260. Besada, V.; Andrade, J.M.; Schultze, F.; González, J.J. Heavy metals in edible seaweeds commercialised for human consumption. *J. Mar. Syst.* **2009**, *75*, 305–313. [CrossRef]
261. Van der Spiegel, M.; Noordam, M.; Fels-Klerx, H. Safety of novel protein sources (insects, microalgae, seaweed, duckweed, and rapeseed) and legislative aspects for their application in food and feed production. *Compr. Rev. Food Sci. Food Saf.* **2013**, *12*, 662–678. [CrossRef]
262. Grobbelaar, J.U. Quality control and assurance: *Crucial* for the sustainability of the applied phycology industry. *J. Appl. Phycol.* **2003**, *15*, 209–215. [CrossRef]
263. Görs, M.; Schumann, R.; Hepperle, D.; Karsten, U. Quality analysis of commercial *Chlorella* products used as dietary supplement in human nutrition. *J. Appl. Phycol.* **2010**, *22*, 265–276. [CrossRef]
264. Harnedy, P.A.; FitzGerald, R.J. Extraction and *Enrichment* of *Protein* from red and green Macroalgae. *Nat. Prod. Mar. Algae Methods Protoc.* **2015**, *1308*, 103–108.

265. Machmudah, S.; Shotipruk, A.; Goto, M.; Sasaki, M.; Hirose, T. Extraction of Astaxanthin from *Haematococcus* pluvialis Using Supercritical CO_2 and Ethanol as Entrainer. *Ind. Eng. Chem. Res.* **2006**, *45*, 3652–3657. [CrossRef]
266. Cuellar-Bermudez, S.P.; Aguilar-Hernandez, I.; Cardenas-Chavez, D.L.; Ornelas-Soto, N.; Romero-Ogawa, M.A.; Parra-Saldivar, R. Extraction and purification of high-value metabolites from microalgae: Essential lipids, astaxanthin and phycobiliproteins. *Microb. Biotechnol.* **2015**, *8*, 190–209. [CrossRef] [PubMed]
267. Marella, C.; Muthukumarappan, K.; Metzger, L. Application of membrane separation technology for developing novel dairy food ingredients. *J. Food Process. Technol.* **2013**, *4*. [CrossRef]
268. Van Hal, J.W.; Huijgen, W.; López-Contreras, A. Opportunities and challenges for seaweed in the biobased economy. *Trends Biotechnol.* **2014**, *32*, 231–233. [CrossRef] [PubMed]

foods

MDPI

Article

Inhibition of *Listeria monocytogenes* on Ready-to-Eat Meats Using Bacteriocin Mixtures Based on Mode-of-Action

Paul Priyesh Vijayakumar [1,†] and Peter M. Muriana [1,2,*,†]

1 Department of Animal and Food Science, University of Kentucky, 213 W.P. Garrigus Building, Lexington, KY 40546-0215, USA; paul.v@uky.edu

2 Robert M. Kerr Food & Agricultural Products Centre, Oklahoma State University, 109 FAPC Building, Monroe Street, Stillwater, OK 74078-6055, USA

* Correspondence: peter.muriana@okstate.edu; Tel.: +1-405-744-5563; Fax: +1-405-744-6313

† These authors contributed equally to this work.

Academic Editor: Maria Hayes

Received: 13 January 2017; Accepted: 9 March 2017; Published: 14 March 2017

Abstract: Bacteriocin-producing (Bac$^+$) lactic acid bacteria (LAB) comprising selected strains of *Lactobacillus curvatus*, *Lactococcus lactis*, *Pediococcus acidilactici*, and *Enterococcus faecium* and *thailandicus* were examined for inhibition of *Listeria monocytogenes* during hotdog challenge studies. The Bac$^+$ strains, or their cell-free supernatants (CFS), were grouped according to mode-of-action (MOA) as determined from prior studies. Making a mixture of as many MOAs as possible is a practical way to obtain a potent natural antimicrobial mixture to address *L. monocytogenes* contamination of RTE meat products (i.e., hotdogs). The heat resistance of the bacteriocins allowed the use of pasteurization to eliminate residual producer cells for use as post-process surface application or their inclusion into hotdog meat emulsion during cooking. The use of Bac$^+$ LAB comprising 3× MOAs directly as co-inoculants on hotdogs was not effective at inhibiting *L. monocytogenes*. However, the use of multiple MOA Bac$^+$ CFS mixtures in a variety of trials demonstrated the effectiveness of this approach by showing a >2-log decrease of *L. monocytogenes* in treatment samples and 6–7 log difference vs. controls. These data suggest that surface application of multiple mode-of-action bacteriocin mixtures can provide for an Alternative 2, and possibly Alternative 1, process category as specified by USDA-FSIS for control of *L. monocytogenes* on RTE meat products.

Keywords: *Listeria monocytogenes*; ready-to-eat meats; bacteriocin; mode-of-action; biopreservatives

1. Introduction

Listeria monocytogenes is a formidable foodborne pathogen that causes listeriosis which results in high hospitalization rates (>90%) and mortalities (20%–30%) in large outbreaks [1]. Vulnerable populations include immuno-compromised, the sick and elderly, pregnant women, and infants. *Listeria monocytogenes* is associated with numerous animals [2] and therefore may be found as a ubiquitous contaminant on many animal-derived raw food products and ingredients that helps the organism find its way into meat and poultry processing facilities. The United States Department of Agriculture's Food Safety and Inspection Service (USDA-FSIS) found incidences as high as 7.24% on small cooked sausages (i.e., hotdogs; 1991) and 7.69% on sliced ham and luncheon meats (1996) in nationwide sampling program for ready-to-eat meats (RTE) in the early 1990's [3]. Their ability to remain as a persistent problem in RTE meat processing plants is a combination of their steady influx on raw ingredients as well as their ability to form biofilms that may resist sanitation efforts and allow the organism to be a persistent contaminant [4,5].

The RTE meat industry has been constantly battling the occurrence of *L. monocytogenes*. While RTE meats primarily rely on salt, curing agents, and refrigerated storage for microbial stability and safety, *Listeria* can capitalize on these conditions by growth at low temperatures and high salt concentrations. The CDC reported that *L. monocytogenes* is responsible for 2500 illness cases and 500 deaths annually (www.cdc.gov/ncidod/disease/foodborn/lister.htm). Hotdogs have maintained a designation as a high-risk RTE meat for *L. monocytogenes* because of high contamination rates [6]. Contamination occurs on the surface of the product during post-process exposure and steps such as peeling and packaging are potential routes for pathogen entry. The primary hurdle against foodborne pathogens and bacterial contamination in the food industry includes preventive measures such as good manufacturing practices (GMPs) and standard operating procedures (SOPs) in addition to a hazard analysis and critical control point (HACCP) food safety plan required for meat and poultry products [7,8]. Though the food industry incorporates a variety of precautionary measures, outbreaks due to foodborne illness continue to occur periodically. Therefore, there is a need for effective antimicrobials that may continue to provide food safety protection during shelf life and distribution of sensitive products.

The lactic acid bacteria (LAB) are well known for producing antimicrobials including organic acids, diacetyl, acetoin, hydrogen peroxide, reuterin, reutericyclin, antifungal peptides, and bacteriocins [9–11]. Although lactic acid is one of the most common acidulants, there has been considerable interest and research in the field of bacteriocins with respect to use of bacteriocinogenic (Bac^+) LAB cultures or bacteriocin-containing culture fermentates as food preservatives [12]. Bacteriocin-producing cultures have been proposed as protective cultures to combat foodborne pathogens and spoilage bacteria in food systems [13–17]. The addition of bacteriocins includes the use of partially purified Bac^+ preparations [18] or pre-cultured bacteriocin-containing (Bac^+) cell-free supernatants (CFS) obtained from Bac^+ LAB [19] as food ingredients. While the addition of purified bacteriocins as food preservatives needs regulatory approval and must be treated as direct food additives, the inclusion of Bac^+ CFS from LAB cultures do not have the same regulatory restrictions [12].

In this study, we examined the effectiveness of Bac^+ LAB and Bac^+ CFS mixtures to prevent the growth of *L. monocytogenes* on RTE meats (hotdogs). Our approach included mixtures of bacteriocins demonstrating different modes-of-action (MOA), or the strains that produce them, that could provide enhanced efficacy against *L. monocytogenes* as opposed to preparations having a single MOA that could allow the development of spontaneous bacteriocin-resistant *L. monocytogenes* [20–22].

2. Materials and Methods

2.1. Bacterial Cultures

Strains of LAB were cultured at 30 °C in Lactobacilli MRS broth (Difco™, Becton-Dickenson Laboratories, Sparks, MD, USA) while *L. monocytogenes* 39-2, an isolate from retail hotdogs [23], was cultured in tryptic soy broth (TSB, Difco™) at 30 °C. Enumeration of LAB from either Bac^+ LAB-*L. monocytogenes* hotdog challenge studies, or as LAB contaminants in Bac^+ CFS-*Listeria* challenge studies, was done using MRS agar adjusted with HCl to pH 5.4–5.5 prior to autoclaving (the pH was found to be ~pH 5.5–5.7 after autoclaving) [24]. Acidified MRS agar inhibited growth of *L. monocytogenes* 39-2 but allowed the growth of LAB as determined from prior studies. *Listeria monocytogenes* 39-2 was selectively enumerated on MOX agar (Modified Oxford agar, Difco™) which was inhibitory to LAB. The *L. monocytogenes* 39-2 strain used in this study is resistant to 50 μg/mL of both streptomycin and rifamycin (Mediatech, Inc., Herndon, VA, USA); plate counts of *L. monocytogenes* 39-2 were occasionally confirmed on TS agar containing antibiotics. Bacterial cultures used in this study are listed in Table 1. Several Bac^+ cultures (FS56-1, FS92) were previously identified as *Lactococcus lactis* and grew well and made bacteriocins in MRS media; however, during the course of our studies they were identified by 16S rRNA PCR/sequencing to be *Enterococcus* sp. (Table 1).

Table 1. Bacterial strains used in this study.

Microorganism	Strain Designation	Source/Reference
Lactobacillus delbrueckii	4797-2	Muriana culture collection
Listeria monocytogenes	39-2 (R0)	[20–23]
Lactobacillus curvatus	FS47	[25]
Lactobacillus curvatus	Beef 3	[26]
Pediococcus acidilactici	Bac 3	[26]
Enterococcus faecium	FS56-1	[21,25,26]
Lactococcus lactis	FLS-1	[26]
Enterococcus thailandicus	RP-1	[21,26]
Enterococcus thailandicus	FS92	[21,25]

2.2. Bacteriocin Preparations

Bacteriocins were prepared by 2× repetitive transfer of individual Bac^+ LAB overnight at 30 °C followed by centrifugation at 20,000× *g* (rcf) for 10 min at 4 °C (Sorvall RC50 Plus, ThermoFisher Scientific, Waltham, MA, USA). The supernatants were carefully decanted to sterile bottles and filter-sterilized through 0.22 μ cellulose acetate syringe filters (VWR, Radnor, PA, USA) or pasteurized at 80 °C for 15 min. Bacteriocin preparations were then stored at 4 °C, or frozen at −20 °C if not expected to be used within a few days. Each of the filter-sterilized or pasteurized Bac^+ CFS preparations were also plated on MRSA plates, or into MRS broth, and incubated (30 °C) in order to check the effectiveness of the pasteurization or filter-sterilization process (i.e., no growth).

2.3. Manufacture of Hotdogs for Bacteriocin Applications

Hotdogs were manufactured in-house for use in shelf life trials (Figure 1). Beef and pork trimmings were used to manufacture hotdogs in the Meat Pilot Plant in the R.M. Kerr Food and Ag Products Center (FAPC) at Oklahoma State University, Stillwater, OK. Hotdogs were manufactured with the following formulation (per 35.52 lbs): beef (81% lean; 4.5 lbs), pork (72% lean; 13.25 lbs), pork (42% lean; 7.25 lbs), water/ice (6.25%; 9.45 lbs), Legg's Bolo seasoning (1.0 lb), cure (6.25% nitrite; 0.06 lb), and sodium erythorbate (0.01 lbs). Antimicrobials such as lactate and diacetate were not added, as is commonly done in commercial frankfurters, so as not to confuse the source of antimicrobial activity during bacteriocin treatments. Emulsions were stuffed into Viscofan 24/USA casings and thermally processed (cooked) in an electric-fired, batch oven (Alkar, DEC International, Washington, DC, USA) to an internal temperature of 88 °C (190 °F). After cooking, hotdogs in casings were chilled with a cold water rinse and then peeled using a peeling machine (PS760L Peeler, Linker Machines, Rockaway, NJ, USA). The formulation above was used for surface-treatment with Bac^+ CFS or Bac^+ LAB applied prior to packaging. Additional hotdog formulation modifications included replacement of the added water with pasteurized Bac^+ CFS, the use of chilled pasteurized Bac^+ CFS spray warm hotdogs still in casings, and surface application of CFS during packaging (Figure 1). The hotdogs manufactured by these different protocols were kept separate from each other, vacuum packaged, and stored frozen until used.

2.4. Hotdog Challenge Studies

2.4.1. Preliminary Treatment of Hotdogs Prior to Challenge Studies

Hotdogs manufactured for use in challenge studies were stored frozen in a blast chiller (−26 °C) in single-layer packages. They were then thawed prior to use and pasteurized by dipping packages into a temperature-controlled, steam-injected 50-gal hot water bath at 82 °C for 5 min in order to eliminate any indigenous bacterial contaminants that could have been acquired during post-process handling. Hotdogs were then aseptically removed from the vacuum packages for use in experimental treatments.

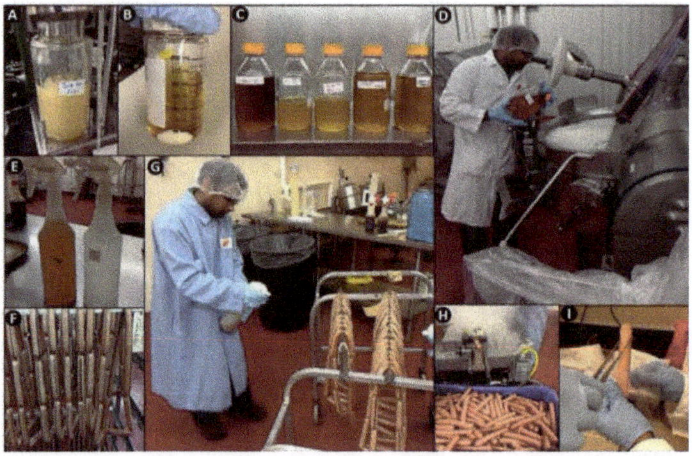

Figure 1. Manufacture of hotdogs for bacteriocin treatment. (**A**) Culturing microorganisms; (**B**) centrifugation of Bac+ supernatants; (**C**) pasteurized Bac+ CFS; (**D**) addition of Bac+ CFS mixture to hotdog meat matrix (Trial #2); (**E**) spray bottles with pasteurized Bac+ CFS; (**F**) pre-cooked hotdogs in casings; (**G**) spraying cooked hotdogs in casings with Bac+ CFS mixture (Trial #2); (**H**) hotdogs after peeling; (**I**) addition of hotdogs to vacuum package bags for addition of Bac+ CFS and the *L. monocytogenes* inoculum (Trials #3, #4, and #5).

2.4.2. Trial #1: Application of Mixed Mode-of-Action (MOA) Bac+ LAB Co-Inoculated with *L. monocytogenes* in Shelf Life Challenge Studies

Selected Bac+ LAB cultures covering 3 different MOA were used in these trials: *Pe. acidilactici* Bac3 (pediocin Bac3), *En. faecium* FS56-1 (enterocin FS56), and *En. thailandicus* FS92 and RP-1 (enterocins FS92 and RP-1). Freshly grown overnight Bac+ LAB cultures were prepared to approximately similar levels by dilution in sterile 0.1% BPW and mixed in equal proportions prior to use. Thawed hotdogs were pasteurized as described above, and then immersed in the Bac+ culture mixture for 30 s using a sterile plastic basket and allowed to drain for 30 s before hotdogs were removed with sterile tongs to sterile vacuum packaging bags. A level of mixed Bac+ culture was used for dipping that would achieve approximately ~10^5 cfu/mL in a recovered minimal hotdog rinse as determined by prior enumeration studies. Similarly, *L. monocytogenes* 39-2 was prepared by dilution in sterile 0.1% BPW and 100 μL was inoculated directly into the bagged hotdogs at a level resulting in recovery of *L. monocytogenes* at approximately ~10^4 cfu/mL from the same minimal rinse recovery solutions. The hotdogs in the vacuum bags were then massaged to distribute *L. monocytogenes* and then vacuum-packaged. Bags were stored at 5 °C and sampled at 0 h, 3 days, and weekly at 1, 2, 3, 4, 5, 6 and 7 weeks. Triplicate replications of each treatment (3 bags/treatment) were sampled at the time intervals mentioned above. During sampling, each package was opened by snipping open the top corner, and a pipette was used to deliver 3 mL of diluent (0.1% BPW). The bags were hand-massaged and then a pipette was used to withdraw the contents into a sterile disposable plastic tube that was kept on ice; this was considered the 10^0 dilution. Further dilutions were made with 0.1% BPW and plated on acidified MRS agar (for LAB) or MOX agar (*L. monocytogenes*). A series of negative control samples containing only *L. monocytogenes* 39-2 were also included.

2.4.3. Trial #2: Application of Mixed MOA Bacteriocin Preparations Added during the Manufacture of Hotdogs

Hotdogs were manufactured in the FAPC Meat Pilot Plant. Four 25-lb batches of hotdogs were manufactured which would be used in a variety of trials involving different formulations or treatments.

Bacteriocin CFS preparations were obtained as described previously (cultured, centrifuged to remove cells, and pasteurized) and mixed in equal volumes: curvaticin FS47, curvaticin Beef3, lacticin FLS1, and pediocin Bac3 representing 3 different MOAs.

Control batches of hotdogs did not receive a bacteriocin application; these were also used in subsequent trials for surface application of a mixed bacteriocin cocktail prior to packaging (see next section). In the current trial, the mixed bacteriocin CFS preparation was added in place of the water component (9.45 lbs) in the raw meat emulsion. In an additional formulation and treatment within this trial, some hotdogs were sprayed after the cook process while still in casings instead of adding the CFS to the meat matrix. Sprayed Bac$^+$ CFS preparation was allowed to absorb onto the permeable casing for up to 30 min after which the hotdog casings were peeled and the hotdogs were vacuum packaged and stored in a blast freezer at −26 °C. Prior to use in experiments, hotdogs were thawed and then pasteurized as described earlier. Hotdogs from these various treatments were then processed identically: they were placed in vacuum packaging bags (2 hotdogs/bag) with sterile tongs, inoculated with 100 µL of *L. monocytogenes* 39-2, hand massaged to evenly distribute the inoculum, and vacuum packaged. The bags were stored at 5 °C and sampled at 0 h, 3 days and weekly at 1, 2, 4, 6, 8, 10, and 12 weeks. Samples were plated on acidified MRS agar (pH 5.5) for enumerating LAB (if present) or MOX agar for enumerating *L. monocytogenes* 39-2.

2.4.4. Trial #3: Application of Mixed Mode-of-Action Bac$^+$ CFS on the Surface of RTE Meats (Hotdogs)

Select Bac$^+$ CFS mixtures comprising 3 MOAs were obtained from *Lb. curvatus* FS47, *Lb. curvatus* Beef3, *Pe. acidilactici* Bac3, *En. faecium* FS56-1, *En. thailandicus* FS92, and/or *Lc. lactis* FLS1. Cultures were propagated, centrifuged, and CFS was processed as described earlier. Equal volumes of each CFS was mixed in a sterile tube to obtain a Bac$^+$ CFS mixture comprising 3 MOA; different bacteriocins were used to obtain the 3 MOA mixture as indicated in the analogous figure legends. As before, hotdogs were pasteurized as described earlier. Using sterile tongs, pasteurized hotdogs were placed in vacuum packaging bags (2 hotdogs/bag) to which 300-µL of sterile water (control) or Bac$^+$ CFS was added, massaged, and then inoculated with 100-µL of *L. monocytogenes* 39-2, hand-massaged again to distribute the inoculum, and then vacuum sealed. Samples were then stored at 5 °C and sampled at 0 h, 3 days and weekly at 1, 2, 4, 6, 8, 10, and 12 weeks. Samples were plated on acidified MRS agar (LAB) or MOX agar (*L. monocytogenes* 39-2).

2.4.5. Trials #4 and #5: Surface Application of Filter vs. Pasteurized Bac+ CFS and Neutralized vs. Non-Neutralized CFS in *L. monocytogenes* Challenge Studies on Hotdogs

Several additional modifications of the above were also examined, including a comparison of filter-sterilized vs. pasteurized Bac$^+$ CFS preparations and pH-neutralized vs. non-neutralized CFS preparations. A summary of the various trials and treatments received are presented in Table 2.

Table 2. Description of trials and treatments used in this study applying bacteriocin-producing cultures (Bac$^+$) or their cell free supernatants (CFS) to inhibit *L. monocytogenes* on RTE meats.

Trial	Description of Treatment	Data
Trial 1	Use of bacteriocin-producing (Bac$^+$) cultures vs. *L. monocytogenes*	Figure 2
Trial 2	Bac$^+$ CFS added into meat matrix before cooking	Figure 3
	Bac+ CFS sprayed onto hotdogs in casings before peeling	
Trial 3	Bac$^+$ CFS as surface treatment (includes CFS from 2 *Enterococcus* strains)	Figure 4A
	Bac$^+$ CFS as surface treatment (includes CFS from 1 *Enterococcus* strain)	Figure 4B
Trial 4	Bac$^+$ CFS as surface treatment: All CFS was from traditional lactic acid bacteria; filter vs. heat-pasteurized Bac$^+$ CFS	Figure 5
Trial 5	Bac$^+$ CFS as surface treatment: Neutralized vs. non-neutralized CFS and Bac$^+$ vs. Bac$^-$ CFS	Figure 6

2.5. Statistical Analysis

Shelf life assays were performed in triplicate and means were plotted versus time. The statistics functions in SigmaPlot 13 (Systat Software, San Jose, CA, USA) were used to perform one-way repeated measures analysis of variance (RM-ANOVA) to determine if significant difference exists between different treatments with level of significance set at 0.05 (p-value).

3. Results and Discussion

3.1. Trial #1: Application of Mixed MOA Bac$^+$ LAB vs. L. monocytogenes on Hotdogs

Trials were performed examining the use of Bac$^+$ LAB cultures as protective co-inoculants that comprised the 3 MOAs described previously [20,22]. The Bac$^+$ LAB were intentionally added at approximately 1-log higher level than the co-inoculated *L. monocytogenes* 39-2. In preliminary co-inoculation challenge studies with individual Bac$^+$ strains, inhibition of *L. monocytogenes* 39-2 was not observed (data not shown). We hoped to demonstrate microbial control of *L. monocytogenes* by mixing cultures comprising the 3 MOAs simultaneously vs. *L. monocytogenes* 39-2. However, we again did not observe any inhibition of *L. monocytogenes* in spite of the additional growth during storage of one or more of the Bac$^+$ strains exceeding that of *L. monocytogenes* by >3 logs (Figure 2). It is likely that the storage conditions were unsuitable for the cultures to produce any, or enough, bacteriocin to be inhibitory to *L. monocytogenes* 39-2. Although others have shown control of *L. monocytogenes* on RTE meats using LAB cultures [27,28], we did not observe this effect and demonstrates the difficulty in relying on the use of live competitive cultures to provide inhibitory protection to food products against potential pathogens. Another potential issue with the use of live protective cultures is that the level required for control, or their potential growth during storage, could be the equivalent of spoilage. It should be noted that lactic acid produced during culture growth may be buffered by the food matrix and that bacteriocins are secondary byproducts and their production is not necessarily concomitant with growth.

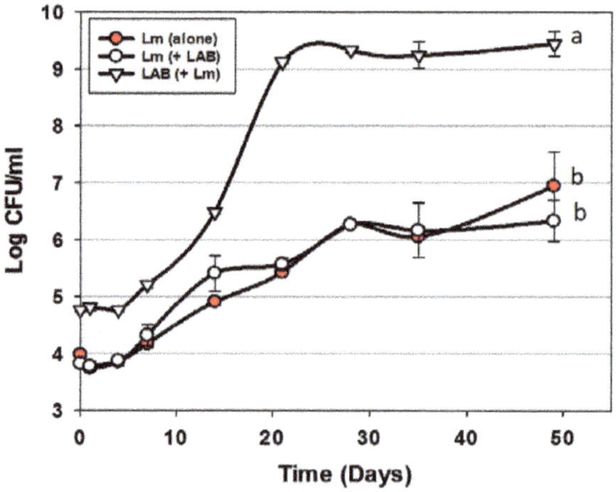

Figure 2. *L. monocytogenes* 39-2 challenge study (hotdogs) with multiple Bac$^+$ LAB cultures held at 5 °C for up to 49 days in vacuum packages (Trial #1). *L. monocytogenes* 39-2 was inoculated alone (Lm alone; ●) or in combination (Lm+LAB; ○) with 5 Bac$^+$ LAB (*En. faecium* FS56-1, *En. thailandicus* FS92, *En. faecium* FS97-2, *En. thailandicus* RP-1, and *Pe. acidilactici* Bac 3). All trials were performed in triplicate replication; data points represent the means and error bars represent standard deviation from the mean. Treatments with different letters are significantly different (repeated measures, $p < 0.05$); those with the same letters are not significantly different ($p > 0.05$).

3.2. Trial #2: Listeria Monocytogenes Challenge Studies Using Hotdogs Made with Bacteriocin Extracts Added during Manufacture or Sprayed Post-Cook onto Encased Products

Additional challenge studies were performed using CFS preparations, either added to the meat emulsion during manufacture or by manual spray onto the encased hotdogs after cooking, but before peeling (Figure 1). In prior testing of the heat stability of our bacteriocins, we found that they were able to tolerate high levels of heating, allowing us to use pasteurization to further insure that extracts were free of producer cells. Moreover, bacteriocins would provide a greater potential for application if their thermal tolerance allowed their inclusion in products that may be heated or cooked.

Application of bacteriocins during the manufacture of hotdogs provided excellent control of *L. monocytogenes* 39-2 during the 12 weeks of challenge showing a slight decline almost immediately that continued slowly through 84 days ending at approximately 1-log lower than was initially added (Figure 3). This level of control is exceptional compared to the >4-log increase observed for the control treatment and resulted in a difference of 5-logs. As with most applications that depend on ingredients added to the entire mass of product volume, this treatment required the highest amount of bacteriocin extract added as an ingredient (i.e., bacteriocin extract was approximately 27% (*w*/*w*) of the total emulsion composition). Another consideration that may occur in such applications is that the active bacteriocin may be reduced due to interaction with food matrix components during cooking, as it is known that bacteriocins have hydrophobic motifs that can partition into the fat phase of certain foods [29,30].

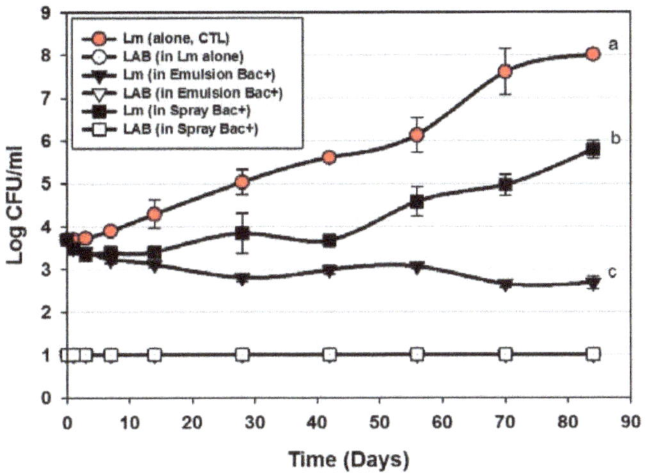

Figure 3. *L. monocytogenes* 39-2 challenge study on hotdogs with multiple MOA Bac$^+$ CFS held at 5 °C for up to 84 days in vacuum packages (Trial #2). *L. monocytogenes* 39-2 was inoculated onto untreated hotdogs (Lm, alone), inoculated onto hotdogs in which the bacteriocin mixture was mixed into the meat emulsion during manufacture (in emulsion Bac+), or inoculated onto hotdogs in which the bacteriocin mixture was previously sprayed while hotdogs were still in casings before peeling (in spray Bac+). The Bac$^+$ CFS was comprised of curvaticin FS47, curvaticin Beef3, pediocin Bac3, and lacticin FLS1. Platings for LAB from all treatments were made on acidified MRS (hollow symbols). All sample treatments were performed in triplicate replication; data points represent the means and error bars represent the standard deviation from the means. Treatments with different letters are significantly different (repeated measures, $p < 0.05$).

In contrast, the cooked and encased product that was sprayed with bacteriocin before peeling showed moderate inhibition relative to the control treatment (Figure 3). Although some bacteriocin probably penetrated the permeable casing, most of it likely washed off. Although a smaller amount was used to spray the encased hotdogs than when included in the meat matrix, one could argue it may be more effective to spray hotdogs after peeling than before peeling, and is reflected in our next approach. In the Bac^+ CFS challenge with *L. monocytogenes*, we also plated the liquid recovered from packages for potential indigenous lactic acid bacteria. We were careful to pasteurize our cooked, frozen, and thawed hotdogs prior to use in challenge studies and did not want production of lactic acid from potential indigenous bacteria to influence interpretation of inhibition from added bacteriocin. The data indicates that indigenous LAB were below our limit of detection and did not contribute to the inhibition observed. In real commercial applications, of course any further contributory inhibition by lactic acid contributed by indigenous LAB would be welcome to help inhibit potential pathogens such as Listeria. The bacteriocin mixture might have also been inhibitory to potential contaminating LAB as well (i.e., sensitive).

3.3. *Trials #3, #4, and #5: Listeria monocytogenes Challenge Studies with Multiple-MOA Bacteriocin Extracts Added after Peeling (During Packaging)*

Several trials were conducted by adding CFS preparations directly to packages prior to vacuum packaging. These applications utilize the least amount of bacteriocin because they are applied on the product surface after cooking and vacuum packaging can provide a tight space for them to perform as a thin film between food product and packaging film where most surface microorganisms may be found. Most post-process contamination of RTE meats usually occurs on the product surface, as *L. monocytogenes* is mostly a surface problem resulting from contact with contaminated food contact surfaces.

During the progression of our studies, we used a variety of strains that were grouped according to MOA as described earlier. We confirmed the identity of our organisms using 16S rRNA PCR amplification followed by sequencing for both old stock cultures and newly identified strains as reported elsewhere [26]. The use of 16S rRNA analysis had shown that some strains previously identified as *Lactococcus* by API metabolic assays were actually *Enterococcus* [25,26]. *Enterococcus* sp. are commonly isolated from foods [31], some are even used as starter cultures [32] and still others as probiotics [33]. However, their use in foods has been challenged because of their involvement as opportunistic human pathogens [34]. We don't feel the use of *Enterococcus* strains are a problem with our work because of the use of cell-free extracts rather than live strains. However, we were still interested in seeing if we could develop a repertoire of strains comprised solely of bacteriocin extracts from what is generally considered traditional 'food-grade' lactic acid bacteria should cell-free extracts from enterococcal strains become a debilitating issue.

Figure 4 represents several hotdog challenge studies whereby we used 2 enterococcal bacteriocins (Figure 4A) and then traded out one of them out for a lactococcal bacteriocin and used only 1 enterococcal bacteriocin (Figure 4B). There was moderate inhibition and control of *L. monocytogenes* when CFS was added to the meat matrix before cooking (Figure 3). However, when Bac^+ CFS was surface-applied, the data shows a significant drop in *L. monocytogenes* within the first few days of storage (~2 logs) and continues until about 7–10 days showing a stable level of *L. monocytogenes* at or near the limit of detection (Figure 4A,B).

The USDA-FSIS regulations for 'control of *L. monocytogenes* in RTE meats' specifies conditions for several risk categories of RTE meats whereby an Alternative 1 process (least risk) is described as possessing a post-process lethality step for *L. monocytogenes* (i.e., $\geq$1-log reduction) and control of *L. monocytogenes* (i.e., $\leq$2-log increase) during shelf life [35]. This is often obtained by two separate mechanisms but can also be achieved by a single treatment. The data presented herein may satisfy those requirements for both post-process lethality and control of *L. monocytogenes* during shelf life using a single surface treatment with these bacteriocin preparations.

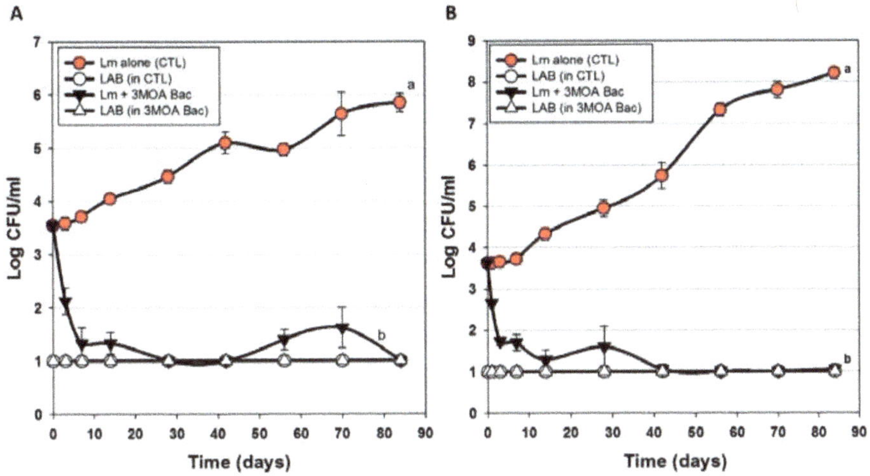

Figure 4. Hotdog challenge study with surface-applied bacteriocin extracts comprising 3 mixed MOAs. *L. monocytogenes* 39-2 was either inoculated onto hotdogs alone (as control) or with added bacteriocin extracts (Trial #3). (**A**) bacteriocin extracts included curvaticin FS47, pediocin Bac3, enterocin FS56-1, and enterocin FS92. (**B**) bacteriocin extracts included curvaticin FS47, pediocin Bac3, enterocin FS56-1, and lacticin FLS1. Plate counts for LAB from these treatments were made on acidified MRS (hollow symbols). All sample treatments were performed in triplicate replication; data points represent the means and error bars represent the standard deviation from the means. Treatments with different letters are significantly different (repeated measures, $p < 0.05$).

We further examined the use of Bac$^+$ CFS preparations as hotdog surface treatments using only traditional lactic acid bacteria from our collection that complied with the 3 mode of actions defined earlier. Using this approach, we also compared the efficacy of filter-sterilized vs. heat-pasteurized bacteriocins (Figure 5). The data appears no different when using filter-sterilized vs. heat-pasteurized bacteriocin extracts demonstrating that heat pasteurization (after centrifugation) imparts no detrimental effect to the bacteriocins and is an easy method of eradicating residual bacteriocin-producer cells (Figure 5). The initial decrease of *L. monocytogenes* was not as dramatic as that observed with the enterococcal bacteriocins (Figure 4). We ascribe this to the moderate production of bacteriocin FLS1 by *Lactococcus lactis* FLS1 compared to that produced by others in the CFS. The enterococcal strains have also been shown to possess genes for multiple enterocins [21]. *L. monocytogenes* increased approximately 5-log in control samples, showing >6.5-log difference between control and treatments. Again, LAB were not detected within the trials and demonstrates that inhibitory action was again solely provided by the added bacteriocins.

Since LAB bacteriocin culture extracts may also contain lactic acid, we further examined the use of neutralized vs. non-neutralized CFS from both Bac$^+$ (bacteriocin treatment) and bacteriocin-negative (Bac$^-$, control) LAB in order to more confidently assert the inhibition to bacteriocin-related antimicrobial activity (Figure 6). The use of Bac$^-$ LAB culture extract (*Lb. delbrueckii* 4797) was an additional control treatment to evaluate whether lactic acid produced by cultures (without the influence of bacteriocin) was contributory to the inhibition observed in these assays as we have observed a contributory effect of lactic acid in culture extracts in microplate *in vitro* assays [22].

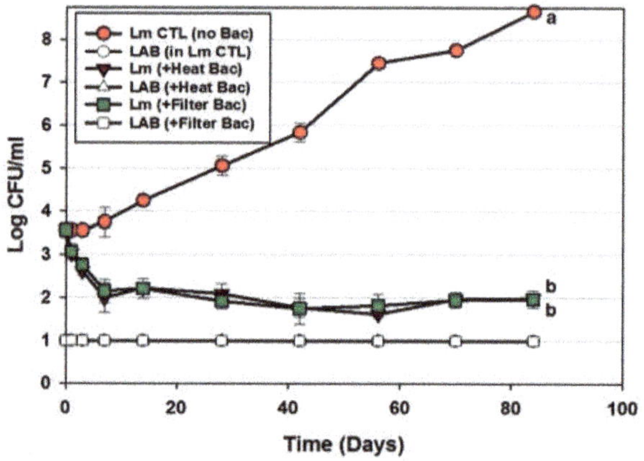

Figure 5. Hotdog challenge study with surface-applied bacteriocin extracts comprising 3 MOAs (curvaticin FS47 and Beef3, pediocin Bac3, and lacticin FLS1; Trial #4). *L. monocytogenes* 39-2 was either inoculated on hotdogs alone, or with added heat-treated or filter-sterilized CFS preparations, vacuum-packaged, and held for up to 12 weeks at 5 °C. Platings for LAB from these 3 treatments were also made on acidified MRS (hollow symbols). All sample treatments were performed in triplicate; data points represent the means and error bars represent the standard deviation from the means. Treatments with different letters are significantly different (repeated measures, $p < 0.05$); those sharing the same letter are not significantly different, $p > 0.05$).

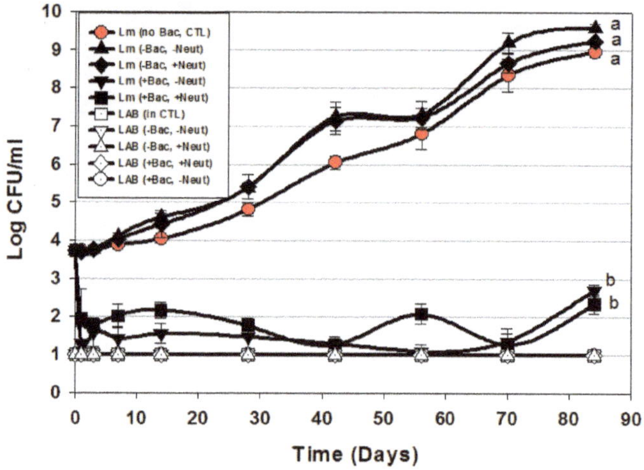

Figure 6. Hotdog challenge study vs. *L. monocytogenes* 39-2 with surface-applied bacteriocin extracts (+Bac) comprising mixed MOAs (curvaticin FS47 and Beef3, pediocin Bac3, and lacticin FLS1) and neutralized (+Neut) vs. non-neutalized (-Neut) culture extracts (Trial #5). *Lb. delbueckii* 4797 was used for bacteriocin-negative (-Bac) CFS extracts that were also used both neutralized (+Neut) and non-neutralized (-Neut). All sample treatments were performed in triplicate replication; data points represent the mean and error bars represent the standard deviation from the mean. Treatments with different letters are significantly different (repeated measures, $p < 0.05$); those sharing the same letter are not significantly different, $p > 0.05$).

The data shows that addition of neutralized (-Bac, +Neut) or non-neutralized (-Bac, -Neut) CFS extracts from a Bac^{-} LAB culture did not show any inhibition of *L. monocytogenes* in comparison to the control treatment in which sterile water was added instead of Bac^- culture extracts (Figure 6). Although the treatments whereby CFS from a Bac^- culture were added showed no significant difference to *L. monocytogenes* 39-2 to which water was added instead of CFS, they did have slightly higher growth levels, perhaps attributed from additional nutrients from the CFS. The main point was that lactic acid in the Bac^- extracts did not inhibit *L. monocytogenes* (Figure 6). In another study using microplate *in vitro* assays, we observed definite lactic acid effects when comparing the effects of neutralized vs. non-neutralized extracts [22]. We suggest that the difference lies in the microplate assay format whereby culture extracts comprised approximately 30% of the assay volume in prior *in vitro* assays [22] whereby in these assays the added culture extracts comprised <0.1% of the package weight which contained substantial organic material (i.e., hotdogs) that can readily buffer the effects of organic acids from even non-neutralized extracts. Addition of the multi-MOA bacteriocin mixture resulted in approximately a 2-log reduction of *L. monocytogenes* within the first 1–3 days of addition that did not show any increase (beyond the initial added level) during the 12-week challenge study, showing ~7-log difference from controls. These data again suggest that these treatments satisfy the requirements for USDA-FSIS Alternative 1 product classification for RTE meat and poultry products [35].

4. Conclusions

The data presented herein shows the efficacious application of a mixture of bacteriocins against *L. monocytogenes*. The bacteriocin-producing strains were isolated from various sources including foods found in supermarkets [25,26], however it was the unique method of procuring and using spontaneous resistant mutants as 'microbial screens' to categorize them into different modes of action (MOA) [20–22] and then using a mixture comprising those different MOAs in order to provide an effective cocktail of natural antimicrobial bacteriocins that active synergistically to inhibit *L. monocytogenes* in RTE meats as observed in this study. Although others are also using surface application of bacteriocins [36], we feel that the multiple MOA approach works well to minimize the possible development of spontaneous bacteriocin-resistant mutants that readily occurs with bacteriocins of similar MOA [20,21]. Since our bacteriocins are heat resistant, they can be added in, or on, foods that may be heated or cooked. The use of culture extracts provides an opportunity for the biological activity to be standardized whereas the use of live cultures to provide antimicrobial protection in non-actively growing situations is tenuous. We feel that such bacteriocin extracts produced by food-grade LAB may be used freely as food ingredients to act as food preservatives (i.e., biopreservatives) in RTE meats and other products where they show proven efficacy against targeted pathogens and susceptible spoilage organisms.

Acknowledgments: This paper was funded in part by an OCAST-OARS grant (#AR12-049), the Nutrition Physiology Co. (Guymon, OK, USA), the Dept. of Animal Science (Advance Foods-Gilliland Professorship), the R.M. Kerr Food & Ag Products Center, the USDA National Institute of Food and Agriculture, [Hatch Project #OKL02885], and the Division of Agricultural Sciences and Natural Resources at Oklahoma State University.

Author Contributions: Paul Priyesh Vijayakumar performed the research as partial fulfillment of the requirements for the PhD degree and wrote the initial draft of the manuscript. Peter M. Muriana was the academic and research advisor on record, the PI of the grant from which funding was obtained to perform the work, and edited the submitted manuscript.

Conflicts of Interest: The authors declare no conflict of interest.

References

1. Hernandez-Milian, A.; Payeras-Cifre, A. What Is new in listeriosis? *BioMed Res. Int.* **2014**, *2014*, 358051. [CrossRef] [PubMed]
2. Leclercq, A. *Chapter 2.9.7, Listeria Monocytogenes*, 2015th ed.; The World Organisation for Animal Health (OIE): Paris, France, 2015; pp. 1–18.
3. Levine, P.; Rose, B.; Green, S.; Ransom, G.; Hill, W. Pathogen testing of ready-to-eat meat and poultry products collected at federally inspected establishments in the United States, 1990 to 1999. *J. Food Prot.* **2001**, *64*, 1188–1193. [CrossRef] [PubMed]

4. Gamble, R.; Muriana, P.M. Microplate fluorescence assay for measurement of the ability of strains of *Listeria monocytogenes* from meat and meat-processing plants to adhere to abiotic surfaces. *Appl. Environ. Microbiol.* **2007**, *73*, 5235–5244. [CrossRef] [PubMed]
5. Kushwaha, K.; Muriana, P.M. Adherence characteristics of *Listeria* strains isolated from three ready-to-eat meat processing plants. *J. Food Prot.* **2009**, *72*, 2125–2131. [CrossRef] [PubMed]
6. U.S. Food and Drug Administration Center for Food Safety and Applied Nutrition (FDA-CFSAN); USDA Food Safety and Inspection Service (USDA-FSIS). *Quantitative Assessment of Relative Risk to Public Health from Foodborne Listeria monocytogenes among Selected Categories of Ready-to-Eat Foods*; U.S. Food and Drug Administration Center for Food Safety and Applied Nutrition: College Park, MD, USA, 2003.
7. Moberg, L. Good manufacturing practices for refrigerated foods. *J. Food Prot.* **1989**, *52*, 363–367. [CrossRef]
8. NACMCF. Hazard analysis and critical control point principles and application guidelines. National Advisory Committee on Microbiological Criteria for Foods. *J. Food Prot.* **1998**, *61*, 762–775.
9. El-Ziney, M.; Debevere, J.; Jakobsen, M.; Reuterin, N.A. *Natural Food Antimicrobial Systems*; CRC Press: London, UK, 2000; pp. 567–587.
10. Holzapfel, W.; Geisen, R.; Schillinger, U. Biological preservation of foods with reference to protective cultures, bacteriocins and food-grade enzymes. *Int. J. Food Microbiol.* **1995**, *24*, 343–362. [CrossRef]
11. Magnusson, J.; Schnürer, J. Lactobacillus coryniformis subsp. coryniformis strain Si3 produces a broad-spectrum proteinaceous antifungal compound. *Appl. Environ. Microbiol.* **2001**, *67*, 1–5. [CrossRef] [PubMed]
12. Deegan, L.H.; Cotter, P.D.; Hill, C.; Ross, P. Bacteriocins: Biological tools for bio-preservation and shelf-life extension. *Int. Dairy J.* **2006**, *16*, 1058–1071. [CrossRef]
13. Papagianni, M.; Anastasiadou, S. Pediocins: The bacteriocins of Pediococci. Sources, production, properties and applications. *Microb. Cell Factories* **2009**, *8*, 3. [CrossRef] [PubMed]
14. Balciunas, E.M.; Castillo Martinez, F.A.; Todorov, S.D.; Franco, B.D.G.D.M.; Converti, A.; Oliveira, R.P.D.S. Novel biotechnological applications of bacteriocins: A review. *Food Control* **2013**, *32*, 134–142. [CrossRef]
15. Cleveland, J.; Montville, T.J.; Nes, I.F.; Chikindas, M.L. Bacteriocins: Safe, natural antimicrobials for food preservation. *Int. J. Food Microbiol.* **2001**, *71*, 1–20. [CrossRef]
16. Daeschel, M. *Applications of Bacteriocins in Food Systems*; Butterworth-Heinemann: Boston, MA, USA, 1990; pp. 91–115.
17. Muriana, P.M. Bacteriocins for control of Listeria spp in food. *J. Food Prot.* **1996**, 54–63.
18. Anonymous. Nisin preparation: Affirmation of GRAS status as a direct human food ingredient. *Fed. Regist.* **1988**, *54*, 11247–11251.
19. Lemay, M.J.; Choquette, J.; Delaquis, P.J.; Claude, G.; Rodrigue, N.; Saucier, L. Antimicrobial effect of natural preservatives in a cooked and acidified chicken meat model. *Int. J. Food Microbiol.* **2002**, *78*, 217–226. [CrossRef]
20. Macwana, S.; Muriana, P.M. Spontaneous bacteriocin resistance in *Listeria monocytogenes* as a susceptibility screen for identifying different mechanisms of resistance and modes of action by bacteriocins of lactic acid bacteria. *J. Microbiol. Methods* **2012**, *88*, 7–13. [CrossRef] [PubMed]
21. Henning, C.; Gautam, D.; Muriana, P. Identification of multiple bacteriocins in *Enterococcus* spp. using an *Enterococcus*-specific bacteriocin PCR array. *Microorganisms* **2015**, *3*, 1–16. [CrossRef] [PubMed]
22. Vijayakumar, P.P.; Muriana, P.M. A microplate growth inhibition assay for screening bacteriocins against *Listeria monocytogenes* to differentiate their mode-of-action. *Biomolecules* **2015**, *5*, 1178–1194. [CrossRef] [PubMed]
23. Chunhua, W.; Muriana, P.M. Incidence of *Listeria monocytogenes* in packages of retail franks. *J. Food Prot.* **1994**, *57*, 382–386.
24. Bouchard, D.S.; Seridan, B.; Saraoui, T.; Rault, L.; Germon, P.; Gonzalez-Moreno, C.; Nader-Macias, F.M.E.; Baud, D.; François, P.; Chuat, V.; et al. Lactic acid bacteria isolated from bovine mammary microbiota: Potential allies against bovine mastitis. *PLoS ONE* **2016**, *10*, e0144831. [CrossRef] [PubMed]
25. Garver, K.I.; Muriana, P.M. Detection, identification and characterization of bacteriocin-producing lactic acid bacteria from retail food products. *Int. J. Food Microbiol.* **1993**, *20*, 241–258. [CrossRef]
26. Henning, C.; Vijayakumar, P.; Adhikari, R.; Jagannathan, B.; Gautam, D.; Muriana, P.M. Isolation and taxonomic identity of bacteriocin-producing lactic acid bacteria from retail foods and animal sources. *Microorganisms* **2015**, *3*, 80–93. [CrossRef] [PubMed]

27. Amezquita, A.; Brashears, M.M. Competitive inhibition of *Listeria monocytogenes* in ready-to-eat meat products by lactic acid bacteria. *J. Food Prot.* **2002**, *65*, 316–325. [CrossRef] [PubMed]
28. Holck, A.; Berg, J. Inhibition of *Listeria monocytogenes* in cooked ham by virulent bacteriophages and protective cultures. *Appl. Environ. Microbiol.* **2009**, *75*, 6944–6946. [CrossRef] [PubMed]
29. Chollet, E.; Sebti, I.; Martial-Gros, A.; Degraeve, P. Nisin preliminary study as a potential preservative for sliced ripened cheese: NaCl, fat and enzymes influence on nisin concentration and its antimicrobial activity. *Food Control* **2008**, *19*, 982–989. [CrossRef]
30. Jung, D.-S.; Bodyfelt, F.W.; Daeschel, M.A. Influence of Fat and Emulsifiers on the Efficacy of Nisin in Inhibiting *Listeria monocytogenes* in Fluid Milk1. *J. Dairy Sci.* **1992**, *75*, 387–393. [CrossRef]
31. Giraffa, G. Enterococci from foods. *FEMS Microbiol. Rev.* **2002**, *26*, 163–171. [CrossRef] [PubMed]
32. Hugas, M.; Garriga, M.; Aymerich, M.T. Functionalty of enterococci in meat products. *Int. J. Food Microbiol.* **2003**, *88*, 223–233. [CrossRef]
33. Franz, C.M.A.P.; Huch, M.; Abriouel, H.; Holzapfel, W.; Gálvez, A. Enterococci as probiotics and their implications in food safety. *Int. J. Food Microbiol.* **2011**, *151*, 125–140. [CrossRef] [PubMed]
34. Franz, C.M.A.P.; Holzapfel, W.H.; Stiles, M.E. Enterococci at the crossroads of food safety? *Int. J. Food Microbiol.* **1999**, *47*, 1–24. [CrossRef]
35. USDA Food Safety and Inspection Service (USDA-FSIS). *Compliance Guidelines to Control Listeria monocytogenes in Post-Lethality Exposed Ready-to-Eat Meat and Poultry Products*; USDA-FSIS: Washington, DC, USA, 2012.
36. Ünlü, G.; Nielsen, B.; Ionita, C. Inhibition of *Listeria monocytogenes* in hot dogs by surface application of freeze-dried bacteriocin-containing powders from lactic acid bacteria. *Probiotics Antimicrob. Proteins* **2016**, *8*, 102–110. [CrossRef] [PubMed]

foods

MDPI

Review

Bioactive Peptides

Eric Banan-Mwine Daliri [1], Deog H. Oh [1] and Byong H. Lee [1,2,*]

[1] Department of Food Science and Biotechnology, Kangwon National University, Chuncheon 24341, Korea; ericdaliri@yahoo.com (E.B.-M.D.); deoghwa@kangwon.ac.kr (D.H.O.)
[2] Department of Microbiology/Immunology, McGill University, Montreal, QC H3A 0G4, Canada
* Correspondence: byong.lee@mail.mcgill.ca; Tel: +82-10-4779-9808

Academic Editor: Maria Hayes
Received: 17 March 2017; Accepted: 20 April 2017; Published: 26 April 2017

Abstract: The increased consumer awareness of the health promoting effects of functional foods and nutraceuticals is the driving force of the functional food and nutraceutical market. Bioactive peptides are known for their high tissue affinity, specificity and efficiency in promoting health. For this reason, the search for food-derived bioactive peptides has increased exponentially. Over the years, many potential bioactive peptides from food have been documented; yet, obstacles such as the need to establish optimal conditions for industrial scale production and the absence of well-designed clinical trials to provide robust evidence for proving health claims continue to exist. Other important factors such as the possibility of allergenicity, cytotoxicity and the stability of the peptides during gastrointestinal digestion would need to be addressed. This review discusses our current knowledge on the health effects of food-derived bioactive peptides, their processing methods and challenges in their development.

Keywords: functional foods; antihypertensive peptides; bioactivity; cytotoxicity

1. Introduction

Proteins in foods do not only serve as nutrients but also perform physiochemical roles that promote health. Most of the physiological activities of proteins are performed by peptide sequences encrypted in the parent protein which become active when cleaved intact [1]. Bioactive peptides are released during enzymatic proteolysis (gastrointestinal digestion, in vitro hydrolysis using proteolytic enzymes) of proteins and also during food processing (cooking, fermentation, ripening). Bioactive peptides are known for their ability to inhibit protein-protein interactions due to their small size and specificity. Nature remains the largest source of bioactive peptides since plants, animals, fungi, microbes and their products contain various proteins in them. Over the years, bioactive peptides in foods have been discovered through a classical approach or a bioinformatic approach. The classical approach involves hydrolyzing proteins with food-grade proteolytic enzymes to release numerous peptide fragments in the hydrolysate [2–4]. Alternatively, proteins could be fermented by bacteria [5–7]. The bacteria proteolytic enzymes hydrolyze the proteins to release peptides into the hydrolysate. The hydrolysates are then tested in vitro for a biological activity. If the hydrolysates show good bioactivity, they are then confirmed through in vivo testing. The biologically active hydrolysate could then be developed into a functional food. The bioactive peptides in the hydrolysates could also be separated and purified into nutraceuticals for nonpharmacological therapy.

On the other hand, the bioinformatic (in silico) approach relies on information available on a database to determine the frequency of occurrence of already identified bioactive peptides in a protein of interest. Specific enzymes that can cleave the identified segments from the parent protein are chosen to hydrolyze the peptides. This strategy enhances the identification of known peptides from unknown proteins. A key challenge in bioactive peptide development for therapeutic purposes has however been the difficulty in establishing a cause and effect relationship between bioactive peptide consumption and their intended health effects in humans. Yet, since studies continue to confirm the therapeutic effects of bioactive peptides, a comprehensive review of recent advances in bioactive peptide research is warranted. In this review, we discuss the recent advances in the production of bioactive peptides from food including antidiabetic peptides, cholesterol-lowering peptides, antihypertensive peptides, anticancer peptides, antimicrobial peptides and multifunctional peptides. We also discuss the effects of processing methods on the bioactivity of peptides and the challenges associated with bioactive peptide development.

2. Production and Processing of Food Protein-Derived Bioactive Peptides

From literature, the most common methods to produce bioactive peptides have been by enzyme hydrolysis of food proteins or by fermentation [8]. In few situations however, water extracts of mushrooms and some plant parts have been found to contain bioactive peptides [9].

2.1. Enzymatic Hydrolysis

In this process, the protein material is subjected to enzyme hydrolysis at a given temperature and pH [10–13]. The use of enzymatic hydrolysis to produce bioactive peptides is preferred than microbial fermentation due to the short reaction time, ease of scalability and predictability. More than a single proteolytic enzyme (whether purified or crude) can be used to hydrolyze the protein to produce the hydrolysate containing short peptide sequences. However, addition of the enzymes (whether simultaneously or sequentially) would depend on the optimal pH and temperature of the enzymes [14–17]. Though no specific proteolytic enzymes are known to produce specific bioactive peptides in foods, subtilisin hydrolysis tends to yield low molecular weight peptides some of which are bioactive. For instance, Huang et al. [18] found that subtilisin hydrolyzed *Achatina fulica* snail foot muscle protein had a higher number of small molecular weight peptides than samples hydrolyzed by papain and trypsin. Zhang et al. [19] also showed that subtilisin hydrolysed rice bran proteins generated the highest number of low molecular weight peptides and showed higher biological activity than samples hydrolyzed with a cysteine endopeptidase, papain and pepsin. However, the enzyme to substrate ratio is an important factor to consider so as to obtain a good degree of hydrolysis. Peptide sequences and their biological activities may differ depending on the type of enzyme used [20]. Low molecular weight peptides (<10 kDa) have been found to be more effective antioxidants and antihypertensive peptides [21–24] than high molecular weight peptides and hence proteases that yield low molecular weight peptides would be helpful for commercial production of antioxidant and antihypertensive peptides.

Since foods may contain several other non-protein bioactive compounds, it is advisable to separate such compounds from the food proteins to avoid interference. For instance, phenolic compounds are known for their antioxidant [25], antihypertensive, antidiabetic [26] and antimicrobial [27] abilities and hence, when present in hydrolysates, can interfere with the biological activities being assayed. Phenolic compounds could be separated from food proteins using methods such as ethanol extraction [24], supercritical carbon dioxide [28] pressurized water extraction [29], ultrasound-assisted extraction [30] and acetone extraction [31] prior to enzymatic hydrolysis.

Protons released during proteolysis result in fluctuations in the pH of the medium and this may affect the efficiency of the hydrolytic process. Though the pH can be adjusted by the addition of acid or alkali solutions [13,32], addition of alkali usually results in high salt concentrations in the hydrolysates. It is therefore advisable to perform the proteolysis in a buffer [24,33,34]. The type of enzyme used, the temperature and the time allowed for hydrolysis affect the extent of hydrolysis and may also affect the type of peptides generated in the hydrolysates. For instance, Rerri et al. [35] observed stronger anti-tyrosinase and anti-inflammatory activities after hydrolyzing rice derived proteins with bacillolysin than samples hydrolyzed with subtilisin while cysteine endopeptidase, papain and leucyl aminopeptidase hydrolyzed samples showed the lowest bioactivity. They also found that, bacillolysin hydrolysed samples had the strongest anti-angiotensin converting enzyme (ACE) activity followed by cysteine endopeptidase, and then subtilisin. However, other studies found that chymotrypsin or cysteine endopeptidase hydrolyzed rice proteins had strong antioxidant abilities [36,37].

After enzymatic hydrolysis, the mixture is centrifuged to separate the supernatant which contains low molecular weight peptides from the precipitates [32,38,39]. The peptides may be recovered by freeze-drying, desalting [19], cross-flow membrane filtration [35], and membrane ultrafiltration or column chromatography. Gel filtration can be used to quickly desalt low molecular weight peptides and separate them based on their sizes.

2.2. Microbial Fermentation

This involves culturing some bacteria or yeast on protein substrates to hydrolyze the proteins with their enzymes as they grow. The growing bacteria or yeast secret their proteolytic enzymes into the protein material to release peptides from the parent proteins. Usually, the bacterium of choice is grown to its exponential phase in a broth at a temperature suitable for the bacterial growth. The cells are then harvested, washed and suspended in sterile distilled water (usually containing glucose) and used as a starter to inoculate a sterilized protein substrate [7,40]. The extent of hydrolysis would depend on the strain used, the type of protein and the fermentation time. We observed that whey fermented by *Lactobacillus brevis* had a stronger ACE inhibitory ability than those fermented with *Lb. acidophilus, Lb. bifermentan, Lb. casei, Lb. helveticus, Lb. lactis, Lb. paracasei, Lb. plantarum and Lb. reuteri* [38]. This shows that the functionality of protein hydrolysates may differ between cultures since microorganisms have different proteolytic systems [41]. Similar results were reported by El-Fattah et al. [42] when they observed that 14 commercial dairy starters exhibited different degrees of proteolysis, ACE inhibition and antioxidant activities after milk fermentation. Bacteria of the same species may also differ in their proteolytic capacities which may result in different bioactivities. For instance, Sanjukta et al. [43] observed that fermenting soybean protein with *B. subtilis* MTCC5480 resulted in a higher degree of hydrolysis and free amino acids than samples fermented with *B. subtilis* MTCC1747. Chen et al. [44] also reported that skim milk fermented with 37 different *Lactobacillus helveticus* strains showed different extents of hydrolysis, titrable acidity, free amino nitrogen and ACE inhibitory abilities.

Apart from bacteria starter, yeast [6,21,45,46] and filamentous fungus [47–49] have also been used in producing bioactive peptides. Proteins can be co-cultured using a combination of different bacteria or even yeast and bacteria to accelerate the proteolytic process [50]. After fermentation, the mixture is centrifuged and the supernatant recovered. The supernatant may then be subjected to further hydrolysis using proteolytic enzymes to obtain shorter peptide sequences [3]. Alternatively, the low molecular weight peptides in the supernatant can be recovered by solvent extraction or other methods, purified and their amino acid sequences determined by mass spectrometry.

3. Food-Derived Bioactive Peptides and Human Health

3.1. Antidiabetic Peptides

Diabetes is a metabolic disease characterized by increased blood sugar level due to insufficiencies in insulin secretion, action, or both. The disease is classified into type I and type II. Type I diabetes (insulin dependent diabetes) is an autoimmune disease that causes the beta cells of the pancreas to secrete little or no insulin. In type 2 diabetes mellitus (T2DM), however, there is an imbalance in insulin secretion and blood sugar absorption [51]. Current synthetic antidiabetic drugs may result in risks of hypoglycemia, weight gain [52], high background risk of pancreatitis [53] and gastrointestinal side effects [54] while some patients may not even tolerate them [55]. For these reasons, the search for food derived anti-diabetic peptides (Table 1) is on the increase. Such alternatives may be safe as they are from food sources and have been consumed over the years without side effects [56]. Many fermented foods such as fermented soybean contain low molecular weight peptides some of which have been shown to induce insulin-stimulated glucose uptake in 3T3-L1 cells [57] and antagonize PPAR-γ activities. The peptides AKSPLF, ATNPLF, FEELN, and LSVSVL isolated from black bean protein hydrolysates effectively inhibited glucose transporter 2 (GLUT2) and sodium-dependent glucose transporter 1 (SGLT1) [58] to reduce blood glucose levels. In another study, peptides in salmon frame protein hydrolysates were found to significantly improve glucose uptake in L6 muscle cells showing their potential to improve blood glucose uptake [59]. Similarly, LPIIDI and APGPAGP from Silver carp (*Hypophthalmichthys molitrix* Val.) protein hydrolysates showed strong competitive/non-competitive mixed-type inhibition against DPP-IV [60].

Table 1. Bioactive peptides and their functions.

Peptide	Function	Reference
Antimicrobial		
LRLKKYKVPQL	Interacts with bacteria to cause inhibition.	[61]
PGTAVFK	Causes bacteria and yeast membrane destruction.	[62]
KVGIN, KVAGT, VRT, PGDL, LPMH, EKF, IRL	Inhibits *Listeria ivanovii* and *E. coli* growth.	[63]
Lp-Def$_1$	Interacts with and impairs mitochondrial functions in *C. albicans*.	[64]
Maize α-hairpinins	Binds to microbial DNA to cause cell death.	[65]
Antihypertensive		
DVWY, FQ, VVG. DVWY, VAE, WTFR	Inhibit ACE in thoracic aorta tissue and suppress angiotensin II-mediated vasoconstriction.	[5,66]
DPYKLRP, PYKLRP, YKLRP, GILRP		[21]
VPP, IPP		[44]
GAAGGAF		[67]
LIVTQ, LIVT		[68]
LLKPY	Competitively bind and inhibit ACE and results in blood pressure reduction	[69]
AHLL		[70]
FISNHAY		[71]
AAATP		[72]
LGL, SFVTT		[73]
IIT		[74]
ADVFNPR, VVLYK, LPILR, VIGPR	Lower endothelia-1 levels significantly	[75]
Anti-type 2 diabetes mellitus		
PPL		[74]
YP, LP, IPI, VPL, IPA, IPAVF		[33]
PGVGGPLGPIGPCTE, CAYNTERPVDRIR, PACCGPTISRPG		[76]
GPAE, GPGA	Inhibits dipeptidyl peptidase-IV	[77]
MHQPPQPL, AWPQYL, SPTVMFPPQSVL, VMFPPQSVL, AWPQYL and INNQFLPYPY		[78]
ILAP, LLAP, MAGVAHI		[79]
IP, MP, VP, LP		[80]
LKPTPEGDL, LPYPY, IPIQY and WR		[81]
Immunomodulatory		
GFLRRIRPKLKT	Significantly inhibits LPS-induced nuclear translocation of NF-κB/p65, inhibits IL-1β and enhances TNF-α release.	[82]
St20	Inhibits human T lymphocyte surface marker CD69 expression and cytokine IL-2 secretion. St20 also inhibits TNF-α and IFN-γ secretion in the activated human T lymphocytes.	[83]
PTGADY	Significantly increases the production of IL-2, IL-4, and IL-6.	[84]

Table 1. *Cont.*

Peptide	Function	Reference
Anti-oxidation		
IP, MP, VP, LP	Scavenge Hydroxyl radicals	[80]
AEERYP, DEDTQAMP	Scavenge reactive oxygen species	[32]
DHTKE, MPDAHL, FFGFN	Oxygen radical scavenging, DPPH radical scavenging.	[85]
RPNYTDA, TSQLLSDQ, TRTGDPFF, NFHPQ	DPPH and ABTS radical scavenging, FRAP-Fe^{3+} reducing ability.	[86]
LANAK, PSLVGRPPVGKLTL, VKVLLEHPVL	DPPH radicals scavenging ability	[87]
WEGPK, GPP and GVPLT	DPPH, ABTS, and hydroxyl radicals	[88]
PYSFK, GFGPEL, GGRP	scavenging ability, inhibiting lipid peroxidation	[89]
LSGYGP	Hydroxyl radicals scavenging.	[90]
GSGGL, GPGGFI, FIGP	DPPH, Hydroxyl and reactive oxygen radical scavenging	[89]
PIIVYWK, TTANIEDRR, FSVVPSPK	Hydrogen peroxide radicals scavenging	[91]
YYIVS		[92]
FIMGPY, GPAGDY and IVAGPQ	DPPH, Hydroxyl and reactive oxygen radical scavenging	[93]
ATSHH	DPPH radicals scavenging	[94]
TPP	Lipid peroxidation, radical scavenging activity.	[95]
WVAPLK	DPPH, Hydroxyl and reactive oxygen radical scavenging	[96]
GASRHWYFL	DPPH, superoxide, ABTS and hydroxyl radical	[97]
PYSFK, GFGPEL, VGGRP	scavenging, lipid peroxidation.	[98]

A = alanine, R = arginine, N = asparagine, D = aspartic acid, C = cysteine, E = glutamic acid, Q = glutamine, G = glycine, H = histidine, I = isoleucine, L = leucine, K = lysine, M = methionine, F = phenylalanine, P = proline, S = serine, T = threonine, W = tryptophan, Y = tyrosine, V = valine. LDL: Low-density lipoprotein, IL: Interleukin, TNFα: tumor necrosis factor alpha, DPPH: 2,2-diphenyl-1-picrylhydrazyl, ABTS: 2,2′-azinobis-(3-ethylbenzothiazoline-6-sulfonic acid), FRAP: ferric reducing antioxidant power, IFNγ: interferon gamma.

α-Amylase inhibitory peptides such as FFRSKLLSNGAAAAKGALLPQYW (CSP1), RCMAFLLSNGAAAAQQLLPQYW (CSP2) and RPAQPNYPYTAVLVFRH (CSP3) obtained from cumin seeds prevented dietary starch absorption by inhibiting the breakdown of complex starches into simpler ones [99] and may therefore have antidiabetic effects. CSP1 directly interacted with α-amylase binding site while CSP2 bound to the enzyme surface. CSP3 however bound to the enzyme interface to inhibit the enzyme activity. Huang and Wu [100] isolated an 8 kDa peptide from shark liver which significantly reduced fasting blood glucose level and caused a significant increase in hepatic glycogen levels in streptozotocin induced diabetic mice. Although many food-derived biopeptides have shown antidiabetic activities in vitro, animal studies are very limited.

3.2. Cholesterol-Lowering

Our bodies require healthy levels of cholesterol for the production of vitamin D and steroid hormones as well as bile acids. Yet, excess cholesterol in the blood could form plaques in arteries resulting in arteriosclerosis. Cholesterol plaques in the coronary artery could reduce oxygen supply to the heart and lead to cardiovascular diseases. Chemical agents for lowering blood cholesterol may result in liver injury or failure, myopathy [101] and diabetes [102,103] whereas other people do not tolerate statins [104]. Therefore, the search for bioactive peptides with cholesterol lowering ability has increased over the years (Table 1).

Cumin seed derived peptides CSP1, CSP 2 and CSP3 have been shown to inhibit cholesterol micelle formation, inhibit lipase activity and bind strongly to bile acids and may therefore lower cholesterol when consumed [105]. Sericin-derived oligopeptides suppressed serum total cholesterol and non-high density lipoprotein (HDL) cholesterol levels in rats fed with high-cholesterol diet. The peptides reduced cholesterol solubility in lipid micelles, and inhibited cholesterol uptake in monolayer Caco-2 cells. They also bound tightly to taurocholate, deoxytaurocholate, and glycodeoxycholate which could lead to a reduced cholesterol absorption in the gut [106]. Soybean peptides LPYP, IAVPGEVA and IAVPTGVA have been reported to effectively activate the LDLR-SREBP 2 pathway and improve LDL uptake. The peptides also inhibited HMGCoA reductase activity in HepG2 cells [107]. Likewise, the consumption of 30 g/mL of lupin protein decreased plasma proprotein convertase subtilisin/kexin type 9 levels in patients with moderate hypercholesterolaemea. The hydrolyzed lupin proteins were found to inhibit HMGCoA reductase activity in HepG2 cells and this may account for its significant hypocholesterolaemic effect [108]. In a similar way, peptides in cowpea inhibited HMGCoA reductase and reduced cholesterol micellar solubilization in vitro [109]. Peptides in rice bran protein hydrolysates were also found to inhibit cholesterol micellar solubilization and may be important in reducing cholesterol [19]. In a recent study, Hernandez et al. [110] observed that black bean and cowpea derived peptide YAAAT could tightly bind to the *N*-terminal domain of Niemann-Pick C1 (NPC1L1) to disrupt interactions between NPC1L1 and membrane proteins to enhance cholesterol absorption. Duranti et al. [111] also observed that consumption of the α' subunit reduced plasma cholesterol levels by 36% and also upregulated liver β-very low density lipoprotein cholesterol receptors in rats. Very little is known about the effects of specific food derived peptides on reducing cholesterol levels in vivo and hence more studies are needed in this area.

3.3. Antihypertensive Peptides

Hypertension (high blood pressure) is characterized by a persistent systolic blood pressure (BP) value of ≥140 mmHg and a diastolic pressure of ≥90 mmHg (140/90). However, BP increases with age and hence only elderly people over 60 years with BPs above 150/90 mmHg may require treatment [41]. Among the physiological mechanisms of hypertension, the renin-angiotensin system has attracted much scientific attention. Renin and angiotensin-converting enzyme (ACE) are the main enzymes involved in the renin-angiotensin system (RAS) [112].

Many synthetic antihypertensive drugs have been reported to cause side effects such as dizziness, dysgeusia, headache, angioedema, and cough [41]. Thus, the search for antihypertensive biopeptides from foods has increased (Table 1). Food-derived antihypertensive peptides are known for their high tissue affinities and hence may be more slowly eliminated from tissues compared to synthetic drugs [5]. To release antihypertensive peptides from whey, we fermented whey from bovine milk with several *Lactobacillus* species and found that *Lactobacillus helvelticus* fermented whey hydrolysates contained peptides AQSAP, IPAVF, APLRV and AHKAL which showed strong angiotensin 1-converting enzyme inhibition. These peptides, at least in part, contributed to the ACE inhibitory effect of the fermentate. Whey fermented with *Lactobacillus brevis* also contained a potent ACE inhibitory peptide identified as AEKTK [38]. Two tripeptides, VPP and IPP from casein have been shown to significantly reduce high blood pressure in humans [113]. The peptides were first reported by Nakamura et al. [114] when they fermented β-casein using *Saccharomyces cerevisiae* and *Lactobacillus helveticus* CP790. Many other food derived antihypertensive peptides have been shown to effectively reduce high blood pressure after a single dose [72,115,116] and after long term administrations [117–119] in animal models.

Recently, a potent ACE inhibitory peptide DPYKLRP was isolated from lactoferrin after *Kluyveromyces marxianus* fermentation. The peptide (10 mg/kg body weight) reduced systolic blood pressure (SBP) in spontaneous hypertensive rats (SHRs) by 27 mmHg relative to control rats which received 650 µL of saline [21]. Also, the dipeptide DY in aqueous extracts from bamboo shoots has been shown to reduce SBP by 18 mmHg in SHRs when administered at 10 mg/kg body weight/day [120]. In another study, Fitzgerald et al. [121] isolated the peptide IRLIIVLMPILMA from hydrolyzed *Palmaria palmata* with papain. The peptide reduced systolic blood pressure by 33 mmHg in SHR. In a double-blind parallel group intervention study, 89 hypertensive subjects who consumed fermented milk containing 5 mg IPP and VPP daily for 12 weeks and a high dose (50 mg/day) for the next 12 weeks experienced a reduction in arterial stiffness leading to reduced blood pressure [122]. Similarly, a spread containing 4.2 mg IPP and VPP significantly reduced SBP by −4.1 mmHg in

104 middle-aged hypertensive subjects after 10 weeks of consumption [123]. Conversely, consumption of a fermented milk containing 5 mg of the tripeptides did not affect the BP of subjects with metabolic syndrome [124]. Over the years, milk-derived antihypertensive peptides are the most studied and several reviews concerning their production, bioavailability and incorporation into foods have been published [125–127]. Also, several fermented milk products such as Evolus® (Valio Ltd., Helsinki, Finland), Danaten®, Ameal® and Calpis® (Calpis Co., Tokyo, Japan) [128] have been developed for managing high blood pressure.

However, the European food safety authority does not consider that a proof of cause and effect relationship between the consumption of these foods has been established [129].

For a strong antihypertensive activity, the position of certain amino acid residues is critical. For instance, the presence of branched amino acids such as valine and isoleucine are important for ACE inhibition [114]. Therefore, hydrolyzing proteins with thermolysin will increase the chances of generating peptides with terminal branched chain amino acids. Also, the presence of proline at the C-terminal has also been shown to enhance ACE inhibition [114] and hence hydrolyzing proteins with prolyl endopeptidases and other proteases that generate proline containing peptides may be helpful in producing antihypertensive peptides. Most of the studies that examined the role of peptide chain amino acid position on the potency of ACE inhibition involved statistical modeling coupled with in vitro experiments and they all agree that the C-terminal sequence are important for ACE inhibition. Yet there is still contradiction about which specific amino acids must be present to enhance C-terminal activity [130].

Several mechanisms may account for the inhibitory ability of antihypertensive peptides. The peptides may inhibit ACE activity by binding to the enzyme. ACE cleaves a dipeptide from angiotensin-I to yield angiotensin II (a vasoactive peptide) which binds with receptors on the vascular wall to cause blood vessel contractions [21]. Therefore, inhibition of ACE reduces high blood pressure. Other peptides such as IRLIIVLMPILMA inhibit renin [121], an enzyme that cleaves a dipeptide from angiotensinogen to yield angiotensin I [131]. Some antihypertensive peptides enhance nitric oxide production [132,133] while other peptides including RVPSL block angiotensin II receptors [134]. IPP has also been proposed to enhance Ang-(1–7) binding with Mas receptors and promotes bradykinin-mediated vasorelaxation which attenuate the development of hypertension [135].

Meanwhile, though many antihypertensive peptides are known and their activities have been confirmed in animal studies, more human intervention studies that use non-invasive techniques to measure hypertension parameters are required to establish the health effects of the peptides [129].

3.4. Anti-Cancer

Most synthetic anticancer agents have been associated with nephrotoxic [136,137], neurotoxic, cardiotoxic [138] and gonadotoxic [139,140] side effects. For this reason, the search for anti-cancer bioactive peptides from food sources has increased.

A cell selective peptide HVLSRAPR isolated from *S. platensis* hydrolysates showed strong inhibitory activity against HT-29 cancer cell proliferation but showed little inhibition against normal liver cells [141]. In another study, the tripeptide WPP isolated from blood clam muscle showed strong cytotoxicity toward PC-3, DU-145, H-1299 and HeLa cell lines [95]. Two peptides from tuna cooking juice KPEGMDPPLSEPEDRRDGAAGPK and KLPPLLLAKLLMSGKLLAEPCTGR have been shown to exhibit strong antiproliferative activity in breast cancer cell line MCF-7. The peptides induced cell arrest in the S phase by increasing p21 and p27 expression while decreasing cyclin A expression. Additionally, the peptides cleaved caspase 3, downregulated Bcl-2, PARP and caspase 9 expression but upregulated p53 and Bax expression [142]. Sepia ink protein hydrolysates contained a peptide QPK which significantly inhibited the proliferation of DU-145, PC-3 and LNCaP cells. The peptide also decreased the expression of the anti-apoptotic protein Bcl-2 and increased the expression of apoptogenic protein Bax [143].

In the same way, the peptide LANAK, isolated from oyster hydrolysate showed anticancer activity against human colon carcinoma (HT-29) cell lines [87]. Another peptide RQSHFANAQP from chickpea hydrolysate increased the level of p53 in breast cancer cell lines and inhibited their proliferation [144] while YALPAH from *Setipinna taty* induced apoptosis in prostate cancer PC-3 cells [145]. Soybean protein hydrolysates contain many anticancer peptides such as Lunasin, RKQLQGVN [146], GLTSK, LSGNK, GEGSGA, MPACGSS and MTEEY [147]. These peptides have been reported to exert strong antiproliferative effects on colorectal cancer HT-29 cells. The peptides RHPFDGPLLPPGD, RCGVNAFLPKSYLVHFGWKLLFHFD and KPEEVGGAGDRWTC obtained from *Dendrobium catenatum* Lindley demonstrated strong antiproliferative activity against HepG-2, SGC-7901 and MCF-7 cancer cells [148]. Peptides from rapeseed protein fermentates have also been shown to inhibit the proliferation of human HepG2 liver cancer, human MCF-7 breast cancer and human MCF-7 breast cancer cell lines [149].

Over the years, only a few studies have tested the potential cytotoxity of anticancer peptides against normal cells. However, more of such studies are needed to confirm the safety of food derived anticancer peptides. To date, most of the anticancer ability of bioactive peptides have only been assessed in vitro while very little is known about the in vivo activities of such peptides. Studies on the anti-cancer bioactive peptides in animal models are therefore warranted.

3.5. *Antimicrobial Peptides*

Antimicrobial peptides (AMP) are known to exert direct effects on a wide range of bacteria, yeast and viruses. Interestingly, many antimicrobial peptides show additional bioactivities such as antioxidant activities [150], immunomodulation [151], and wound healing activity [152]. These properties of AMPs make them better alternatives for conventional antibiotics which have recorded much resistance among pathogenic bacteria. AMPs vary in length (between 12–50 amino acids), amino acid composition, charge and position of disulphide bonds [153]. The presence of a positive charge or the presence of both hydrophilic and hydrophobic amino acids at the terminals (amphipathic) are recognized as major structural motifs by which AMPs interact with microbes. It has been shown that the antimicrobial potency of a cationic AMP is directly related to the product of its charge, hydrophobicity and length of the peptide [153]. AMPs may directly kill bacteria either by making pores through the bacteria cell membrane [154,155] or by interacting with macromolecules inside the microbial cells [156,157]. It has been reported that AMPs rich in positively charged amino acids such as arginine and lysine enter into cells by inducing energy dependent endocytic pathway such as micropinocytosis [158]. Many AMPs have been identified over the years and are available on several databases such as APD3 [159], CAMPR3 [160], DRAMP [161] and YADAMP [162]. Extensive studies have been done on the antimicrobial activity of milk derived peptides and AMPs have been mostly identified in peptides fragments from casein, and β-lactoglobulin and α-lactalbumin [162]. Recently, Zhang et al. [155] found that the peptide ELLLNPTHQIYPVTQPLAPV isolated from human colostrum killed bacteria by cell wall and cytoplasmic membrane destruction. In another study, an AMP SSSEESII from α_{s2}–casein was found to inhibit the growth of *Listeria innocua, Micrococcus luteus, Salmonella enteritidis* as well as *E. coli*. The nanopeptide, IKHQGLPQE in casein hydrolysates effectively reduced the number of pathogenic bacteria spiked in infant formula [163]. Potent AMPs have also been isolated from fish and fish products. Hydrolysis of Mackerel by-products yielded SIFIQRFTT which inhibited *Listeria innocua* and *Escherichia coli* [164]. When anchovy cooking waste water was treated with protamix, the AMP GLSRLFTALK was isolated. The peptide showed strong inhibition against *S. aureus, B. subtilis, S. pneumoniae, E. coli, S. dysenteriae, P. aeruginosa* and *S. typhimurium* [165]. Hydrolysis of *Scorpaena notata* (Small red scorpionfish) viscera protein with a neutral protease from *Trichoderma harzianum* showed remarkable antimicrobial activities. The peptide FPIGMGHGSRPA was isolate from the hydrolysate and was found to inhibit *Bacillus subtilis, Bacillus cereus, Listeria innocua, Salmonella* sp. and *E. coli* [166]. In addition to AMPs from fish and fish products, AMPs were identified in bovine blood. The peptide VNFKLLSHSLLVTLASHL isolated from bovine haemoglobin strongly inhibited the growth of *Candida albicans, Escherichia coli* and *Staphylococcus aureus* [167]. Controlled pepsin hydrolysis of haemoglobin yielded the AMPs VLSAADKGNVKAAWGKVGGHAAEYGAEALERMF, ASHLPSDFTPAVHASLDKFLANVSTVLTSKYR and VLSAADKGNVKAAWGKVGGHAAEYGAEAL ERMFLSF. The peptides showed strong inhibition against *Salmonella enteritidis, Escherichia coli, Shigella sonnei, Micrococcus luteus, Enterococcus faecalis, Listeria innocua, Staphylococcus saprophyticus, Bacillus cereus* and *Staphylococcus simulants* [168,169].

3.6. Multifunctional Peptides

Food-derived bioactive peptides with single activities have been well documented, yet only few peptides with multiple functions have been reported. Meanwhile, single peptides with multifunctional bioactivities will be more preferred over single activity peptides as the former would simultaneously elicit multiple health benefits. For this reason, García-Mora et al. [170] hydrolyzed lentil proteins using Savinase® (Novozymes, Bagsvaerd, Denmark) to search for peptides with multiple functions. They observed that peptides LLSGTQNQPSFLSGF, NSLTLPILRYL and TLEPNSVFLPVLLH present in the hydrolysate had strong antioxidant and antihypertensive effects. YSK from rice bran protein [4], WVYY and PSLPA from hemp seed protein hydrolysates [171] also showed both antioxidant and antihypertensive effects. Cummin seed peptides CSP1, CSP2 and CSP3 have been reported to exhibit cholesterol lowering activity [105], anti-oxidant and anti-amylase activities in vitro [99]. Four other peptides, YINQMPQKSRE, YINQMPQKSREA, VTGRFAGHPAAQ and YIEAVNKVSPRAGQF isolated from egg yolk have been reported to show antidiabetic, ACE inhibitory and antioxidant activities [172]. These peptides could be important in managing metabolic diseases such as diabetes and hypercholesterolemia.

Another peptide WPP isolated from *Tegillarca granosa* hydrolysates is reported to have strong antioxidant and anticancer activities [95]. Also, RQSHFANAQP peptide from chickpea albumin showed strong antioxidative and anticancer effects in MCF-7 and MDA-MB-231 cells [144].

A peptide that may prevent reactive oxygen species-induced cancer has been isolated recently. The polypeptide was obtained from *Pleurotus eryngii* mycelium and has been reported to possess a strong reducing power and a strong oxygen radical scavenging ability. The peptide also showed strong antitumor effect against stomach, cervical and breast cancer cell lines and stimulated the immune system [173]. Another recent study has shown that lunasin isolated from quinoa has strong anti-inflammatory and antioxidant abilities [174]. Since reactive oxygen species can cause inflammation [175], lunasin could be important in treating or preventing inflammatory diseases. In a similar way, NTDGSTDYGILQINSR from egg white lysozyme hydrolysates scavenge DPPH and ABTS radicals and also inhibits the growth of both Gram-negative and Gram-positive bacteria [150]. The functions of this peptide make it a potential agent for food preservation.

Low molecular weight peptides from various foods have also shown important multifunctional abilities. For instance, Aguilar-Toalá et al. [7] fermented milk with *Lactobacillus plantarum* strains and observed that the crude fermented milk extracts had strong antihemolytic, antioxidant, anti-inflammatory, antimicrobial and antimutagenic activities. Also, low molecular weight peptides from hydrolyzed brewer's yeast showed strong antiulcer and antiproliferative activity against leukemia cell lines [176]. Low molecular weight peptides from *I. badionotus* hydrolysates have also been reported to inhibit ACE activity, kill colorectal cancer cells and scavenge free radicals [177]. Similarly, peptides in cowpea hydrolysates were found to exhibit antioxidant activities and inhibit HMGCoA reductase activity [109]. Likewise, Rohu roe hydrolysates showed strong ACE inhibitory ability as well as antiproliferative activity against Caco-2 cell lines [178].

These observations could however be due the different bioactive peptides released during the fermentation and hydrolysis process and not necessarily the presence of single multifunctional peptides. Though the isolation of single multifunctional peptides from the hydrolysates would be of much interest, consumption of the hydrolysates may yield multiple health effects.

4. Effects of Processing Methods on Bioactivity

Food processing methods can significantly affect the biological activity of bioactive peptides. Physical processing methods such as ultrasound, heat and irradiation may affect the protein structure and functions. Processing could also result in maillard reactions and lead to the production of allergenic compounds [179]. Processing may increase the susceptibility of peptides to gastrointestinal digestion, absorption and response to the immune system. Therefore, it is important to determine the optimal conditions within which proteins (and peptides) could be processed to maintain or enhance their bioactivities. However, processing methods that may reduce the activity of one peptide may enhance the activities of others. For instance, the antibacterial activity of α-lactalbumin [180] and lysozyme [181] increased after they were denatured by heat while other peptides may lose their activity after heating.

The effects of boiling on the activity of biopeptides may depend on the enzyme formulation as well as the treatment conditions of the parent protein. It has been shown that hydrolyzing raw casein using chymotrypsin yielded antidiabetic peptides with higher bioactivity that those release from boiled casein [182] and hence, raw milk hydrolysates may be more effective in diabetes management than pasteurized milk. Similarly, thermal processing has been reported to significantly decrease the antioxidant activity of cowpea derived peptides. However, the thermal treatment increased the ability of the peptides to inhibit the micellar solubility of cholesterol relative to raw samples [109].

In another study, though the DPP-IV, α-glucosidase inhibitory activities and ACE inhibitory potencies of Navy beans hydrolysates did not change significantly after cooking, α-amylase inhibitory ability reduced significantly [183]. ACE inhibitory peptides tend to retain their bioactivity after thermal processing. For instance, the ACE inhibitory peptides YAGGS and YAAGS from beans maintained their bioactivities even after precooking, and this indicates that precooked beans could still provide beneficial health effects [183]. The peptides KAAAAP, AAPLAP, KPVAAP, IAGRP, and KAAAATP from ham also retained their ACE inhibitory abilities after heat treatment (at 117 °C for 6 min) [184]. Similarly, collagen derived ACE inhibitory peptides retained their bioactivity between a pH range of 2–6 and after 2 h of heating at 100 °C. Their bioactivity however significantly reduced in alkaline conditions [185].

In another study, pulsed electric field (PEF) was used to improve the ability of antioxidant peptides KCHQP (from pine nut) [186] and SHCMN (from soybean protein) [187]. Though PEF did not change the basic structures of the peptides, their zwitter potentials were significantly reduced. It is therefore important to ascertain the optimal conditions within which bioactive peptides could be processed to retain their bioactivity.

5. Challenges in Bioactive Peptide Development

Since bioactive peptides are encrypted in foods, the extent of protein hydrolysis during their production is an indispensable factor worth considering. Yet, peptides released during fermentation and enzymatic proteolysis remain susceptible to further hydrolysis as long as the enzymatic reaction or the fermentation process goes on. Such a situation may result in a decrease or loss of bioactivity due to continuous degradation of the peptides. This makes designing of a kinetic model for protein hydrolysis very challenging [180].

Also, peptides produced by microbial fermentation using wild microbes are not reproducible. This is because, the microbes are live cells and their metabolic activities, the type of enzymes and enzyme levels cannot be controlled. Therefore, the quantities of specific bioactive peptides released after fermentation cannot be guaranteed. Therefore, improved strains or genetically recombinant strains or pure enzymes may help mitigate this challenge. Meanwhile, the use of enzymes for protein hydrolysis is more expensive than microbial fermentation. Also, bioactive peptides in food hydrolysates sometimes have improved activity due to their synergistic effects with other components in the hydrolysate. Therefore, some single isolated peptides may demonstrate reduced bioactivities when tested alone as nutraceuticals.

Another important challenge in bioactive peptide production is the stability of the generated peptides. Most food derived peptides are easily degraded in the gut and therefore do no exhibit any resultant activity in the body when tested in vivo. In this case, pure isolated peptides with good activities could be stabilized against digestion by inserting a structure inducing probe tail [188] and also by clipping the peptide sequence. Such stabilization strategies would improve the bioavailability of the peptides after consumption.

6. Conclusions and Future Perspectives

Bioactive peptides from foods are valuable functional agents in healthy diets that can prevent and treat diseases. Consumer awareness of the effects of functional foods on health is a strong drive for the search and production of bioactive peptides in foods. Milk derived tripeptides IPP and VPP are the most studied food derived antihypertensive peptides and have shown positive effects in human studies. This warrants the confirmation of other food derived bioactive peptides in human studies. Many food hydrolysates have also shown multifunctional bioactive effects [7,176,177], however, their components are unknown. Identifying the peptides in such hydrolysates will be important in studying the mechanisms by which they exert their health effects.

The search of bioactive peptides through microbial fermentation will remain a promising and a cheap strategy for generating bioactive peptides in foods as generally regarded as safe microbial proteolytic systems yield several peptides of diverse potentials during fermentation. In the near future, development and use of genetically improved strains will become important as they would release large amounts of proteolytic enzymes to hydrolyze food proteins. Also, pure food derived bioactive peptides would soon be abundant on the market and sold as nutraceuticals. Such peptides could be regulated as drugs since they would be well characterized and their properties and mechanisms of action established.

Conflicts of Interest: The authors declare no conflict of interest.

References

1. Rizzello, C.G.; Tagliazucchi, D.; Babini, E.; Rutella, G.S.; Saa, D.L.T.; Gianotti, A. Bioactive peptides from vegetable food matrices: Research trends and novel biotechnologies for synthesis and recovery. *J. Funct. Foods* **2016**, *27*, 549–569. [CrossRef]
2. Abdel-Hamid, M.; Otte, J.; De Gobba, C.; Osman, A.; Hamad, E. Angiotensin I-converting enzyme inhibitory activity and antioxidant capacity of bioactive peptides derived from enzymatic hydrolysis of buffalo milk proteins. *Int. Dairy J.* **2017**, *66*, 91–98. [CrossRef]
3. Babini, E.; Tagliazucchi, D.; Martini, S.; Dei Più, L.; Gianotti, A. LC-ESI-QTOF-MS identification of novel antioxidant peptides obtained by enzymatic and microbial hydrolysis of vegetable proteins. *Food Chem.* **2017**, *228*, 186–196. [CrossRef] [PubMed]
4. Wang, X.; Chen, H.; Fu, X.; Li, S.; Wei, J. A novel antioxidant and ace inhibitory peptide from rice bran protein: Biochemical characterization and molecular docking study. *LWT-Food Sci. Technol.* **2017**, *75*, 93–99. [CrossRef]
5. Koyama, M.; Hattori, S.; Amano, Y.; Watanabe, M.; Nakamura, K. Blood pressure-lowering peptides from neo-fermented buckwheat sprouts: A new approach to estimating ACE-inhibitory activity. *PLoS ONE* **2014**, *9*, e105802. [CrossRef] [PubMed]
6. García-Tejedor, A.; Sánchez-Rivera, L.; Recio, I.; Salom, J.B.; Manzanares, P. Dairy *Debaryomyces hansenii* strains produce the antihypertensive casein-derived peptides LHLPLP and HLPLP. *LWT-Food Sci. Technol.* **2015**, *61*, 550–556. [CrossRef]
7. Aguilar-Toalá, J.; Santiago-López, L.; Peres, C.; Peres, C.; Garcia, H.; Vallejo-Cordoba, B.; González-Córdova, A.; Hernández-Mendoza, A. Assessment of multifunctional activity of bioactive peptides derived from fermented milk by specific *Lactobacillus plantarum* strains. *J. Dairy Sci.* **2017**, *100*, 65–75. [CrossRef] [PubMed]
8. Lee, S.Y.; Hur, S.J. Antihypertensive peptides from animal products, marine organisms, and plants. *Food Chem.* **2017**, *228*, 506–517. [CrossRef] [PubMed]

9. Geng, X.; Tian, G.; Zhang, W.; Zhao, Y.; Zhao, L.; Wang, H.; Ng, T.B. A *Tricholoma matsutake* peptide with angiotensin converting enzyme inhibitory and antioxidative activities and antihypertensive effects in spontaneously hypertensive rats. *Sci. Rep.* **2016**, *6*, 24130. [CrossRef] [PubMed]
10. Norris, R.; FitzGerald, R.J. Antihypertensive peptides from food proteins. In *Bioactive Food Peptides in Health and Disease*; Hernández-Ledesma, B., Hsieh, C.-C., Eds.; InTech: Rijeka, Croatia, 2013.
11. He, R.; Girgih, A.T.; Rozoy, E.; Bazinet, L.; Ju, X.-R.; Aluko, R.E. Selective separation and concentration of antihypertensive peptides from rapeseed protein hydrolysate by electrodialysis with ultrafiltration membranes. *Food Chem.* **2016**, *197*, 1008–1014. [CrossRef] [PubMed]
12. Huang, J.; Liu, Q.; Xue, B.; Chen, L.; Wang, Y.; Ou, S.; Peng, X. Angiotensin-I-converting enzyme inhibitory activities and in vivo antihypertensive effects of Sardine protein hydrolysate. *J. Food Sci.* **2016**, *81*, H2831–H2840. [CrossRef] [PubMed]
13. E Silva, F.G.D.; Hernández-Ledesma, B.; Amigo, L.; Netto, F.M.; Miralles, B. Identification of peptides released from flaxseed (*Linum usitatissimum*) protein by alcalase® hydrolysis: Antioxidant activity. *LWT-Food Sci. Technol.* **2017**, *76*, 140–146. [CrossRef]
14. Sangsawad, P.; Roytrakul, S.; Yongsawatdigul, J. Angiotensin converting enzyme (ACE) inhibitory peptides derived from the simulated *in vitro* gastrointestinal digestion of cooked chicken breast. *J. Funct. Foods* **2017**, *29*, 77–83. [CrossRef]
15. Zhang, Y.; Ma, L.; Otte, J. Optimization of hydrolysis conditions for production of angiotensin-converting enzyme inhibitory peptides from Basa fish skin using response surface methodology. *J. Aquat. Food Prod. Technol.* **2016**, *25*, 684–693. [CrossRef]
16. Cai, S.; Ling, C.; Lu, J.; Duan, S.; Wang, Y.; Zhu, H.; Lin, R.; Chen, L.; Pan, X.; Cai, M.; et al. Food protein-derived tetrapeptide, reduces seizure activity in pentylenetetrazole-induced epilepsy models through α-amino-3-hydroxy-5-methyl-4-isoxazole propionate receptors. *Neurotherapeutics* **2017**, *14*, 212–226. [CrossRef] [PubMed]
17. Khiari, Z.; Ndagijimana, M.; Betti, M. Low molecular weight bioactive peptides derived from the enzymatic hydrolysis of collagen after isoelectric solubilization/precipitation process of Turkey by-products. *Poult. Sci.* **2014**, *93*, 2347–2362. [CrossRef] [PubMed]
18. Huang, Y.-L.; Ma, M.-F.; Chow, C.-J.; Tsai, Y.-H. Angiotensin I-converting enzyme inhibitory and hypocholesterolemic activities: Effects of protein hydrolysates prepared from *Achatina fulica* snail foot muscle. *Int. J. Food Prop.* **2017**. [CrossRef]
19. Zhang, H.; Yokoyama, W.H.; Zhang, H. Concentration-dependent displacement of cholesterol in micelles by hydrophobic rice bran protein hydrolysates. *J. Sci. Food Agric.* **2012**, *92*, 1395–1401. [CrossRef] [PubMed]
20. Mojica, L.; de Mejía, E.G. Optimization of enzymatic production of anti-diabetic peptides from black bean (*Phaseolus vulgaris L.*) proteins, their characterization and biological potential. *Food Funct.* **2016**, *7*, 713–727. [CrossRef] [PubMed]
21. García-Tejedor, A.; Sánchez-Rivera, L.; Castelló-Ruiz, M.; Recio, I.; Salom, J.B.; Manzanares, P. Novel antihypertensive lactoferrin-derived peptides produced by *Kluyveromyces marxianus*: Gastrointestinal stability profile and in vivo angiotensin I-converting enzyme (ACE) inhibition. *J. Agric. Food Chem.* **2014**, *62*, 1609–1616. [CrossRef] [PubMed]
22. Fernández-Musoles, R.; Manzanares, P.; Burguete, M.C.; Alborch, E.; Salom, J.B. In vivo angiotensin I-converting enzyme inhibition by long-term intake of antihypertensive lactoferrin hydrolysate in spontaneously hypertensive rats. *Food Res. Int.* **2013**, *54*, 627–632. [CrossRef]
23. Ruiz-Ruiz, J.; Dávila-Ortíz, G.; Chel-Guerrero, L.; Betancur-Ancona, D. Angiotensin I-converting enzyme inhibitory and antioxidant peptide fractions from hard-to-cook bean enzymatic hydrolysates. *J. Food Biochem.* **2013**, *37*, 26–35. [CrossRef]
24. Wattanasiritham, L.; Theerakulkait, C.; Wickramasekara, S.; Maier, C.S.; Stevens, J.F. Isolation and identification of antioxidant peptides from enzymatically hydrolyzed rice bran protein. *Food Chem.* **2016**, *192*, 156–162. [CrossRef] [PubMed]
25. Singh, B.; Sharma, P.; Kumar, A.; Chadha, P.; Kaur, R.; Kaur, A. Antioxidant and in vivo genoprotective effects of phenolic compounds identified from an endophytic *Cladosporium velox* and their relationship with its host plant *Tinospora cordifolia*. *J. Ethnopharmacol.* **2016**, *194*, 450–456. [CrossRef] [PubMed]

26. Alu'datt, M.H.; Rababah, T.; Johargy, A.; Gammoh, S.; Ereifej, K.; Alhamad, M.N.; Brewer, M.S.; Saati, A.A.; Kubow, S.; Rawshdeh, M. Extraction, optimisation and characterisation of phenolics from *Thymus vulgaris* L.: Phenolic content and profiles in relation to antioxidant, antidiabetic and antihypertensive properties. *Int. J. Food Sci. Technol.* **2016**, *51*, 720–730. [CrossRef]
27. Maddox, C.E.; Laur, L.M.; Tian, L. Antibacterial activity of phenolic compounds against the phytopathogen *Xylella fastidiosa*. *Curr. Microbiol.* **2010**, *60*, 53. [CrossRef] [PubMed]
28. Da Porto, C.; Natolino, A. Supercritical fluid extraction of polyphenols from grape seed (*Vitis vinifera*): Study on process variables and kinetics. *J. Supercrit. Fluids* **2017**. [CrossRef]
29. Uzel, R.A. A practical method for isolation of phenolic compounds from black carrot utilizing pressurized water extraction with in-site particle generation in hot air assistance. *J. Supercrit. Fluids* **2017**, *120*, 320–327. [CrossRef]
30. Kumari, B.; Tiwari, B.K.; Hossain, M.B.; Rai, D.K.; Brunton, N.P. Ultrasound-assisted extraction of polyphenols from potato peels: Profiling and kinetic modelling. *Int. J. Food Sci. Technol.* **2017**. [CrossRef]
31. Do, Q.D.; Angkawijaya, A.E.; Tran-Nguyen, P.L.; Huynh, L.H.; Soetaredjo, F.E.; Ismadji, S.; Ju, Y.-H. Effect of extraction solvent on total phenol content, total flavonoid content, and antioxidant activity of limnophila aromatica. *J. Food Drug Anal.* **2014**, *22*, 296–302. [CrossRef]
32. Nimalaratne, C.; Bandara, N.; Wu, J. Purification and characterization of antioxidant peptides from enzymatically hydrolyzed chicken egg white. *Food Chem.* **2015**, *188*, 467–472. [CrossRef] [PubMed]
33. Nongonierma, A.B.; FitzGerald, R.J. Susceptibility of milk protein-derived peptides to dipeptidyl peptidase IV (DPP-IV) hydrolysis. *Food Chem.* **2014**, *145*, 845–852. [CrossRef] [PubMed]
34. Rahimi, M.; Ghaffari, S.M.; Salami, M.; Mousavy, S.J.; Niasari-Naslaji, A.; Jahanbani, R.; Yousefinejad, S.; Khalesi, M.; Moosavi-Movahedi, A.A. ACE-inhibitory and radical scavenging activities of bioactive peptides obtained from camel milk casein hydrolysis with proteinase K. *Dairy Sci. Technol.* **2016**, *96*, 489–499. [CrossRef]
35. Ferri, M.; Graen-Heedfeld, J.; Bretz, K.; Guillon, F.; Michelini, E.; Calabretta, M.M.; Lamborghini, M.; Gruarin, N.; Roda, A.; Kraft, A. Peptide fractions obtained from rice by-products by means of an environment-friendly process show in vitro health-related bioactivities. *PLoS ONE* **2017**, *12*, e0170954. [CrossRef] [PubMed]
36. Dei Piu, L.; Tassoni, A.; Serrazanetti, D.I.; Ferri, M.; Babini, E.; Tagliazucchi, D.; Gianotti, A. Exploitation of starch industry liquid by-product to produce bioactive peptides from rice hydrolyzed proteins. *Food Chem.* **2014**, *155*, 199–206. [CrossRef] [PubMed]
37. Zhang, J.; Zhang, H.; Wang, L.; Guo, X.; Wang, X.; Yao, H. Isolation and identification of antioxidative peptides from rice endosperm protein enzymatic hydrolysate by consecutive chromatography and MALDI-TOF/TOF MS/MS. *Food Chem.* **2010**, *119*, 226–234. [CrossRef]
38. Ahn, J.; Park, S.; Atwal, A.; Gibbs, B.; Lee, B. Angiotensin I-converting enzyme (ACE) inhibitory peptides from whey fermented by Lactobacillus species. *J. Food Biochem.* **2009**, *33*, 587–602. [CrossRef]
39. Zhang, N.; Zhang, C.; Chen, Y.; Zheng, B. Purification and characterization of antioxidant peptides of pseudosciaena crocea protein hydrolysates. *Molecules* **2016**, *22*, 57. [CrossRef] [PubMed]
40. Rizzello, C.G.; Lorusso, A.; Russo, V.; Pinto, D.; Marzani, B.; Gobbetti, M. Improving the antioxidant properties of quinoa flour through fermentation with selected autochthonous lactic acid bacteria. *Int. J. Food Microbiol.* **2017**, *241*, 252–261. [CrossRef] [PubMed]
41. Daliri, E.B.-M.; Lee, B.H.; Oh, D.H. Current perspectives on antihypertensive probiotics. *Probiotics Antimicrob. Proteins* **2016**. [CrossRef] [PubMed]
42. El-Fattah, A.A.; Sakr, S.; El-Dieb, S.; Elkashef, H. Angiotensin-converting enzyme inhibition and antioxidant activity of commercial dairy starter cultures. *Food Sci. Biotechnol.* **2016**, *25*, 1745–1751. [CrossRef]
43. Sanjukta, S.; Rai, A.K.; Muhammed, A.; Jeyaram, K.; Talukdar, N.C. Enhancement of antioxidant properties of two soybean varieties of sikkim himalayan region by proteolytic *Bacillus subtilis* fermentation. *J. Funct. Foods* **2015**, *14*, 650–658. [CrossRef]
44. Chen, Y.; Liu, W.; Xue, J.; Yang, J.; Chen, X.; Shao, Y.; Kwok, L.-Y.; Bilige, M.; Mang, L.; Zhang, H. Angiotensin-converting enzyme inhibitory activity of *Lactobacillus helveticus* strains from traditional fermented dairy foods and antihypertensive effect of fermented milk of strain H9. *J. Dairy Sci.* **2014**, *97*, 6680–6692. [CrossRef] [PubMed]

45. Chaves-López, C.; Tofalo, R.; Serio, A.; Paparella, A.; Sacchetti, G.; Suzzi, G. Yeasts from *Colombian kumis* as source of peptides with angiotensin I-converting enzyme (ACE) inhibitory activity in milk. *Int. J. Food Microbiol.* **2012**, *159*, 39–46. [CrossRef] [PubMed]
46. Rai, A.K.; Kumari, R.; Sanjukta, S.; Sahoo, D. Production of bioactive protein hydrolysate using the yeasts isolated from soft chhurpi. *Bioresour. Technol.* **2016**, *219*, 239–245. [CrossRef] [PubMed]
47. Lima, C.A.; Campos, J.F.; Lima Filho, J.L.; Converti, A.; da Cunha, M.G.C.; Porto, A.L. Antimicrobial and radical scavenging properties of bovine collagen hydrolysates produced by *Penicillium aurantiogriseum* URM 4622 collagenase. *J. Food Sci. Technol.* **2015**, *52*, 4459–4466. [CrossRef] [PubMed]
48. Hou, Y.; Liu, W.; Cheng, Y.; Zhou, J.; Wu, L.; Yang, G. Production optimization and characterization of immunomodulatory peptides obtained from fermented goat placenta. *Food Sci. Technol.* **2014**, *34*, 723–729. [CrossRef]
49. Giri, A.; Nasu, M.; Ohshima, T. Bioactive properties of Japanese fermented fish paste, fish miso, using koji inoculated with *Aspergillus oryzae*. *Int. J. Nutr. Food Sci.* **2012**, *1*, 13–22. [CrossRef]
50. Chaves-López, C.; Serio, A.; Paparella, A.; Martuscelli, M.; Corsetti, A.; Tofalo, R.; Suzzi, G. Impact of microbial cultures on proteolysis and release of bioactive peptides in fermented milk. *Food Microbiol.* **2014**, *42*, 117–121. [CrossRef] [PubMed]
51. Chaudhury, A.; Duvoor, C.; Dendi, V.S.R.; Kraleti, S.; Chada, A.; Ravilla, R.; Marco, A.; Shekhawat, N.S.; Montales, M.T.; Kuriakose, K. Clinical review of antidiabetic drugs: Implications for type 2 diabetes mellitus management. *Front. Endocrinol.* **2017**, *8*, 6. [CrossRef] [PubMed]
52. Thulé, P.M.; Umpierrez, G. Sulfonylureas: A new look at old therapy. *Curr. Diabetes Rep.* **2014**, *14*, 1–8. [CrossRef] [PubMed]
53. Meier, J.J.; Nauck, M.A. Risk of pancreatitis in patients treated with incretin-based therapies. *Diabetologia* **2014**, *57*, 1320–1324. [CrossRef] [PubMed]
54. Thong, K.; Gupta, P.S.; Blann, A.; Ryder, R. The influence of age and metformin treatment status on reported gastrointestinal side effects with liraglutide treatment in Type 2 diabetes. *Diabetes Res. Clin. Pract.* **2015**, *109*, 124–129. [CrossRef] [PubMed]
55. Dujic, T.; Causevic, A.; Bego, T.; Malenica, M.; Velija-Asimi, Z.; Pearson, E.; Semiz, S. Organic cation transporter 1 variants and gastrointestinal side effects of metformin in patients with Type 2 diabetes. *Diabet. Med.* **2015**, *4*, 511–514. [CrossRef] [PubMed]
56. Chakrabarti, S.; Jahandideh, F.; Wu, J. Food-derived bioactive peptides on inflammation and oxidative stress. *BioMed Res. Int.* **2014**, *2014*, 608979. [CrossRef] [PubMed]
57. Kwon, D.Y.; Hong, S.M.; Ahn, I.S.; Kim, M.J.; Yang, H.J.; Park, S. Isoflavonoids and peptides from meju, long-term fermented soybeans, increase insulin sensitivity and exert insulinotropic effects in vitro. *Nutrition* **2011**, *27*, 244–252. [CrossRef] [PubMed]
58. Mojica, L.; de Mejia, E.G.; Granados-Silvestre, M.Á.; Menjivar, M. Evaluation of the hypoglycemic potential of a black bean hydrolyzed protein isolate and its pure peptides using in silico, in vitro and in vivo approaches. *J. Funct. Foods* **2017**, *31*, 274–286. [CrossRef]
59. Roblet, C.; Akhtar, M.J.; Mikhaylin, S.; Pilon, G.; Gill, T.; Marette, A.; Bazinet, L. Enhancement of glucose uptake in muscular cell by peptide fractions separated by electrodialysis with filtration membrane from Salmon frame protein hydrolysate. *J. Funct. Foods* **2016**, *22*, 337–346. [CrossRef]
60. Zhang, Y.; Chen, R.; Chen, X.; Zeng, Z.; Ma, H.; Chen, S. Dipeptidyl peptidase IV-inhibitory peptides derived from Silver carp (*Hypophthalmichthys molitrix* val.) proteins. *J. Agric. Food Chem.* **2016**, *64*, 831–839. [CrossRef] [PubMed]
61. Tang, W.; Yuan, H.; Zhang, H.; Wang, L.; Qian, H.; Qi, X. An antimicrobial peptide screened from casein hydrolyzate by *Saccharomyces cerevisiae* cell membrane affinity method. *Food Control* **2015**, *50*, 413–422. [CrossRef]
62. McClean, S.; Beggs, L.B.; Welch, R.W. Antimicrobial activity of antihypertensive food-derived peptides and selected alanine analogues. *Food Chem.* **2014**, *146*, 443–447. [CrossRef] [PubMed]
63. Théolier, J.; Hammami, R.; Labelle, P.; Fliss, I.; Jean, J. Isolation and identification of antimicrobial peptides derived by peptic cleavage of whey protein isolate. *J. Funct. Foods* **2013**, *5*, 706–714. [CrossRef]
64. Taveira, G.B.; Carvalho, A.O.; Rodrigues, R.; Trindade, F.G.; Da Cunha, M.; Gomes, V.M. Thionin-like peptide from *Capsicum annuum* fruits: Mechanism of action and synergism with fluconazole against *Candida* species. *BMC Microbiol.* **2016**, *16*, 12. [CrossRef] [PubMed]

65. Sousa, D.A.; Porto, W.F.; Silva, M.Z.; da Silva, T.R.; Franco, O.L. Influence of cysteine and tryptophan substitution on DNA-binding activity on maize α-hairpinin antimicrobial peptide. *Molecules* **2016**, *21*, 1062. [CrossRef] [PubMed]
66. Koyama, M.; Naramoto, K.; Nakajima, T.; Aoyama, T.; Watanabe, M.; Nakamura, K. Purification and identification of antihypertensive peptides from fermented buckwheat sprouts. *J. Agric. Food Chem.* **2013**, *61*, 3013–3021. [CrossRef] [PubMed]
67. Li, B.; Qiao, L.; Li, L.; Zhang, Y.; Li, K.; Wang, L.; Qiao, Y. A novel antihypertensive derived from Adlay (*Coix larchryma-jobi* L. Var. ma-yuen Stapf) Glutelin. *Molecules* **2017**, *22*, 123. [CrossRef] [PubMed]
68. Vallabha, V.S.; Tiku, P.K. Antihypertensive peptides derived from soy protein by fermentation. *Int. J. Pept. Res. Ther.* **2014**, *20*, 161–168. [CrossRef]
69. Xu, X.; Gao, Y. Purification and identification of angiotensin I-converting enzyme-inhibitory peptides from apalbumin 2 during simulated gastrointestinal digestion. *J. Sci. Food Agric.* **2015**, *95*, 906–914. [CrossRef] [PubMed]
70. Li, Y.; Zhou, J.; Huang, K.; Sun, Y.; Zeng, X. Purification of a novel angiotensin I-converting enzyme (ACE) inhibitory peptide with an antihypertensive effect from loach (*Misgurnus anguillicaudatus*). *J. Agric. Food Chem.* **2012**, *60*, 1320–1325. [CrossRef] [PubMed]
71. Castellano, P.; Aristoy, M.-C.; Sentandreu, M.Á.; Vignolo, G.; Toldrá, F. Peptides with angiotensin I converting enzyme (ACE) inhibitory activity generated from porcine skeletal muscle proteins by the action of meat-borne lactobacillus. *J. Proteom.* **2013**, *89*, 183–190. [CrossRef] [PubMed]
72. Escudero, E.; Mora, L.; Fraser, P.D.; Aristoy, M.-C.; Arihara, K.; Toldrá, F. Purification and identification of antihypertensive peptides in spanish dry-cured ham. *J. Proteom.* **2013**, *78*, 499–507. [CrossRef] [PubMed]
73. Dellafiora, L.; Paolella, S.; Dall'Asta, C.; Dossena, A.; Cozzini, P.; Galaverna, G. Hybrid in silico/in vitro approach for the identification of angiotensin I converting enzyme inhibitory peptides from parma dry-cured ham. *J. Agric. Food Chem.* **2015**, *63*, 6366–6375. [CrossRef] [PubMed]
74. Lafarga, T.; O'Connor, P.; Hayes, M. Identification of novel dipeptidyl peptidase-IV and angiotensin-I-converting enzyme inhibitory peptides from meat proteins using in silico analysis. *Peptides* **2014**, *59*, 53–62. [CrossRef] [PubMed]
75. Zheng, Y.; Li, Y.; Zhang, Y.; Ruan, X.; Zhang, R. Purification, characterization, synthesis, in vitro ACE inhibition and in vivo antihypertensive activity of bioactive peptides derived from oil palm kernel glutelin-2 hydrolysates. *J. Funct. Foods* **2017**, *28*, 48–58. [CrossRef]
76. Huang, S.-L.; Jao, C.-L.; Ho, K.-P.; Hsu, K.-C. Dipeptidyl-peptidase IV inhibitory activity of peptides derived from tuna cooking juice hydrolysates. *Peptides* **2012**, *35*, 114–121. [CrossRef] [PubMed]
77. Li-Chan, E.C.; Hunag, S.-L.; Jao, C.-L.; Ho, K.-P.; Hsu, K.-C. Peptides derived from atlantic salmon skin gelatin as dipeptidyl-peptidase IV inhibitors. *J. Agric. Food Chem.* **2012**, *60*, 973–978. [CrossRef] [PubMed]
78. Zhang, Y.; Chen, R.; Ma, H.; Chen, S. Isolation and identification of dipeptidyl peptidase IV-inhibitory peptides from trypsin/chymotrypsin-treated goat milk casein hydrolysates by 2D-TLC and LC–MS/MS. *J. Agric. Food Chem.* **2015**, *63*, 8819–8828. [CrossRef] [PubMed]
79. Harnedy, P.A.; O'Keeffe, M.B.; FitzGerald, R.J. Purification and identification of dipeptidyl peptidase (DPP) IV inhibitory peptides from the macroalga *Palmaria palmata*. *Food Chem.* **2015**, *172*, 400–406. [CrossRef] [PubMed]
80. Hatanaka, T.; Uraji, M.; Fujita, A.; Kawakami, K. Anti-oxidation activities of rice-derived peptides and their inhibitory effects on dipeptidylpeptidase-IV. *Int. J. Pept. Res. Ther.* **2015**, *21*, 479–485. [CrossRef]
81. Lacroix, I.M.; Chen, X.-M.; Kitts, D.; Li-Chan, E.C. Investigation into the bioavailability of milk protein-derived peptides with dipeptidyl-peptidase IV inhibitory activity using Caco-2 cell monolayers. *Food Funct.* **2017**, *8*, 701–709. [CrossRef] [PubMed]
82. Rahiman, S.S.F.; Morgan, M.; Gray, P.; Shaw, P.N.; Cabot, P.J. Inhibitory effects of dynorphin 3–14 on the lipopolysaccharide-induced toll-like receptor 4 signalling pathway. *Peptides* **2017**, *90*, 48–54. [CrossRef] [PubMed]
83. Xiao, M.; Ding, L.; Yang, W.; Chai, L.; Sun, Y.; Yang, X.; Li, D.; Zhang, H.; Li, W.; Cao, Z. St20, a new venomous animal derived natural peptide with immunosuppressive and anti-inflammatory activities. *Toxicon* **2017**, *127*, 37–43. [CrossRef] [PubMed]

84. Hou, H.; Fan, Y.; Wang, S.; Si, L.; Li, B. Immunomodulatory activity of Alaska pollock hydrolysates obtained by glutamic acid biosensor—Artificial neural network and the identification of its active central fragment. *J. Funct. Foods* **2016**, *24*, 37–47. [CrossRef]
85. Liu, J.; Jin, Y.; Lin, S.; Jones, G.S.; Chen, F. Purification and identification of novel antioxidant peptides from egg white protein and their antioxidant activities. *Food Chem.* **2015**, *175*, 258–266. [CrossRef] [PubMed]
86. Yan, Q.-J.; Huang, L.-H.; Sun, Q.; Jiang, Z.-Q.; Wu, X. Isolation, identification and synthesis of four novel antioxidant peptides from rice residue protein hydrolyzed by multiple proteases. *Food Chem.* **2015**, *179*, 290–295. [CrossRef] [PubMed]
87. Umayaparvathi, S.; Meenakshi, S.; Vimalraj, V.; Arumugam, M.; Sivagami, G.; Balasubramanian, T. Antioxidant activity and anticancer effect of bioactive peptide from enzymatic hydrolysate of oyster (*Saccostrea cucullata*). *Biomed. Prev. Nutr.* **2014**, *4*, 343–353. [CrossRef]
88. Chi, C.-F.; Wang, B.; Wang, Y.-M.; Zhang, B.; Deng, S.-G. Isolation and characterization of three antioxidant peptides from protein hydrolysate of Bluefin leatherjacket (*Navodon septentrionalis*) heads. *J. Funct. Foods* **2015**, *12*, 1–10. [CrossRef]
89. Chi, C.-F.; Wang, B.; Hu, F.-Y.; Wang, Y.-M.; Zhang, B.; Deng, S.-G.; Wu, C.-W. Purification and identification of three novel antioxidant peptides from protein hydrolysate of Bluefin leatherjacket (*Navodon septentrionalis*) skin. *Food Res. Int.* **2015**, *73*, 124–129. [CrossRef]
90. Sun, L.; Zhang, Y.; Zhuang, Y. Antiphotoaging effect and purification of an antioxidant peptide from Tilapia (*Oreochromis niloticus*) gelatin peptides. *J. Funct. Foods* **2013**, *5*, 154–162. [CrossRef]
91. Park, S.Y.; Kim, Y.-S.; Ahn, C.-B.; Je, J.-Y. Partial purification and identification of three antioxidant peptides with hepatoprotective effects from blue mussel (*Mytilus edulis*) hydrolysate by peptic hydrolysis. *J. Funct. Foods* **2016**, *20*, 88–95. [CrossRef]
92. Zhang, M.; Mu, T.H. Optimisation of antioxidant hydrolysate production from sweet potato protein and effect of in vitro gastrointestinal digestion. *Int. J. Food Sci. Technol.* **2016**, *51*, 1844–1850. [CrossRef]
93. Pan, X.; Zhao, Y.-Q.; Hu, F.-Y.; Wang, B. Preparation and identification of antioxidant peptides from protein hydrolysate of skate (*Raja porosa*) cartilage. *J. Funct. Foods* **2016**, *25*, 220–230. [CrossRef]
94. Jang, H.L.; Liceaga, A.M.; Yoon, K.Y. Purification, characterisation and stability of an antioxidant peptide derived from sandfish (*Arctoscopus japonicus*) protein hydrolysates. *J. Funct. Foods* **2016**, *20*, 433–442. [CrossRef]
95. Chi, C.-F.; Hu, F.-Y.; Wang, B.; Li, T.; Ding, G.-F. Antioxidant and anticancer peptides from the protein hydrolysate of blood clam (*Tegillarca granosa*) muscle. *J. Funct. Foods* **2015**, *15*, 301–313. [CrossRef]
96. Song, R.; Zhang, K.-Q.; Wei, R.-B. In vitro antioxidative activities of squid (*Ommastrephes bartrami*) viscera autolysates and identification of active peptides. *Process Biochem.* **2016**, *51*, 1674–1682. [CrossRef]
97. Shabestarian, H.; Asoodeh, A.; Homayouni-Tabrizi, M.; Hossein-Nejad-Ariani, H. Antioxidant and angiotensin I converting enzyme (ACE) inhibitory properties of GL-9 peptide. *J. Food Process. Preserv.* **2016**. [CrossRef]
98. Cai, L.; Wu, X.; Zhang, Y.; Li, X.; Ma, S.; Li, J. Purification and characterization of three antioxidant peptides from protein hydrolysate of grass carp (*Ctenopharyngodon idella*) skin. *J. Funct. Foods* **2015**, *16*, 234–242. [CrossRef]
99. Siow, H.-L.; Gan, C.-Y. Extraction, identification, and structure—Activity relationship of antioxidative and α-amylase inhibitory peptides from cumin seeds (*Cuminum cyminum*). *J. Funct. Foods* **2016**, *22*, 1–12. [CrossRef]
100. Huang, F.; Wu, W. Antidiabetic effect of a new peptide from *Squalus mitsukurii* liver (S-8300) in streptozocin-induced diabetic mice. *J. Pharm. Pharmacol.* **2005**, *57*, 1575–1580. [CrossRef] [PubMed]
101. Mancini, G.J.; Baker, S.; Bergeron, J.; Fitchett, D.; Frohlich, J.; Genest, J.; Gupta, M.; Hegele, R.A.; Ng, D.; Pearson, G.J. Diagnosis, prevention, and management of statin adverse effects and intolerance: Canadian consensus working group update (2016). *Can. J. Cardiol.* **2016**, *32*, S35–S65. [CrossRef] [PubMed]
102. Carter, A.A.; Gomes, T.; Camacho, X.; Juurlink, D.N.; Shah, B.R.; Mamdani, M.M. Risk of incident diabetes among patients treated with statins: Population based study. *BMJ* **2013**, *346*, f2610. [CrossRef] [PubMed]
103. Katsiki, N.; Banach, M. Statin use and risk of diabetes mellitus in postmenopausal women. *Clin. Lipidol.* **2012**, *7*, 267–270. [CrossRef]
104. Ahmad, Z. Statin intolerance. *Am. J. Cardiol.* **2014**, *113*, 1765–1771. [CrossRef] [PubMed]

105. Siow, H.-L.; Choi, S.-B.; Gan, C.-Y. Structure—Activity studies of protease activating, lipase inhibiting, bile acid binding and cholesterol-lowering effects of pre-screened Cumin seed bioactive peptides. *J. Funct. Foods* **2016**, *27*, 600–611. [CrossRef]
106. Lapphanichayakool, P.; Sutheerawattananonda, M.; Limpeanchob, N. Hypocholesterolemic effect of sericin-derived oligopeptides in high-cholesterol fed rats. *J. Nat. Med.* **2017**, *71*, 208–215. [CrossRef] [PubMed]
107. Lammi, C.; Zanoni, C.; Arnoldi, A. IAVPGEVA, IAVPTGVA, and LPYP, three peptides from soy glycinin, modulate cholesterol metabolism in HepG2 cells through the activation of the LDLR-SREBP2 pathway. *J. Funct. Foods* **2015**, *14*, 469–478. [CrossRef]
108. Lammi, C.; Zanoni, C.; Calabresi, L.; Arnoldi, A. Lupin protein exerts cholesterol-lowering effects targeting PCSK9: From clinical evidences to elucidation of the in vitro molecular mechanism using HepG2 cells. *J. Funct. Foods* **2016**, *23*, 230–240. [CrossRef]
109. Marques, M.R.; Freitas, R.A.M.S.; Carlos, A.C.C.; Siguemoto, É.S.; Fontanari, G.G.; Arêas, J.A. Peptides from cowpea present antioxidant activity, inhibit cholesterol synthesis and its solubilisation into micelles. *Food Chem.* **2015**, *168*, 288–293. [CrossRef] [PubMed]
110. Hernandez, L.M.R.; de Mejia, E.G. Bean peptides have higher in silico binding affinities than Ezetimibe for the N-terminal domain of cholesterol receptor Niemann-Pick C1 Like-1. *Peptides* **2017**, *17*, 30052–30059. [CrossRef] [PubMed]
111. Duranti, M.; Lovati, M.R.; Dani, V.; Barbiroli, A.; Scarafoni, A.; Castiglioni, S.; Ponzone, C.; Morazzoni, P. The α' subunit from soybean 7S globulin lowers plasma lipids and upregulates liver β-VLDL receptors in rats fed a hypercholesterolemic diet. *J. Nutr.* **2004**, *134*, 1334–1339. [PubMed]
112. Guang, C.; Phillips, R.D.; Jiang, B.; Milani, F. Three key proteases—Angiotensin-I-converting enzyme (ACE), ACE2 and Renin–within and beyond the renin-angiotensin system. *Arch. Cardiovasc. Dis.* **2012**, *105*, 373–385. [CrossRef] [PubMed]
113. Fekete, Á.A.; Givens, D.I.; Lovegrove, J.A. Casein-derived lactotripeptides reduce systolic and diastolic blood pressure in a meta-analysis of randomised clinical trials. *Nutrients* **2015**, *7*, 659–681. [CrossRef] [PubMed]
114. Nakamura, Y.; Yamamoto, N.; Sakai, K.; Okubo, A.; Yamazaki, S.; Takano, T. Purification and characterization of Angiotensin I-converting enzyme inhibitors from sour milk. *J. Dairy Sci.* **1995**, *78*, 777–783. [CrossRef]
115. Bernabucci, U.; Catalani, E.; Basiricò, L.; Morera, P.; Nardone, A. In vitro ACE-inhibitory activity and in vivo antihypertensive effects of water-soluble extract by Parmigiano reggiano and Grana padano cheeses. *Int. Dairy J.* **2014**, *37*, 16–19. [CrossRef]
116. García-Tejedor, A.; Castelló-Ruiz, M.; Gimeno-Alcañíz, J.V.; Manzanares, P.; Salom, J.B. In vivo antihypertensive mechanism of Lactoferrin-derived peptides: Reversion of Angiotensin I-and Angiotensin II-induced hypertension in wistar rats. *J. Funct. Foods* **2015**, *15*, 294–300. [CrossRef]
117. Sipola, M.; Finckenberg, P.; Korpela, R.; Vapaatalo, H.; Nurminen, M. Effect of long-term intake of milk products on blood pressure in hypertensive rats. *J. Dairy Res.* **2002**, *6*, 103–111. [CrossRef]
118. Sipola, M.; Finckenberg, P.; Santisteban, J.; Korpela, R.; Vapaatalo, H.; Nurminen, M. Long term intake of milk peptides attenuates development of hypertension in spontaneously hypertensive rats. *J. Physiol. Pharmacol.* **2001**, *52*, 745–754. [PubMed]
119. Jauhiainen, T.; Collin, M.; Narva, M.; Paussa, T.; Korpela, R. Effect of long-term intake of milk peptides and minerals on blood pressure and arterial function in spontaneously hypertensive rats. *Milchwissenschaft* **2005**, *60*, 358–362.
120. Liu, L.; Liu, L.; Lu, B.; Chen, M.; Zhang, Y. Evaluation of bamboo shoot peptide preparation with Angiotensin converting enzyme inhibitory and antioxidant abilities from byproducts of canned bamboo shoots. *J. Agric. Food Chem.* **2013**, *61*, 5526–5533. [CrossRef] [PubMed]
121. Fitzgerald, C.; Aluko, R.E.; Hossain, M.; Rai, D.K.; Hayes, M. Potential of a renin inhibitory peptide from the red seaweed *Palmaria palmata* as a functional food ingredient following confirmation and characterization of a hypotensive effect in spontaneously hypertensive rats. *J. Agric. Food Chem.* **2014**, *62*, 8352–8356. [CrossRef] [PubMed]
122. Jauhiainen, T.; Rönnback, M.; Vapaatalo, H.; Wuolle, K.; Kautiainen, H.; Groop, P.H.; Korpela, R. Long-term intervention with *Lactobacillus helveticus* fermented milk reduces augmentation index in hypertensive subjects. *Eur. J. Clin. Nutr.* **2010**, *64*, 424–431. [CrossRef] [PubMed]

123. Turpeinen, A.M.; Ikonen, M.; Kivimäki, A.S.; Kautiainen, H.; Vapaatalo, H.; Korpela, R. A spread containing bioactive milk peptides Ile–Pro–Pro and Val–Pro–Pro, and plant sterols has antihypertensive and cholesterol-lowering effects. *Food Funct.* **2012**, *3*, 621–627. [CrossRef] [PubMed]
124. Hautaniemi, E.J.; Tikkakoski, A.J.; Tahvanainen, A.; Nordhausen, K.; Kähönen, M.; Mattsson, T.; Luhtala, S.; Turpeinen, A.M.; Niemelä, O.; Vapaatalo, H.; et al. Effect of fermented milk product containing lactotripeptides and plant sterol esters on haemodynamics in subjects with the metabolic syndrome—A randomised, double-blind, placebo-controlled study. *Br. J. Nutr.* **2015**, *114*, 376–386. [CrossRef] [PubMed]
125. Pihlanto, A.; Korhonen, H. Bioactive peptides and proteins. *Adv. Food Nutr. Res.* **2003**, *47*, 175–276. [PubMed]
126. Jäkälä, P.; Vapaatalo, H. Antihypertensive peptides from milk proteins. *Pharmaceuticals* **2010**, *3*, 251–272. [CrossRef] [PubMed]
127. Beltrán-Barrientos, L.M.; Hernández-Mendoza, A.; Torres-Llanez, M.J.; González-Córdova, A.F.; Vallejo-Córdoba, B. Invited review: Fermented milk as antihypertensive functional food. *J. Dairy Sci.* **2016**, *99*, 4099–4110. [CrossRef] [PubMed]
128. Ricci-Cabello, I.; Olalla, M.; Artacho, R. Possible role of milk-derived bioactive peptides in the treatment and prevention of metabolic syndrome. *Nutr. Rev.* **2012**, *70*, 241–255. [CrossRef] [PubMed]
129. European Food Safety Authority (EFSA). Scientific Opinion of the Panel on Dietetic Products, Nutrition and Allergies on a request from Valio Ltd. on the scientific substantiation of a health claim related to *Lactobacillus helveticus* fermented Evolus® low-fat milk products and reduction of arterial stiffness. *EFSA J.* **2008**, *824*, 1–12.
130. Aluko, R.E. Antihypertensive peptides from food proteins. *Annu. Rev. Food Sci. Technol.* **2015**, *6*, 235–262. [CrossRef] [PubMed]
131. Malomo, S.A.; Onuh, J.O.; Girgih, A.T.; Aluko, R.E. Structural and antihypertensive properties of enzymatic hemp seed protein hydrolysates. *Nutrients* **2015**, *7*, 7616–7632. [CrossRef] [PubMed]
132. Wei, Y.; Jiehua, W.; Fengjuan, Z. In vivo hypotensive and physiological effects of a silk fibroin hydrolysate on spontaneously hypertensive rats. *Biosci. Biotechnol. Biochem.* **2012**, *76*, 1987–1989.
133. Majumder, K.; Chakrabarti, S.; Morton, J.S.; Panahi, S.; Kaufman, S.; Davidge, S.T.; Wu, J. Egg-derived tri-peptide IRW exerts antihypertensive effects in spontaneously hypertensive rats. *PLoS ONE* **2013**, *8*, e82829. [CrossRef] [PubMed]
134. Yu, Z.; Yin, Y.; Zhao, W.; Chen, F.; Liu, J. Antihypertensive effect of angiotensin-converting enzyme inhibitory peptide RVPSL on spontaneously hypertensive rats by regulating gene expression of the Renin–Angiotensin System. *J. Agric. Food Chem.* **2014**, *62*, 912–917. [CrossRef] [PubMed]
135. Ehlers, P.I.; Nurmi, L.; Turpeinen, A.M.; Korpela, R.; Vapaatalo, H. Casein-derived tripeptide Ile–Pro–Pro improves angiotensin-(1–7)-and bradykinin-induced rat mesenteric artery relaxation. *Life Sci.* **2011**, *88*, 206–211. [CrossRef] [PubMed]
136. Van Acker, T.; Van Malderen, S.J.; Van Heerden, M.; McDuffie, J.E.; Cuyckens, F.; Vanhaecke, F. High-resolution laser ablation-inductively coupled plasma-mass spectrometry imaging of Cisplatin-induced nephrotoxic side effects. *Anal. Chim. Acta* **2016**, *945*, 23–30. [CrossRef] [PubMed]
137. Kamisli, S.; Ciftci, O.; Kaya, K.; Cetin, A.; Kamisli, O.; Ozcan, C. Hesperidin protects brain and sciatic nerve tissues against cisplatin-induced oxidative, histological and electromyographical side effects in rats. *Toxicol. Ind. Health* **2015**, *31*, 841–851. [CrossRef] [PubMed]
138. Oun, R.; Plumb, J.; Rowan, E.; Wheate, N. Encapsulation of cisplatin by cucurbit[7] uril decreases the neurotoxic and cardiotoxic side effects of cisplatin. *Toxicol. Lett.* **2013**, S92. [CrossRef]
139. Gutierrez, K.; Glanzner, W.G.; Chemeris, R.O.; Rigo, M.L.; Comim, F.V.; Bordignon, V.; Gonçalves, P.B. Gonadotoxic effects of busulfan in two strains of mice. *Reprod. Toxicol.* **2016**, *59*, 31–39. [CrossRef] [PubMed]
140. Ahar, N.H.; Khaki, A.; Akbari, G.; Novin, M.G. The effect of busulfan on body weight, testis weight and mda enzymes in male rats. *Int. J. Women's Health Reprod. Sci.* **2014**, *2*, 316–319. [CrossRef]
141. Wang, Z.; Zhang, X. Isolation and identification of anti-proliferative peptides from spirulina platensis using three-step hydrolysis. *J. Sci. Food Agric.* **2017**, *97*, 918–922. [CrossRef] [PubMed]
142. Hung, C.-C.; Yang, Y.-H.; Kuo, P.-F.; Hsu, K.-C. Protein hydrolysates from tuna cooking juice inhibit cell growth and induce apoptosis of human breast cancer cell line MCF-7. *J. Funct. Foods* **2014**, *11*, 563–570. [CrossRef]

143. Huang, F.; Yang, Z.; Yu, D.; Wang, J.; Li, R.; Ding, G. Sepia ink oligopeptide induces apoptosis in prostate cancer cell lines via caspase-3 activation and elevation of Bax/Bcl-2 ratio. *Mar. Drugs* **2012**, *10*, 2153–2165. [CrossRef] [PubMed]
144. Xue, Z.; Wen, H.; Zhai, L.; Yu, Y.; Li, Y.; Yu, W.; Cheng, A.; Wang, C.; Kou, X. Antioxidant activity and anti-proliferative effect of a bioactive peptide from chickpea (*Cicer arietinum* L.). *Food Res. Int.* **2015**, *77*, 75–81. [CrossRef]
145. Song, R.; Wei, R.-B.; Luo, H.-Y.; Yang, Z.-S. Isolation and identification of an antiproliferative peptide derived from heated products of peptic hydrolysates of half-fin anchovy (*Setipinna taty*). *J. Funct. Foods* **2014**, *10*, 104–111. [CrossRef]
146. Fernández-Tomé, S.; Sanchón, J.; Recio, I.; Hernández-Ledesma, B. Transepithelial transport of lunasin and derived peptides: Inhibitory effects on the gastrointestinal cancer cells viability. *J. Food Compos. Anal.* **2017**. [CrossRef]
147. Vital, D.A.L.; de Mejía, E.G.; Dia, V.P.; Loarca-Piña, G. Peptides in common bean fractions inhibit human colorectal cancer cells. *Food Chem.* **2014**, *157*, 347–355. [CrossRef] [PubMed]
148. Zheng, Q.; Qiu, D.; Liu, X.; Zhang, L.; Cai, S.; Zhang, X. Antiproliferative effect of dendrobium catenatum lindley polypeptides against human liver, gastric and breast cancer cell lines. *Food Funct.* **2015**, *6*, 1489–1495. [CrossRef] [PubMed]
149. Xie, H.; Wang, Y.; Zhang, J.; Chen, J.; Wu, D.; Wang, L. Study of the fermentation conditions and the antiproliferative activity of rapeseed peptides by bacterial and enzymatic cooperation. *Int. J. Food Sci. Technol.* **2015**, *50*, 619–625. [CrossRef]
150. Memarpoor-Yazdi, M.; Asoodeh, A.; Chamani, J. A novel antioxidant and antimicrobial peptide from hen egg white lysozyme hydrolysates. *J. Funct. Foods* **2012**, *4*, 278–286. [CrossRef]
151. Mansour, S.C.; Pena, O.M.; Hancock, R.E. Host defense peptides: Front-line immunomodulators. *Trends Immunol.* **2014**, *35*, 443–450. [CrossRef] [PubMed]
152. Tomioka, H.; Nakagami, H.; Tenma, A.; Saito, Y.; Kaga, T.; Kanamori, T.; Tamura, N.; Tomono, K.; Kaneda, Y.; Morishita, R. Novel anti-microbial peptide SR-0379 accelerates wound healing via the PI3 Kinase/Akt/mTOR pathway. *PLoS ONE* **2014**, *9*, e92597. [CrossRef] [PubMed]
153. Pane, K.; Durante, L.; Crescenzi, O.; Cafaro, V.; Pizzo, E.; Varcamonti, M.; Zanfardino, A.; Izzo, V.; Di Donato, A.; Notomista, E. Antimicrobial potency of cationic antimicrobial peptides can be predicted from their amino acid composition: Application to the detection of "cryptic" antimicrobial peptides. *J. Theor. Biol.* **2017**, *419*, 254–265. [CrossRef] [PubMed]
154. Farkas, A.; Maróti, G.; Kereszt, A.; Kondorosi, É. Comparative analysis of the bacterial membrane disruption effect of two natural plant antimicrobial peptides. *Front. Microbiol.* **2017**, *8*, 51. [CrossRef] [PubMed]
155. Zhang, F.; Cui, X.; Fu, Y.; Zhang, J.; Zhou, Y.; Sun, Y.; Wang, X.; Li, Y.; Liu, Q.; Chen, T. Antimicrobial activity and mechanism of the human milk-sourced peptide casein201. *Biochem. Biophys. Res. Commun.* **2017**, *485*, 698–704. [CrossRef] [PubMed]
156. Shah, P.; Hsiao, F.S.H.; Ho, Y.H.; Chen, C.S. The proteome targets of intracellular targeting antimicrobial peptides. *Proteomics* **2016**, *16*, 1225–1237. [CrossRef] [PubMed]
157. Taniguchi, M.; Ochiai, A.; Kondo, H.; Fukuda, S.; Ishiyama, Y.; Saitoh, E.; Kato, T.; Tanaka, T. Pyrrhocoricin, a proline-rich antimicrobial peptide derived from insect, inhibits the translation process in the cell-free *Escherichia coli* protein synthesis system. *J. Biosci. Bioeng.* **2016**, *121*, 591–598. [CrossRef] [PubMed]
158. Guterstam, P.; Madani, F.; Hirose, H.; Takeuchi, T.; Futaki, S.; Andaloussi, S.E.; Gräslund, A.; Langel, Ü. Elucidating cell-penetrating peptide mechanisms of action for membrane interaction, cellular uptake, and translocation utilizing the hydrophobic counter-anion pyrenebutyrate. *Biochim. Biophys. Acta (BBA)-Biomembr.* **2009**, *1788*, 2509–2517. [CrossRef] [PubMed]
159. Wang, G.; Li, X.; Wang, Z. APD3: The antimicrobial peptide database as a tool for research and education. *Nucleic Acids Res.* **2016**, *44*, D1087–D1093. [CrossRef] [PubMed]
160. Waghu, F.H.; Barai, R.S.; Gurung, P.; Idicula-Thomas, S. CAMPR3: A database on sequences, structures and signatures of antimicrobial peptides. *Nucleic Acids Res.* **2016**, *44*, D1094–D1097. [CrossRef] [PubMed]
161. Fan, L.; Sun, J.; Zhou, M.; Zhou, J.; Lao, X.; Zheng, H.; Xu, H. Dramp: A comprehensive data repository of antimicrobial peptides. *Sci. Rep.* **2016**, *6*, 24482. [CrossRef] [PubMed]
162. Piotto, S.P.; Sessa, L.; Concilio, S.; Iannelli, P. Yadamp: Yet another database of antimicrobial peptides. *Int. J. Antimicrob. Agents* **2012**, *39*, 346–351. [CrossRef] [PubMed]

163. Kamali Alamdari, E.; Ehsani, M. Antimicrobial peptides derived from milk: A review. *J. Food Biosci. Technol.* **2017**, *7*, 49–56.
164. Guinane, C.M.; Kent, R.M.; Norberg, S.; O'Connor, P.M.; Cotter, P.D.; Hill, C.; Fitzgerald, G.F.; Stanton, C.; Ross, R.P. Generation of the antimicrobial peptide caseicin a from casein by hydrolysis with thermolysin enzymes. *Int. Dairy J.* **2015**, *49*, 1–7. [CrossRef]
165. Ennaas, N.; Hammami, R.; Beaulieu, L.; Fliss, I. Purification and characterization of four antibacterial peptides from protamex hydrolysate of Atlantic mackerel (*Scomber scombrus*) by-products. *Biochem. Biophys. Res. Commun.* **2015**, *462*, 195–200. [CrossRef] [PubMed]
166. Tang, W.; Zhang, H.; Wang, L.; Qian, H.; Qi, X. Targeted separation of antibacterial peptide from protein hydrolysate of anchovy cooking wastewater by equilibrium dialysis. *Food Chem.* **2015**, *168*, 115–123. [CrossRef] [PubMed]
167. Aissaoui, N.; Chobert, J.-M.; Haertlé, T.; Marzouki, M.N.; Abidi, F. Purification and biochemical characterization of a neutral serine protease from *Trichoderma harzianum*. Use in antibacterial peptide production from a fish by-product hydrolysate. *Appl. Biochem. Biotechnol.* **2016**. [CrossRef] [PubMed]
168. Hu, J.; Xu, M.; Hang, B.; Wang, L.; Wang, Q.; Chen, J.; Song, T.; Fu, D.; Wang, Z.; Wang, S. Isolation and characterization of an antimicrobial peptide from bovine hemoglobin α-subunit. *World J. Microbiol. Biotechnol.* **2011**, *27*, 767–771. [CrossRef]
169. Adje, E.Y.; Balti, R.; Lecouturier, D.; Kouach, M.; Dhulster, P.; Guillochon, D.; Nedjar-Arroume, N. Controlled enzymatic hydrolysis: A new strategy for the discovery of antimicrobial peptides. *Probiotics Antimicrob. Proteins* **2013**, *5*, 176–186. [CrossRef] [PubMed]
170. García-Mora, P.; Martín-Martínez, M.; Bonache, M.A.; González-Múniz, R.; Peñas, E.; Frias, J.; Martinez-Villaluenga, C. Identification, functional gastrointestinal stability and molecular docking studies of Lentil peptides with dual antioxidant and Angiotensin I converting enzyme inhibitory activities. *Food Chem.* **2017**, *221*, 464–472. [CrossRef] [PubMed]
171. Girgih, A.T.; He, R.; Malomo, S.; Offengenden, M.; Wu, J.; Aluko, R.E. Structural and functional characterization of hemp seed (*Cannabis sativa* L.) protein-derived antioxidant and antihypertensive peptides. *J. Funct. Foods* **2014**, *6*, 384–394. [CrossRef]
172. Zambrowicz, A.; Pokora, M.; Setner, B.; Dąbrowska, A.; Szołtysik, M.; Babij, K.; Szewczuk, Z.; Trziszka, T.; Lubec, G.; Chrzanowska, J. Multifunctional peptides derived from an egg yolk protein hydrolysate: Isolation and characterization. *Amino Acids* **2015**, *47*, 369–380. [CrossRef] [PubMed]
173. Sun, Y.; Hu, X.; Li, W. Antioxidant, antitumor and immunostimulatory activities of the polypeptide from *Pleurotus eryngii* mycelium. *Int. J. Biol. Macromol.* **2017**, *97*, 323–330. [CrossRef] [PubMed]
174. Ren, G.; Zhu, Y.; Shi, Z. Detection of lunasin in quinoa (*Chenopodium quinoa*, willd) and the in vitro evaluation of its antioxidant and anti-inflammatory activities. *J. Sci. Food Agric.* **2017**. [CrossRef] [PubMed]
175. Blaser, H.; Dostert, C.; Mak, T.W.; Brenner, D. TNF and ROS crosstalk in inflammation. *Trends Cell Biol.* **2016**, *26*, 249–261. [CrossRef] [PubMed]
176. Amorim, M.M.; Pereira, J.O.; Monteiro, K.M.; Ruiz, A.L.; Carvalho, J.E.; Pinheiro, H.; Pintado, M. Antiulcer and antiproliferative properties of spent brewer's yeast peptide extracts for incorporation into foods. *Food Funct.* **2016**, *7*, 2331–2337. [CrossRef] [PubMed]
177. Pérez-Vega, J.A.; Olivera-Castillo, L.; Gómez-Ruiz, J.Á.; Hernández-Ledesma, B. Release of multifunctional peptides by gastrointestinal digestion of sea cucumber (*Isostichopus badionotus*). *J. Funct. Foods* **2013**, *5*, 869–877. [CrossRef]
178. Chalamaiah, M.; Jyothirmayi, T.; Diwan, P.V.; Kumar, B.D. Antiproliferative, ace-inhibitory and functional properties of protein hydrolysates from Rohu (*Labeo rohita*) roe (egg) prepared by gastrointestinal proteases. *J. Food Sci. Technol.* **2015**, *52*, 8300–8307. [CrossRef] [PubMed]
179. Davis, P.J.; Smales, C.M.; James, D.C. How can thermal processing modify the antigenicity of proteins? *Allergy* **2001**, *56*, 56–60. [CrossRef] [PubMed]
180. Agyei, D.; Ongkudon, C.M.; Wei, C.Y.; Chan, A.S.; Danquah, M.K. Bioprocess challenges to the isolation and purification of bioactive peptides. *Food Bioprod. Process.* **2016**, *98*, 244–256. [CrossRef]
181. Takahashi, H.; Tsuchiya, T.; Takahashi, M.; Nakazawa, M.; Watanabe, T.; Takeuchi, A.; Kuda, T.; Kimura, B. Viability of murine norovirus in salads and dressings and its inactivation using heat-denatured lysozyme. *Int. J. Food Microbiol.* **2016**, *233*, 29–33. [CrossRef] [PubMed]

182. Jan, F.; Kumar, S.; Jha, R. Effect of boiling on the antidiabetic property of enzyme treated sheep milk casein. *Vet. World* **2016**, *9*, 1152. [CrossRef] [PubMed]
183. Mojica, L.; Chen, K.; Mejía, E.G. Impact of commercial precooking of common bean (*Phaseolus vulgaris*) on the generation of peptides, after pepsin—Pancreatin hydrolysis, capable to inhibit dipeptidyl peptidase-IV. *J. Food Sci.* **2015**, *80*, H188–H198. [CrossRef] [PubMed]
184. Escudero, E.; Mora, L.; Toldrá, F. Stability of ACE inhibitory ham peptides against heat treatment and in vitro digestion. *Food Chem.* **2014**, *161*, 305–311. [CrossRef] [PubMed]
185. Fu, Y.; Young, J.F.; Dalsgaard, T.K.; Therkildsen, M. Separation of angiotensin I-converting enzyme inhibitory peptides from bovine connective tissue and their stability towards temperature, pH and digestive enzymes. *Int. J. Food Sci. Technol.* **2015**, *50*, 1234–1243. [CrossRef]
186. Lin, S.; Liang, R.; Xue, P.; Zhang, S.; Liu, Z.; Dong, X. Antioxidant activity improvement of identified pine nut peptides by pulsed electric field (PEF) and the mechanism exploration. *LWT-Food Sci. Technol.* **2017**, *75*, 366–372. [CrossRef]
187. Lin, S.; Liang, R.; Li, X.; Xing, J.; Yuan, Y. Effect of pulsed electric field (PEF) on structures and antioxidant activity of soybean source peptides-SHCMN. *Food Chem.* **2016**, *213*, 588–594. [CrossRef] [PubMed]
188. Kaspar, A.A.; Reichert, J.M. Future directions for peptide therapeutics development. *Drug Discov. Today* **2013**, *18*, 807–817. [CrossRef] [PubMed]

foods

MDPI

Review

Bioactive Peptides in Animal Food Products

Marzia Albenzio *, Antonella Santillo, Mariangela Caroprese, Antonella della Malva and Rosaria Marino

Department of Agricultural Food and Environmental Sciences (SAFE), University of Foggia, Via Napoli 25, 71122 Foggia, Italy; antonella.santillo@unifg.it (A.S.); mariangela.caroprese@unifg.it (M.C.); antonella.dellamalva@unifg.it (A.d.M.); rosaria.marino@unifg.it (R.M.)

* Correspondence: marzia.albenzio@unifg.it

Academic Editor: Maria Hayes
Received: 1 March 2017; Accepted: 5 May 2017; Published: 9 May 2017

Abstract: Proteins of animal origin represent physiologically active components in the human diet; they exert a direct action or constitute a substrate for enzymatic hydrolysis upon food processing and consumption. Bioactive peptides may descend from the hydrolysis by digestive enzymes, enzymes endogenous to raw food materials, and enzymes from microorganisms added during food processing. Milk proteins have different polymorphisms for each dairy species that influence the amount and the biochemical characteristics (e.g., amino acid chain, phosphorylation, and glycosylation) of the protein. Milk from other species alternative to cow has been exploited for their role in children with cow milk allergy and in some infant pathologies, such as epilepsy, by monitoring the immune status. Different mechanisms concur for bioactive peptides generation from meat and meat products, and their functionality and application as functional ingredients have proven effects on consumer health. Animal food proteins are currently the main source of a range of biologically-active peptides which have gained special interest because they may also influence numerous physiological responses in the organism. The addition of probiotics to animal food products represent a strategy for the increase of molecules with health and functional properties.

Keywords: milk; cheese; meat; bioactive peptides; human health

1. Introduction

The general consensus on the impact of lifestyle on human health considers that diet represents a crucial factor in terms of human health status. Proteins of animal origin have been recognized for their nutritional properties as an essential source of amino acids upon digestion, but both digestion and industrial processing may liberate peptides from the parent protein which have biological functions. Animal food products, in particularly dairy foods, were characterized by genetic polymorphisms of the main proteins that impact on protein hydrolysis during food processing prior to consumption and digestion in the human organism. Biologically-active peptides can be produced from milk proteins through different pathways involving milk secretion, milk storage, milk processing, and milk digestion due to enzymatic hydrolyses by indigenous enzymes, digestive enzymes, and microbial enzymes from starter and non-starter cultures. The integrity and structure of meat proteins undergo changes during rigor mortis, the resolution of rigor mortis, and long-term frozen storage. Particularly, a large number of peptides showing important physiological activities are released during meat processing. Dietary supplements allow the delivery of positive molecules in dosages that exceed those obtained from conventional food products. However, great interest has been observed regarding the bioactive components naturally contained in foods which have an impact on biological processes. Bioactive components in foods represent dietary elements that impart a measurable biological effect that affect health in a beneficial way, such as immune-modulating, antihypertnesive, osteoprotective, antilipemic, opiate, antioxidative, and antimicrobial activities [1]. The in vitro bioactivities of food components have been widely explored and the present effort is to study their effects in vivo on healthy subjects or patients with different pathologies. This review provides an integrated overview on the occurrence of bioactive peptides in animal food products, and the role on human health of milk and meat peptides is also discussed.

2. Role of Dairy Proteins on Human Health

Protein is a very heterogeneous component in cow, sheep, and goat milk, mainly influenced by genetic variants. The genetic polymorphisms of milk proteins are of importance as they are associated with quantitative and qualitative parameters in milk. In particular, genetic polymorphism was associated with different levels of protein synthesis in milk, different rates of phosphorylation and glycosilation of the peptide chain, and amino acid sequences of the protein [2]. In cattle, the six main milk proteins are encoded by highly-polymorphic genes with up to 47 protein variants identified, affecting not only the specific protein expression and, as a consequence, milk composition and cheesemaking, but they are also involved in various aspects of human nutrition [3,4]. A recent review paper [5] deepened the presence of a complex polymorphism at casein loci levels in small ruminant species and its role on the nutritional properties on milk and dairy products. In this contest the polymorphisms of milk proteins from small ruminant species have significant potential in human pathology. Genetic polymorphisms of milk proteins also play an important role in eliciting different degrees of allergic reaction [6–8]. Caseins, and especially α-CN, are among the most important milk allergens [9–11]. Milk from other species alternative to cow has been investigated for its role in children with cow milk allergy (CMA); higher TNF-α levels were indeed found after exposure to cow milk casein and β-Lg than after exposure to the same fractions from goat milk [12]. Some studies have indicated an unusually high incidence of allergenic illnesses in those suffering with epilepsy [13]. Most of the authors examined the relationship between food allergy and epilepsy by comparing groups of adult patients with healthy control subjects [14]. Cytokine productions by cultured PBMCs from infants with generalized epilepsy was influenced by protein fractions of milk from bovine, caprine, and ovine species. PBMC's ability to secrete cytokines in response to milk and protein fraction stimulation may be a predictor of the secretion of pro- and anti-inflammatory cytokines in the bloodstream of the challenged patients [14].

2.1. Occurrence of Bioactive Peptides in Milk

Biologically-active peptides can be produced from milk proteins through different pathways involving the action of indigenous enzymes, digestive enzymes, and microbial enzymes from starter and non-starter cultures acting during milk secretion, milk storage, milk processing, and milk digestion. Proteolytic activity in fresh raw milk is attributed to indigenous and microbial enzymes. Among the indigenous enzymes, milk contains at least two main proteinase systems, the plasmin-plasminogen system and lysosomal enzymes, as well as possibly other proteolytic enzymes. Plasmin is the principal proteolytic enzyme in raw milk and is associated with casein micelles. The second proteinase in milk is cathepsin D, activity of which is significantly correlated with somatic cell count, which contains several proteinases, including cathepsin B, L, and G, and elastase [15]. The principal indigenous proteolytic enzymes were investigated and characterized in ovine and caprine milk [16–20]. Some bioactive peptides found in milk and dairy products and their functionality have been reported in Table 1. Indigenous enzymes play a role in the liberation of bioactive peptides during milk secretion and storage. A great number of peptides were found in goat milk incubated up to seven days without any protease inhibitors; plasmin was shown to play a major role in the hydrolysis of casein and high numbers of peptides were derived from the hydrolysis of β-casein. Almost 90% of the peptides identified shared a structural homology with previously-described bioactive peptides in caprine and bovine milk and dairy products showing encrypted sequences of bioactive peptides able to exert ACE-inhibitory activity [21,22]; antihypertensive activity [22,23], and antioxidant activity [24].

Table 1. Bioactive peptides in milk and dairy products.

Product	Carrier/Regulation	Peptide Sequence/Protein Fragment	Functionality	References
Milk	Endogenous enzymes	PYVRYL, LVYPFTGPIPN	ACE-I activity	[21,22]
	Protease from *Enterococcus faecalis*, enzymatic hydrolysis	LHLPLPL,αs1-CN f(90–94) (RYLGY), αs1-CN f(143–149) (AYFYPEL), and αS2-CN f(89–95) (YQKFPQY)	Antihypertensive activity	[22,23,25]
	endogenous enzymes	VLPVPQK	Antioxidant activity	[24]
	Proteinase of *Lactobacillus helveticus* PR4	Bovine αS1-casein; (αS1-CN) 24–47 fragment (f24–47), f(169–193), and β-CN f(58–76); ovine αS1-CN f(1–6) and αS2-CN f(182–185) and f(186–188); caprine β-CN f(58–65) and αS2-CN f(182–187); buffalo β-CN f(58–66);	ACE-I activity	[26]
		Ovine as2-CN fragments; f(165–170) LKKISQ, f(165–181) LKKISQYYYQKFAWPQYL, f(184–208) VDQHQKAMKPWTQPKTNAIPYVRYL, f(203–208) PYVRYL.	Antibacterial activity	[27]
		as1-casein f(1–23)	Immunomodulating activity	[28]
Cheese		Bovine β-CN f(13–28), αS2-CN f(5–21)	Mineral binding	[29]
Manchego		Ovine β-CN, fragment (199–204); α_{s1}-CN f(102–109) KKYNVPQL; αs2-CN f(205–208) VRYL	ACE-I activity	[30,31]
Emmental		Fragments from as1_CN and β-CN	Immunostimulator, antimicrobial and ACE-I activity	[32]
Gouda		α_{s1}-CN f (1–9), α_{s1}-C f (1–13), β-CN f (60–68), β-CN f (109–111)	ACE-I activity	[33]
Crescenza		b-CN f(58–72)	ACE-I activity	[34]

In sheep milk, several peptides with functional activity were found deriving from the action of peptidases of different origins on casein fractions. At least three ACE inhibitory peptides were liberated by purified proteinase of *Lb helveticus* [26] from αs1- and αs2-caseins, and antihypertensive and antioxidant peptides were found in ovine sodium caseinate incubated with *Bacillus* sp. P7 [35]. Four antibacterial peptides were identified from a pepsin hydrolysate of ovine αs2-casein [27], corresponding to αs2-casein fragments f(165–170), f(165–181), f(184–208), and f(203–208), with the former being most effective against Gram-negative bacteria. The peptide corresponding to ovine αs2-casein f(203–208) is a good example for a multifunctional peptide, because it exhibited not only antimicrobial activity, but also potent antihypertensive and antioxidant activity [36].

The most common way to produce bioactive peptides is through enzymatic hydrolyses of whole protein molecules: digestive enzymes and different enzyme combinations of proteinases, including alcalase, chymotripsin, pancreatin, pepsin, and thermolysin have been utilized to generate bioactive peptides from various proteins [37]. Ingested proteins undergo different stages of gastrointestinal hydrolysis in the stomach and intestinal lumen due to proteinases, such as pepsin, trypsin, and chymotripsin. Finally, these peptides are further digested by brush border peptidases at the surface of intestinal epithelial cells to produce amino acids and oligopeptides able to undergo the absorption process. For example, β-casomorphins and phosphopeptides derived from casein (CPPs) are produced in vivo during digestion of dairy products, including milk, fermented milk, cheese, and yogurt [38]. The quantity of peptides released upon digestion is hardly predictable and, consequently, the beneficial effects of human health. Peptide bioavailability is dependent on the resistance of the peptide to hydrolysis in the gastrointestinal tract and serum and its ability to be absorbed across the intestinal epithelium [39]. However, some authors report that the potential yield of bioactive peptides, during the digestion of the major dairy proteins, is relatively high. Meisel and Fitzgerald [40] estimated the theoretical yield of opioid peptides encrypted in milk proteins ranged between 2% and 6%.

2.2. Occurrence of Bioactive Peptides in Dairy Products

The ripening process in cheese encompasses several biochemical pathways dealing with the proteolytic, lipolytic, and glicolytic processes. Many dairy cultures are highly proteolytic, leading to bioactive peptide accumulation in ripened dairy products. Depending on the type of dairy products the level of peptides naturally formed in the matrix varies along with the equilibrium between the liberation and the further hydrolysis during ripening. However, the bioactive peptides have been characterized in a wide variety of dairy products distinguished on the basis of the time of ripening in fresh, short, and long ripened cheese, and on the basis of the technological process of fermented cheese, pasta filata cheese, and cooked cheese.

In long-ripened Gruyere de Comté and Cheddar cheese CPPs naturally occurred due to the primary action of chymosin and plasmin and further hydrolysis of endopeptidases from non-starter lactic acid bacteria [29,41]. The maximum ACE-inhibitor activities were found in Gouda cheese ripened for three months than in short- and long-ripened cheese. On the contrary, in Manchego cheese, from ovine milk, the ACE-inhibitory activity showed a different and complex evolution along with the ripening time decreasing in the first four months, with a subsequent increase and then decreasing again in twelve-month cheese [30]. In Emmental cheese, different bioactivities were detected as mineral-carrying, antimicrobial, antihypertensive, and immunestimulatory due to both the action of plasmin and cathepsin D and to proteinases associated with microbial starter [32]. In Cheddar cheese, the sequence RPKHPIK was found in Festivo and Iberian ovine cheeses [42–44] and was also found when the cheeses were subjected to a hydrolysis process that simulated gastric digestion and reported antimicrobial activity. The sequence RPKHPIKHQ was found in water-soluble peptide preparation isolated from Gouda ripened for eight months, showing a potent antihypertensive activity tested in spontaneously-hypertensive rats [33]. Furthermore, the fragment 1–23 of αs_1-CN, known as Isracidin, originated from the proteolytic activity of chymosin and exerted antimicrobial activity on several microorganisms [45]. The sequence PQEVLNENLLRF was referenced by Minkiewicz et al. [28] as an immunomodulating and antimicrobial peptide sequence in the primary structure of αs_1-CN freed by chymosin activity. Furthermore, antimicrobial peptides were isolated from Mozzarella, Italico, Crescenza, and Gorgonzola cheeses [34] with a specific inhibitory action towards endopeptidase from *Pseudomonas fluorescens*. Such a microorganism is responsible for the impairment of technological and organoleptic features of dairy products. The fermented milks are a source of bioactive peptides with anticariogenic, antihypertensive, mineral binding, and stress relieving activities due to the action of probiotic strains such as *Lb. casei, Lb. helveticus*, and *S. cerevisiae* [46–48].

The development of probiotic cheeses regarded Cheddar cheese [42,49–54], Gouda cheese [55], Cottage cheese [56], Pategrás cheese [57], Crescenza cheese [58], Minas fresh cheese [59,60], and Turkish white cheese [61]. Few studies have been conducted on the production of functional cheeses made from ewe milk; the first research was performed on PDO Canestrato pugliese cheese using *B. bifidum* and *B. longum* [62] as a starter adjunct. Probiotics added to cheese yield a wide spectrum of enzymes able to influence the biochemical events involving the protein and lipid fractions in cheese during ripening. These events have an impact on the development of texture, flavor and health components of cheese. The use of lamb rennet paste containing probiotics is a suitable strategy for innovation in traditional ovine cheese without modification of the production procedures [19,63–67]. This could provide a spin-off for health properties of cheese and for its ripening features, such as an acceleration of the ripening process with economic advantages to producers. It was found that using starter cultures and *L. acidophilus* and *Bifidobacteria* spp. Produced ACE-inhibitory activity peptides in Festivo cheese [43] and Manchego cheese [31]; peptides with antimicrobial activity were found in Cottage cheese produced with *Bifidobacteria* [68]. In functional Scamorza cheese made from ovine milk, containing a mix of *B. longum-, B. lactis-*, and *L. acidophilus*-specific peptides deriving from microbial enzymes were found in cheese at fifteen days of ripening. Several fragments were identified which shared structural homology with previously-reported peptides with ACE-inhibitory activity, antimicrobial activity, antihypertensive activity, and immunomodulating activity. Specific peptides deriving from microbial enzymes may be regarded as tracing fragments and may represent a tool to verify the presence and activity of probiotic cultures in cheese. In functional Scamorza cheese fragments were identified deriving from β-galactosidase and from endonuclease associated to *B. longum*, or deriving from enzymes yielded by *Lactobacillus acidophilus*.

3. Bioactive Peptides in Meat and Meat Products

3.1. Bioactive Peptide Generation

Due to the presence of high-quality proteins, meat represents the most investigated source for the isolation of novel bioactive peptides. Different mechanisms concur for bioactive peptide generation from meat and meat products (Table 2). During meat post-mortem aging, the proteolytic activity due to endogenous enzymes (calpains and cathepsins) is a key process that affects the destructuration of proteins and, consequently, the production and release of a large number of peptides and free amino acids [69,70]. Bauchart et al. [71], in a study on aged beef, found an increase of bioactive peptides in meat after 14 days of post mortem storage than in fresh meat. In a recent study, Fu et al. [72], also demonstrated that post-mortem aging can generate bioactive peptides of about 3 kDa in longissimus dorsi and in semitendinosus muscles after 20 days of extensive proteolysis. During post-mortem meat storage the generation of peptides may also be driven by oxidation processes [73]. An oxidative status could regulate the endogenous enzymatic activity and, consequently, the myofibrillar and sarcoplasmic protein degradation [74]. Changes of temperature and pH can affect the content of bioactive peptides during meat storage due to the variation in the activity of endogenous enzymes and the destruction of pH or heat-sensitive amino acids [75,76].

It is known that bioactive peptides are generated naturally in mammals within the gastrointestinal tract during the metabolisms of dietary meat proteins [77,78]. During gastrointestinal proteolysis, ingested meat-derivative proteins are attacked by stomach-secreted digestive enzymes, such as pepsin, followed by trypsin, chymotrypsin, elastase, and carboxypeptidase secreted in the small intestine with a consequent generation of biological peptides [79]. For this reason, in order to generate potentially-functional peptides from meat products, the gastrointestinal digestive system has been simulated to generate peptides similar to those released in a physiological digestion process. The process that simulates the gastrointestinal digestion is based on an enzymatic hydrolysis using different commercial exogenous proteinases obtained from animal tissues (pepsin and tripsin), plants (papain, ficin, and bromelain), and microbial sources (alcalase®, flavourzyme®, neutrase®, collagenase, or proteinase K) [79–81]. Enzymatic hydrolysis is a widespread method selected by food and pharmaceutical industries to produce bioactive peptides. In addition to meat sources, several bioactive peptides have been obtained through enzymatic hydrolysis from meat collagen or slaughtered by-products (trimmings, organs, hemoglobin), as reported in many studies [73,82].

Table 2. Schematic representation of processes generated for obtaining meat bioactive peptides.

Product	Process	Carrier/Regulation	Functionality	Peptide Sequence	References
Meat	Proteolysis, oxidation	Endogenous enzymes	ACE-I activity	APPPPAEVPEVHEEVH, PPPAEVPEVHEEVH, IPITAAKASRNIA, LPLGG, FAGGRGG, APPPPAEVP	[71,72,74]
	Enzymatic hydrolysis	Exogenous enzymes	ACE-I, antioxidant, antithrombotic, antimicrobial, and anticancer activity	KRQKYD, EKERERQ, KAPVA, PTPVT, RPR, GLSDGEWQ, GFHI, DFHING, FHG	[83–91]
	Cooking	High temperature	ACE-I activity	SPLPPPE, EGPQGPPGPVG, PGLIGARGPPGP	[72]
Collagen	Enzymatic hydrolysis	Bacterial collagenase, exogenous enzymes, protease from *Aspergillus oryzae*	ACE-I and antioxidant activity	AKGANGAPGIAGAPGFPGARGPSGPQGPSGPP, PAGNPGADGQPGAKGANGAP, GAXGLXGP, GPRGF, VGPV, QGAR, LQGM, LQGMH, LC	[92–95]
Cured products	Proteolysis	Endogenous enzymes	Antioxidant activity	DSGVT, IEAEGE, EELDNALN, VPSIDDQEELM, DAQEKLE, ALTA, SLTA, VT, SAGNPN, GLAGA, DLEE	[96,97]
Fermented products	Proteolysis	Presence of starter cultures	Antioxidant activity	FGG, DM	[98]

Other mechanisms, such as freezing and cooking processes, can affect the isolation and availability of bioactive peptides from meat. Freezing can denature proteins due to different chemical and physical stress mechanisms, including ice formation, pH variations, and cold temperature [99], leading to an increase of bioactive peptides. Cooking can affect the generation of peptides and their related bioactivities [72,76] due to changes in the native conformation (denaturation) and rupture of intramolecular forces of proteins caused by heat [100].

A number of bioactive peptides were shown to be released, also, from meat products during curing or ripening processes [101]. The proteolytic degradation that occurs during the ripening of dry-cured ham or during fermentation of sausages, responsible also for flavor and texture, lead to a production of small peptides and free amino acids [83,102]. In particular, in fermented meat products the protein degradation is influenced by different variables as product formulation, processing conditions, and the presence of starter cultures. The content of peptides is influenced by proteolytic degradation of endogenous enzymes together with lactic acid bacteria. In particular, the presence of lactic acid bacteria induces a decrease of pH resulting in a greater activity of endogenous muscle proteases [103].

3.2. Functionality of Meat Bioactive Peptides

Meat peptides have proven effects on consumer health due to different types of bioactivity, including antihypertensive, antioxidant, antithrombotic, antimicrobial, or anticancer activities [104]. Bioactivities of peptides depend on the sequence, amino acid composition, and molecular mass [105]. Furthermore, Vermeirssen et al. [39] reported that the length of peptides could affect the intensity of the bioactivity, with smaller peptides characterized by greater bioactivity.

The most extensively-studied meat bioactive peptides are the angiotensin I-converting enzyme inhibitory (ACE-I) peptides, probably due to their implication in the regulation of blood pressure. ACE is a dipeptidylcarboxypeptidase enzyme that convert angiotensin I (decapeptide) into angiotensin II (octapeptide) resulting in a vasoconstriction of the arteries and, consequently, an increase of blood pressure. Therefore, the inhibition of ACE could be linked to the prevention of cardiovascular disease [106]. Meat proteins are a good source of ACE-I peptides with in vitro and in vivo bioactivities. In recent years, several bioactive peptides have been isolated through the hydrolysis of meat proteins with gastrointestinal enzymes, like pepsin, trypsin, chymotrypsin, or pancreatin. Katayama et al. [84] found two different ACE-I peptides from pork meat (KRQKYD, EKERERQ) through pepsin treatment. Both isolated peptides were studied in vivo in rats showing a hypotensive activity after three and six hours of oral administration.

Twenty-two ACE-I peptides from pork meat using pepsin and pancreatin proteases were isolated in vitro. Among these, KAPVA and PTPVP peptide sequences showed the highest antihypertensive activity [85]. Subsequently, in 2012, the same authors [86] investigated, in vivo, the bioactivity of KAPVA, PTPVP, and RPR peptides in rats, highlighting a major decrease of blood pressure by KAPVA and PTPVP peptides than RPR sequence in rats after eight hours of oral administration.

Peptides extracted from connective tissue were also identified as inhibitors of ACE [92,93,107]. Gómez-Guillén et al. [108] reported that the bioactivities of collagen-derived peptides depends on the amount of Gly and Pro amino acids. In vitro and in vivo ACE-I properties were found in peptides isolated from hydrolysate of bovine Achilles tendon collagen with bacterial collagenase [92]. After hydrolysis, samples were purified, sequenced, and identified as AKGANGA PGIAGAPGFPGARGPSGPQGPSGPP and PAGNPGADGQPGAKGANGAP. Both peptides showed ACE-I activity after an oral administration in rats. In recent years, Fu et al. [72,107] also found bioactive peptides from collagen extracted derived both from nuchal ligament of bovine carcasses (GPRGF) and from cooked semitendinosus muscle (SPLPPPE, EGPQGPPGPVG, and PGLIGARGPPGP) showing greater ACE and renin-inhibitory activities. In addition, Saiga et al. [94] isolated peptides with in vivo ACE-I activity from chicken collagen after hydrolysis with a protease from *Aspergillus oryzae*.

Several peptides isolated from meat are characterized by an antioxidant activity due to their capability to inhibit lipid peroxidation, chelate metal ions, and remove free radicals and ROS [109,110].

The most important antioxidants naturally present in meat are carnosine and anserine dipeptides, which explicate their antioxidant activity chelating pro-oxidative metals [87]. In addition to the peptides that are naturally present in meat, peptides with antioxidant activity were also generated through the hydrolysis with specific proteases. Saiga et al. [87], in an in vitro study on porcine myofibrillar proteins hydrolyzed with papain and actinase E, found five peptides (DSGVT, IEAEGE, EELDNALN, VPSIDDQEELM, and DAQEKLE) that exhibited an antioxidant activity using the linolenic acid peroxidation system. The same authors suggested that the highest antioxidant activity was reached by the DAQEKLE peptide obtained by actinase E, corresponding to a part of the tropomyosin alpha-1 chain. Thus, the type and specificity of proteases used play an important role in determining the antioxidative properties of peptides. Furthermore, three peptides (ALTA, SLTA, and VT) obtained from porcine skeletal muscle actomyosin showed antioxidative activity not only in vitro, but also in vivo in rats [88]. Four antioxidant peptides were also obtained from porcine collagen by Li et al. [95] using three different protease treatments (pepsin and papain, protease from bovine pancreas, and a cocktail of protease from bovine pancreas, bacterial proteases from *Streptomyces*, and *Bacillus polymyxa*). Results of this study showed that collagen treated with the cocktail of three enzymes demonstrate higher antioxidant activity and a major number of peptides (QGAR, LQGM, LQGMH, and LC) rather than the other treatments. In recent years, Banerjee and Shanthi [92] isolated a 36-amino acid residue peptide with free radical scavenging and metal chelating properties from bovine tendon collagen α1. Peptides with antioxidant activity can be produced during meat processing. Twenty-seven antioxidant peptides were sequenced using LC-MS/MS in samples of Spanish dry-cured ham [96]; in this study the highest scavenging activity was identified in the two different peptides (SAGNPN and GLAGA). Broncano et al. [98] also isolated two peptides (FGG and DM) with antioxidant activity in pork Chorizo sausages. Recently, Xing et al. [97] purified several antioxidant peptides from dry-cured Xuanwei ham, highlighting the highest antioxidant activity in DLEE peptide.

Peptides with antithrombotic properties were also isolated from meat. Morimatsu et al. [89] and Shimizu et al. [90] isolated peptides that exhibited antithrombotic activity from porcine longissimus dorsi muscle hydrolyzed with papain. Particularly, Shimizu et al. [90] tested the antithrombotic activity both in vitro, by a platelet function test using rat blood, and in vivo, by oral administration to mice (dose 70 mg/kg of body weight). In vivo results showed that the meat-derived peptide significantly reduced carotid artery thrombosis and decrease platelet activity with a comparable effect to aspirin treatment (at a dose of 50 mg/kg of body weight).

Although a number of peptides with antimicrobial activity have been isolated from bovine blood, only one study showed the presence of antimicrobial peptides derived from bovine meat [91]. In this study, Jang et al. [91] isolated four peptides (GLSDGEWQ, GFHI, DFHING, and FHG) after the hydrolysis with commercial enzymes of beef sarcoplasmic proteins. All peptides were subsequently tested for antimicrobial activity against six pathogens (*Escherichia coli, Pseudomonas aeruginosa, Salmonella typhimurium, Staphylococcus aureus, Bacillus cereus*, and *Listeria monocytogenes*). Results showed a different antimicrobial effect against one, or more, bacteria. In particular, GLSDGEWQ peptide showed an inhibition effect on *Escherichia coli, Salmonella typhimurium, Bacillus cereus*, and *Listeria monocytogenes*, while all tested peptides were found to be active against *Pseudomonas aeruginosa*.

It is known that some peptides can also exhibit anti-cancer activity, inhibit cell proliferation and have cytotoxic effects against cancer cells [111]. Jang et al. [91], investigated four peptides extracted from bovine sarcoplasmic proteins against breast, gastric, and lung adenocarcinoma. Results showed that the GFHI peptide had a greater cytotoxic effect against cancer cells of the breast and decreased the viability of gastric cells. In addition, an inhibitory effect on the proliferation of gastric cells has been found for the GLSDGEWQ peptide.

It is known that, after oral intake, bioactive peptides need to be absorbed intact to ensure their bioactivity within the cellular environments. In this regard, it is important that peptides enter the circularly system intact and remain active during the digestive process [112]. Small-sized peptides are more resistant to degradation by the intestinal enzymes and more easily absorbed to the circularly system [113]. Ohara et al. [114] detected small peptides derived from collagen in blood after oral ingestion of protein hydrolysate products. In recent years, nutrient absorption at the intestinal level is studied using an experimental model involving cultures of colon Caco-2 cells. Shimizu et al. [115] reported that chicken collagen octapeptide (GAXGLXGP) can be transported across a human intestinal epithelium. Recently, Fu et al. [107] also identified two peptides derived from bovine collagen (VGPV and GPRGF) with ACE-inhibitory activity into Caco-2 cells in the human intestinal epithelium, highlighting the bioavailability of these peptides.

Meat-derived bioactive peptides, due to their biological properties, are promising candidates as ingredients of functional or health-promoting foods [116]. Although the meat functional peptide-based products have not yet been commercialized by the industry, meat functional products could open a new market. In particular, development of functional fermented meat products could be a strategy to introduce to the market products with high nutritional value.

4. Occurrence of Bioactive Peptides in Egg

The avian egg is an important source of nutrients, containing all of the proteins, lipids, vitamins, minerals, and growth factors required by the developing embryo, as well as a number of defense factors to protect against bacterial and viral infection [117]. Especially, egg white contains a number of proteins with antimicrobial activities, including bacterial cell lysis, metal binding, and vitamin binding.

Lysozyme is well known to exert antimicrobial activity and, more recently, enzymatic hydrolysis of lysozyme has been found to enhance its activity by exposing antibacterial portions of the protein and producing peptides with antibacterial activity. Peptides corresponding to amino acid residues 98–112 [118], 98–108, and 15–21 [119] possessed antimicrobial activity against *E. coli* and *S. aureus*. Furthermore, peptides produced by the enzymatic digestion of ovalbumin, and their synthetic counterparts, were found to be strongly active against *Bacillus subtilus* and, to a lesser extent, against *E. coli*, *Bordetella bronchiseptica*, *Pseudomonas aeruginosa*, and *Serratia marcescens*, as well as *Candida albicans* [120].

Several egg white proteins and peptides have demonstrated immunomodulating activity. Tezuka and Yoshikawa [121] found that the phagocytic activity of macrophages was increased by the addition of ovalbumin peptides, OA 77-84 and OA 126-134, derived from peptic and chymotryptic digestions, respectively.

It has been reported that certain egg white-derived peptides can play a role in controlling the development of hypertension by exerting vasorelaxing effects [122]; a vasorelaxing peptide, ovokinin (OA 358-365), was isolated by the peptic digestion of ovalbumin. Additionally, a peptide produced by chymotrypsin digestion and corresponding to OA 359-364, was found to possess vasorelaxing activity. Both peptides were administered orally in spontaneously hypertensive rats and were found to significantly lower the systolic blood pressure. The replacement of amino acids in the ovokinin (2–7) peptide has resulted in enhanced antihypertensive activity, with the most potent derivative resulting in a 100-fold more potent antihypertensive activity [123]. Two angiotensin I converting enzyme (ACE)-inhibitory peptides were also identified in ovalbumin by peptic (OA 183-184) and tryptic (OA 200-218) digestions. Miguel et al. [124] examined peptides with ACE-inhibitory properties produced by enzymatic hydrolysis of crude egg white, which were mainly derived from ovalbumin. Among these peptides, two novel peptides with potent ACE-inhibitory activity were found, with amino acid sequences Arg-Ala-Asp-His-Pro-Phe-Leu and Tyr-Ala-Glu-Glu-Arg-Tyr-Pro-Ile-Leu.

Purified fractions from egg white protein hydrolysate showed several peptides identified as RVPSLM, TPSPR, DLQGK, AGLAPY, RVPSL, DHPFLF, HAEIN, QIGLF, HANENIF, VKELY, and TNGIIR, and investigated for angiotensin I-converting enzyme inhibitory activity, antioxidant properties, and anticoagulation activity [125]. In particular, the sequences ascribed to RVPSL, QIGLF, and TNGIIR exhibited high ACE inhibitory activity in vitro, with the IC50 value 20 lM, 75 lM, and 70 lM, respectively.

Hen's egg white lysozyme-derived peptides showed moderate inhibitory activities against calmodulin-dependent phosphodiesterase (CaMPDE) and free-radical scavenging properties [126]. Egg lysozyme hydrolysates have potential as functional foods and nutraceuticals, although bioavailability studies are required to confirm their health benefits in humans.

Author Contributions: Authors equally contributed to literature search and wrote the manuscript.

Conflicts of Interest: The authors declare no conflict of interest.

References

1. Möller, N.P.; Scholz-Ahrens, K.E.; Roos, N.; Schrezenmeier, J. Bioactive peptides and proteins from foods: Indication for health effects. *Eur. J. Nutr.* **2008**, *47*, 171–182. [CrossRef] [PubMed]
2. Albenzio, M.; Santillo, A. Biochemical characteristics of ewe and goat milk: Effect on the quality of dairy products. *Small Rumin. Res.* **2011**, *101*, 33–40. [CrossRef]
3. Kamiński, S.; Cieślińska, A.; Kostyra, E. Polymorphism of bovine-casein and its potential effect on human health. *J. Appl. Genet.* **2007**, *48*, 189–198. [CrossRef] [PubMed]
4. Caroli, A.M.; Chessa, S.; Erhardt, G.J. Invited review: Milk protein polymorphisms in cattle: Effect on animal breeding and human nutrition. *J. Dairy Sci.* **2009**, *92*, 5335–5352. [CrossRef] [PubMed]
5. Albenzio, M.; Santillo, A.; Avondo, M.; Nudda, A.; Chessa, S.; Pirisi, A.; Banni, S. Nutritional properties of small ruminant food products and their role on human health. *Small Rumin. Res.* **2016**, *135*, 3–12. [CrossRef]
6. El-Agamy, E.I. The Challenge of Cow Milk Protein Allergy. *Small Rumin. Res.* **2007**, *68*, 64–72. [CrossRef]
7. Park, Y.W. Hypo-Allergenic and Therapeutic Significance of Goat Milk. *Small Rumin. Res.* **1994**, *14*, 151–159. [CrossRef]
8. Saini, A.L.; Gill, L.S. Goat Milk: An Attractive Alternative. *Indian Dairym.* **1991**, *42*, 562–564.
9. Ballabio, C.; Chessa, S.; Rignanese, D.; Gigliotti, C.; Pagnacco, G.; Terracciano, L.; Fiocchi, A.; Restani, P.; Caroli, A.M. Goat Milk Allergenicity as a Function of α_{S1}-casein Genetic Polymorphism. *J. Dairy Sci.* **2011**, *94*, 998–1004. [CrossRef] [PubMed]
10. Bevilacqua, C.; Martin, P.; Candalh, C.; Fauquant, J.; Piot, M.; Roucayrol, A.M.; Pilla, F.; Heyman, M. Goat's Milk of Defective Alpha(s1)-Casein Genotype Decreases Intestinal and Systemic Sensitization to Beta-Lactoglobulin in Guinea Pigs. *J. Dairy Res.* **2001**, *68*, 217–227. [CrossRef] [PubMed]
11. Slačanac, V.; Božanić, R.; Hardi, J.; Szabó, J.R.; Lučan, M.; Krstanović, V. Nutritional and therapeutic value of fermented caprine milk. *Int. J. Dairy Technol.* **2010**, *63*, 171–189. [CrossRef]
12. Albenzio, M.; Campanozzi, A.; D'Apolito, M.; Santillo, A.; Pettoello Mantovani, M.; Sevi, A. Differences in protein fraction from goat and cow milk and their role on cytokine production in children with cow's milk protein allergy. *Small Rumin. Res.* **2012**, *105*, 202–205. [CrossRef]
13. Frediani, T.; Lucarelli, S.; Pelliccia, A.; Vagnucci, B.; Cerminara, C.; Barbato, M.; Cardi, E. Allergy and Childhood Epilepsy: A Close Relationship? *Acta Neurol. Scand.* **2001**, *104*, 349–352. [CrossRef] [PubMed]
14. Albenzio, M.; Santillo, A.; Ciliberti, M.G.; Figliola, L.; Caroprese, M.; Marino, R.; Polito, A.N. Milk from different species: Relationship between protein fractions and inflammatory response in infant affected by generalized epilepsy. *J. Dairy Sci.* **2016**, *99*, 5032–5038. [CrossRef] [PubMed]
15. Fox, P.F.; Kelly, A.L. Indigenous enzymes in milk: Overview and historical aspects-part 1. *Int. Dairy J.* **2006**, *16*, 500–516. [CrossRef]
16. Albenzio, M.; Caroprese, M.; Santillo, A.; Marino, R.; Taibi, L.; Sevi, A. Effects of somatic cell count and stage of lactation on the plasmin activity and cheese-making properties of ewe milk. *J. Dairy Sci.* **2004**, *87*, 533–542. [CrossRef]
17. Albenzio, M.; Caroprese, M.; Santillo, A.; Marino, R.; Muscio, A.; Sevi, A. Proteolytic Patterns and Plasmin Activity in Ewe Milk with High Somatic Cell Count. *J. Dairy Res.* **2005**, *72*, 86–92. [CrossRef] [PubMed]
18. Albenzio, M.; Santillo, A.; d'Angelo, F.; Sevi, A. Focusing on casein gene cluster and protein profile in Garganica goat milk. *J. Dairy Res.* **2009**, *76*, 83–89. [CrossRef] [PubMed]
19. Santillo, A.; Kelly, A.L.; Palermo, C.; Sevi, A.; Albenzio, M. Role of indigenous enzymes in proteolysis of casein in caprine milk. *Int. Dairy J.* **2009**, *19*, 655–660. [CrossRef]
20. Kelly, A.L.; O'Flaherty, F.; Fox, P.F. Indigenous proteolytic enzymes in milk: A brief overview of the present state of knowledge. *Int. Dairy J.* **2006**, *16*, 563–572. [CrossRef]

21. Quirós, A.; Hernandez-Ledesma, B.; Ramos, M.; Amigo, L.; Recio, I. Angiotensin-converting enzyme inhibitory activity of peptides derived from caprine kefir. *J. Dairy Sci.* **2005**, *88*, 3480–3487. [CrossRef]
22. Yamamoto, N.; Akino, A.; Takano, T. Antihypertensive effect of the peptides derived from casein by an extracellular proteinase from Lactobacillus helveticus CP790. *J. Dairy Sci.* **1994**, *77*, 917–922. [CrossRef]
23. Quirós, A.; Ramos, M.; Muguerza, B.; Delgado, M.A.; Miguel, M.; Aleixandre, M.; Recio, I. Identification of novel antihipertensive peptides in milk fermented with *Enterococcus faecalis*. *Int. Dairy J.* **2007**, *17*, 33–41. [CrossRef]
24. Rival, S.G.; Fornaroli, S.; Boeriu, C.G.; Wichers, H.J. Caseins and casein hydrolysates. 1. Lipoxygenase inhibitory properties. *J. Agric. Food Chem.* **2001**, *49*, 287–294. [CrossRef] [PubMed]
25. del Mar Contreras, M.; Sanchez, D.; Amigo, L.; Sevilla, M.A.; Recio, I. Resistance of casein-derived bioactive peptides to simulated gastrointestinal digestion. *Int. Dairy J.* **2013**, *32*, 71–78. [CrossRef]
26. Minervini, F.; Algaron, F.; Rizzello, C.G.; Fox, P.F.; Monnet, V.; Gobbetti, M. Angiotensin I-converting-enzyme-inhibitory and antibacterial peptides from *Lactobacillus helveticus* PR-4 proteinase-hydrolyzed casein of milk from 6 species. *Appl. Environ. Microbiol.* **2003**, *69*, 5297–5305. [CrossRef] [PubMed]
27. Lopez-Exposito, I.; Gomez-Ruiz, J.A.; Amigo, L.; Recio, I. Identification of antibacterial peptides from ovine alpha-s2 casein. *Int. Dairy J.* **2006**, *16*, 1072–1080. [CrossRef]
28. Minkiewicz, P.; Slangen, C.J.; Dziuba, J.; Visser, S.; Mioduszewska, H. Identification of peptides obtained via hydrolis of bovine casein using HPLC and mass spectrometry. *Milkchwissenschaft* **2000**, *55*, 14–17.
29. Roudot-Algaron, F.; LeBars, D.; Kerhoas, L.; Einhorn, J.; Gripon, J.C. Phosphopeptides from Comté cheese: Nature and origin. *J. Food Sci.* **1994**, *59*, 544–547. [CrossRef]
30. Gómez-Ruiz, J.A.; Ramos, M.; Recio, I. Angiotensin-converting enzyme-inhibitory peptides in Manchego cheeses manufactured with different starter cultures. *Int. Dairy J.* **2002**, *12*, 697–706. [CrossRef]
31. Gomez-Ruiz, J.A.; Ramos, M.; Recio, I. Angiotensin converting enzyme-inhibitory activity of peptides isolated from Manchego cheese. Stability under simulated gastrointestinal digestion. *Int. Dairy J.* **2004**, *14*, 1075–1080. [CrossRef]
32. Gagnaire, V.; Mollé, M.; Herrouin, M.; Léonil, J. Peptides identified during Emmental cheese ripening: Origin and proteolytic systems involved. *J. Agric. Food Chem.* **2001**, *49*, 4402–4413. [CrossRef] [PubMed]
33. Saito, R.; Nakamura, T.; Kitazawa, H.; Kawai, Y.; Itoh, T. Isolation and structural analysis of antihypertensive peptides that exist naturally in Gouda cheese. *J. Dairy Sci.* **2000**, *83*, 1434–1440. [CrossRef]
34. Smacchi, E.; Gobbetti, M. Peptides from several italian cheeses inhibitory to proteolytic enzymes of lactic acid bacteria, *Pseudomonas fluorescens* ATCC 948 and to the angiotensin I-converting enzyme. *Enz. Micr. Tech.* **1998**, *22*, 687–694. [CrossRef]
35. Corrêa, A.P.F.; Dariot, D.J.; Coelho, J.; Meira, S.M.M.; Lopes, F.C.; Segalin, J. Antioxidant, antihypertensive and antimicrobial properties of ovine milk caseinate hydrolyzed with a microbial proteinase. *J. Sci. Food Agric.* **2011**, *91*, 2247–2254. [PubMed]
36. Recio, I.; Quiros, A.; Hernandes-Ledesma, B.; Gomez-Ruiz, J.A.; Miguel, M.; Amigo, L.; Lopez-Exposito, I.; Ramos, M.; Alexandre, A. Bioactive Peptides Identified in Enzyme Hydrolysates from Milk Caseins and Procedure for Their Obtention. European Patent 2005011373, 2005.
37. Korhonen, H.; Pihlanto, A. Bioactive peptides: Production and functionality. *Int. Dairy J.* **2006**, *16*, 945–960. [CrossRef]
38. Phelan, M.; Aherne, A.; FitzGerald, R.J.; O'Brien, N.M. Casein-derived peptides: Biological effects, industrial uses, safety aspects and regulatory status. *Int. Dairy J.* **2009**, *19*, 643–654. [CrossRef]
39. Vermeirssen, V.; van Camp, J.; Verstraete, W. Bioavailability of angiotensin I-converting enzyme inhibitory peptides. *Br. J. Nutr.* **2004**, *92*, 357–366. [CrossRef] [PubMed]
40. Meisel, H.; FitzGerald, R.J. Opioid peptides encrypted in milk proteins. *Br. J. Nutr.* **2000**, *84*, S27–S31. [CrossRef] [PubMed]
41. Singh, T.K.; Fox, P.F.; Healy, A. Isolation and identification of further peptides in the diafiltration retentate of the water-soluble fraction of Cheddar cheese. *J. Dairy Res.* **1997**, *64*, 433–443. [CrossRef] [PubMed]
42. Ong, L.; Henriksson, A.; Shah, N.P. Chemical analysis and sensory evaluation of Cheddar cheese produced with *Lactobacillus acidophilus, Lb. casei, Lb. paracasei* or *Bifidobacterium*. sp. *Int. Dairy J.* **2007**, *17*, 937–945. [CrossRef]

43. Ryhänen, E.L.; Pihlanto, L.A.; Pahkala, E. A new type of ripened; low-fat cheese with bioactive properties. *Int. Dairy J.* **2001**, *11*, 441–447. [CrossRef]
44. Recio, I. Health and Nutritional aspects of cheese with a focus on the bioactive peptides. In Proceedings of the 8th Cheese Symposium, Moorepark, Cork, Ireland, 28–29 September 2011; Special Issue of Dairy Science & Technology. Volume 91, p. 31.
45. Gobbetti, M.; Minervini, F.; Rizzello, C.G. Angiotensin Iconverting- enzyme-inhibitory and antimicrobial bioactive peptides. *Appl. Environ. Microbiol.* **2004**, *57*, 172–188.
46. Nakamura, Y.; Yamamoto, M.; Sakai, K.; Okubo, A.; Yamazaki, S.; Takano, T. Purification and characterization of angiotensin Iconverting enzyme inhibitors from sour milk. *J. Dairy Sci.* **1995**, *78*, 777–783. [CrossRef]
47. Meisel, H. Biochemical prperties of bioactive peptides derived from milk proteins: Potential nutraceuticals for food and pharmaceutical applications. *Livestock Prod. Sci.* **1997**, *50*, 125–138. [CrossRef]
48. Ashar, M.N.; Chand, R. Fermented milk containing ACE-inhibitory peptides reduces blood pressure in middle aged hypertensive subjects. *Milchwissenschaft* **2004**, *59*, 363–366.
49. Dinakar, P.; Mistry, V.V. Growth and viability of *Bifidobacterium. bifidum* in Cheddar cheese. *J. Dairy Sci.* **1994**, *77*, 2854–2864. [CrossRef]
50. Ong, L.; Henriksson, A.; Shah, N.P. Development of probiotic Cheddar cheese containing *Lb. acidophilus, Lb. paracasei, Lb. casei* and *Bifidobacterium.* spp. and the influence of these bacteria on proteolytic patterns and production of organic acid. *Int. Dairy J.* **2006**, *16*, 446–456. [CrossRef]
51. Daigle, A.; Roy, D.; Vuillemand, J.C. Production of probiotic cheese (Cheddar-like cheese) using enriched cream fermented by *Bifidobacterium infantis*. *J. Dairy Sci.* **1999**, *82*, 1081–1091. [CrossRef]
52. Gardiner, G.; Ross, R.P.; Collins, J.K.; Fitzgerald, G.; Stanton, C. Development of a probiotic Cheddar cheese containing human-derived *Lactobacillus paracasei* strains. *Appl. Environ. Microbiol.* **1998**, *6*, 2192–2199.
53. Gardiner, G.; Stanton, C.; Lynch, P.B.; Collins, K.; Fitzgerald, G.; Ross, R.P. Evaluation of Cheddar cheese as a food carrier for delivery of a probiotic strain to the gastrointestinal tract. *J. Dairy Sci.* **1999**, *82*, 1379–1387. [CrossRef]
54. McBrearty, S.; Ross, R.P.; Fitzgerald, G.F.; Collins, J.K.; Wallace, J.M.; Stanton, C. Influence of two commercially available bifidobacteria cultures on Cheddar cheese quality. *Int. Dairy J.* **2001**, *11*, 599–610. [CrossRef]
55. Gomes, A.M.P.; Malcata, F.X.; Klaver, F.A.M.; Grande, H.J. Incorporation and survival of *bifidobacterium* spp. Strain Bo and *Lactobacillus acidophilus* strain Ki in a cheese product. *Neth. Milk Dairy J.* **1995**, *49*, 71–95.
56. Blanchette, L.; Roy, D.; Bèlanger, G.; Gauthier, S.F. Production of Cottage cheese using dressing fermented by bifidobacteria. *J. Dairy Sci.* **1996**, *79*, 8–15. [CrossRef]
57. Bergamini, C.V.; Hynes, E.R.; Palma, S.B.; Sabbag, N.G.; Zalazar, C.A. Proteolytic activity of the three probiotic strains in semi-hard cheese as single and mixed cultures: *Lactobacillus acidophilus, Lactobcillus. paracasei*, and *Bifidabacterium. lactis*. *Int. Dairy J.* **2009**, *19*, 467–475. [CrossRef]
58. Gobbetti, M.; Corsetti, A.; Smacchi, E.; Zocchetti, A.; De Angelis, M. Production of Crescenza cheese by incorporation of bifidobacteria. *J. Dairy Sci.* **1998**, *81*, 37–47. [CrossRef]
59. Buriti, F.; da Rocha, J.S.; Assis, E.G.; Saad, S.M.I. Probiotic potential of Minas fresh cheese prepared with the addition of *Lactobacillus paracasei*. *LWT—Food Sci. Technol.* **2005**, *38*, 173–180. [CrossRef]
60. Souza, C.H.B.; Saad, S.M.I. Viability of *Lactobacillus acidophilus* La-5 added solely or in co-culture with a yoghurt starter culture and implications on physico-chemical and related properties of Minas fresh cheese during storage. *LWT—Food Sci. Technol.* **2009**, *42*, 633–640. [CrossRef]
61. Kasimoğlu, A.; Göncüoğlu, M.; Akgün, S. Probiotic white cheese with *Lactobacillus acidophilus*. *Int. Dairy J.* **2004**, *14*, 1067–1073. [CrossRef]
62. Corbo, M.R.; Albenzio, M.; De Angelis, M.; Sevi, A.; Gobbetti, M. Microbiological and biochemical properties of Canestrato pugliese hard cheese supplemented with Bifidobacteria. *J. Dairy Sci.* **2001**, *84*, 551–561. [CrossRef]
63. Santillo, A.; Caroprese, M.; Marino, R.; Muscio, A.; Sevi, A.; Albenzio, M. Influence of lamb rennet paste on composition and proteolysis during ripening of Pecorino foggiano cheese. *Int. Dairy J.* **2007**, *17*, 535–546. [CrossRef]
64. Santillo, A.; Albenzio, M. Influence of lamb rennet paste containing probiotic on proteolysis and rheological properties of pecorino cheese. *J. Dairy Sci.* **2008**, *91*, 1733–1742. [CrossRef] [PubMed]
65. Albenzio, M.; Santillo, A.; Caroprese, M.; Marino, R.; Trani, A.; Faccia, M. Biochemical patterns in ovine cheese: Influence of probiotic strains. *J. Dairy Sci.* **2010**, *93*, 3487–3496. [CrossRef] [PubMed]

66. Santillo, A.; Albenzio, M. Focusing on Lamb Rennet Paste: Combining Tradition and Innovation in Cheese Production. In *Food Engineering*; Siegler, B.C., Ed.; Nova Publishers, Inc.: New York, NY, USA, 2010; Chapter 17; ISBN: 978-1-61728-913-2.
67. Santillo, A.; Albenzio, M.; Bevilacqua, A.; Corbo, M.R.; Sevi, A. Encapsulation of probiotic bacteria in lamb rennet paste: Effects on the quality of Pecorino cheese. *J. Dairy Sci.* **2012**, *95*, 3489–3500. [CrossRef] [PubMed]
68. O'Riordan, K.; Fitzgerald, G.F. Evaluation of bifidobacteria for the production of antimicrobial compounds and assessment of performance in Cottage cheese at refrigeration temperature. *J. Appl. Microbiol.* **1998**, *85*, 103–114. [CrossRef] [PubMed]
69. Sentandreu, M.A.; Coulis, G.; Ouali, A. Role of muscle endopeptidases and their inhibitors in meat tenderness. *Trends Food Sci. Technol.* **2002**, *13*, 400–421. [CrossRef]
70. Toldrá, F.; Aristoy, M.C.; Mora, L.; Reig, M. Innovations in value-addition of edible meat by-products. *Meat Sci.* **2012**, *92*, 290–296. [CrossRef] [PubMed]
71. Bauchart, C.; Remond, D.; Chambon, C.; Mirand, P.P.; Savary-Auzeloux, I.; Reynes, C.; Morzel, M. Small peptides (<5 kDa) found in ready-to-eat beef meat. *Meat Sci.* **2006**, *74*, 658–666.
72. Fu, Y.; Jette, F.Y.; Therkildsen, M. Bioactive peptides in beef: Endogenous generation through postmortem aging. *Meat Sci.* **2017**, *123*, 134–142. [CrossRef] [PubMed]
73. Lafarga, T.; Hayes, M. Bioactive peptides from meat muscle and by-products: Generation, functionality and application as functional ingredients. *Meat Sci.* **2014**, *98*, 227–239. [CrossRef] [PubMed]
74. Zhang, W.G.; Xiao, S.; Ahn, D.U. Protein oxidation: Basic principles and implications for meat quality. *Crit. Rev. Food Sci. Nutr.* **2013**, *53*, 1191–1201. [CrossRef] [PubMed]
75. Korhonen, H.; Pihlanto-Leppäla, A.; Rantamäki, P.; Tupasela, T. Impact of pro- cessing on bioactive proteins and peptides. *Trends Food Sci. Technol.* **1998**, *9*, 307–319. [CrossRef]
76. Leygonie, C.; Britz, T.J.; Hoffman, L.C. Impact of freezing and thawing on the quality of meat: Review. *Meat Sci.* **2012**, *91*, 93–98. [CrossRef] [PubMed]
77. Bauchart, C.; Morzel, M.; Chambon, C.; Mirand, P.P.; Reynès, C.; Buffère, C.; Rémond, D. Peptides reproducibly released by in vivo digestion of beef meat and trout flesh in pigs. *Br. J. Nutr.* **2007**, *98*, 1187–1195. [CrossRef] [PubMed]
78. Adje, E.; Balti, R.; Kouach, M.; Guillochon, D.; Nedjar-Arroume, N. α 67–106 of bovine hemoglobin: A new family of antimicrobial and angiotensin I-converting en- zyme inhibitory peptides. *Eur. Food Res. Technol.* **2011**, *232*, 637–646. [CrossRef]
79. Pihlanto, A.; Korhonen, H. Bioactive peptides and proteins. *Adv. Food Nutr. Res.* **2003**, *47*, 175–276. [PubMed]
80. Lafarga, T.; O'Connor, P.; Hayes, M. In silico methods to identify meat-derived prolyl endopeptidase inhibitors. *Food Chem.* **2015**, *175*, 337–343. [CrossRef] [PubMed]
81. Cheung, I.W.Y.; Nakayama, S.; Hsu, M.N.K.; Samaranayaka, A.G.P.; Li-Chan, E.C.Y. Angiotensin-I converting enzyme inhibitory activity of hydrolysates from Oat (avena sativa) proteins by in silico and in vitro analyses. *J. Agric. Food Chem.* **2009**, *57*, 9234–9242. [CrossRef] [PubMed]
82. Vercruysse, L.; Van Camp, J.; Smagghe, G. ACE inhibitory peptides derived from enzymatic hydrolysate of animal protein: A review. *J. Agric. Food Chem.* **2005**, *53*, 8106–8115. [CrossRef] [PubMed]
83. Arihara, K. Strategies for designing novel functional meat products. *Meat Sci.* **2006**, *74*, 219–229. [CrossRef] [PubMed]
84. Katayama, K.; Anggraeni, H.E.; Mori, T.; Ahhmed, A.M.; Kawahara, S.; Sugiyama, M.; Nakayama, T.; Maruyama, M.; Muguruma, M. Porcine skeletal muscle tro- ponin is a good source of peptides with angiotensin-I converting enzyme inhibitory activity and antihypertensive effects in spontaneously hypertensive rats. *J. Agric. Food Chem.* **2008**, *56*, 355–360. [CrossRef] [PubMed]
85. Escudero, E.; Sentandreu, M.A.; Arihara, K.; Toldra, F. Angiotensin I-converting enzyme inhibitory peptides generated from in vitro gastrointestinal digestion of pork meat. *J. Agric. Food Chem.* **2010**, *58*, 2895–2901. [CrossRef] [PubMed]
86. Escudero, E.; Toldrá, F.; Sentandreu, M.A.; Nishimura, H.; Arihara, K. Anti- hypertensive activity of peptides identified in the in vitro gastrointestinal digest of pork meat. *Meat Sci.* **2012**, *91*, 382–384. [CrossRef] [PubMed]
87. Saiga, A.; Tanabe, S.; Nishimura, T. Antioxidant activity of peptides obtained from porcine myofibrillar proteins by protease treatment. *J. Agric. Food Chem.* **2003**, *51*, 3661–3667. [CrossRef] [PubMed]

88. Arihara, K.; Ohata, M. Functional Properties of Bioactive Peptides Derived from meat Proteins. In *Advanced Technologies for Meat Processing*; Toldra, F., Ed.; Springer: New York, NY, USA, 2006; pp. 245–274.
89. Morimatsu, F.; Ito, M.; Budijanto, S.; Watanabe, I.; Furukawa, Y.; Kimura, S. Plasma cholesterol-suppressing effect of papain-hydrolyzed pork meat in rats fed hypercholesterolemic diet. *J. Nutr. Sci. Vitam.* **1996**, *42*, 145–153. [CrossRef]
90. Shimizu, M.; Sawashita, N.; Morimatsu, F.; Ichikawa, J.; Taguchi, Y.; Ijiri, Y.; Yamamoto, J. Antithrombotic papain-hydrolyzed peptides isolated from pork meat. *Thromb. Res.* **2009**, *123*, 753–757. [CrossRef] [PubMed]
91. Jang, A.; Jo, C.; Kang, K.S.; Lee, M. Antimicrobial and human cancer cell cytotoxic effect of synthetic angiotensin-converting enzyme (ACE) inhibitory peptides. *Food Chem.* **2008**, *107*, 327–336. [CrossRef]
92. Banerjee, P.; Shanthi, C. Isolation of novel bioactive regions from bovine Achilles tendon collagen having angiotensin I-converting enzyme-inhibitory properties. *Process. Biochem.* **2012**, *47*, 2335–2346. [CrossRef]
93. Fu, Y.; Young, J.F.; Løkke, M.M.; Lametsch, R.; Aluko, R.E.; Therkildsen, M. Re- valorisation of bovine collagen as a potential precursor of angiotensin I-converting enzyme (ACE) inhibitory peptides based on in silico and in vitro protein digestions. *J. Funct. Foods* **2016**, *24*, 196–206. [CrossRef]
94. Saiga, A.; Iwai, K.; Hayakawa, T.; Takahata, Y.; Kitamura, S.; Nishimura, T.; Morimatsu, F. Angiotensin I-converting enzyme-inhibitory peptides obtained from chicken collagen hydrolysate. *J. Agric. Food Chem.* **2008**, *56*, 9586–9591. [CrossRef] [PubMed]
95. Li, B.; Chen, F.; Wang, X.; Ji, B.; Wu, Y. Isolation and identification of antioxidative peptides from porcine collagen hydrolysate by consecutive chromatography and electrospray ionization–mass spectrometry. *Food Chem.* **2007**, *102*, 1135–1143. [CrossRef]
96. Escudero, E.; Mora, L.; Fraser, P.D.; Aristoy, M.C.; Toldrá, F. Identification of novel antioxidant peptides generated in Spanish dry-cured ham. *Food Chem.* **2013**, *138*, 1282–1288. [CrossRef] [PubMed]
97. Xing, L.; Hu, Y.; Hu, H.; Ge, Q.; Zhou, G.; Zhang, W. Purification and identification of antioxidative peptides from dry-cured Xuanwei ham. *Food Chem.* **2016**, *194*, 951–958. [CrossRef] [PubMed]
98. Broncano, J.M.; Otte, J.; Petrón, M.J.; Parra, V.; Timón, M.L. Isolation and identification of low molecular weight antioxidant compounds from fermented "chorizo" sausages. *Meat Sci.* **2012**, *90*, 494–501. [CrossRef] [PubMed]
99. Christensen, L.; Ertbjerg, P.; Løje, H.; Risbo, J.; van den Berg, F.W.; Christensen, M. Relationship between meat toughness and properties of connective tissue from cows and young bulls heat treated at low temperatures for prolonged times. *Meat Sci.* **2013**, *93*, 787–795. [CrossRef] [PubMed]
100. Yu, T.; Morton, J.D.; Clerens, S.; Dyer, M.J. Cooking-induced protein modifications in meat. *Compr. Rev. Food Sci. Food Saf.* **2017**, *16*, 141–159. [CrossRef]
101. Stadnik, J.; Keska, P. Meat and fermented meat products as a source of bioactive peptides. *Acta Sci. Pol. Technol. Aliment.* **2015**, *14*, 181–190. [CrossRef] [PubMed]
102. Toldra, F. Dry. In *Encyclopedia of Meat Sciences*; Jensen, W.K., Devine, C., Dikeman, M., Eds.; Elsevier: Oxford, UK, 2004; pp. 360–365.
103. Kato, T.; Matsuda, T.; Tahara, T.; Sugimoto, M.; Sato, Y.; Nakamura, R. Effects of meat conditioning and lactic fermentation on pork muscle protein degradation. *Biosci. Biotechnol. Biochem.* **1994**, *58*, 408–410. [CrossRef]
104. Udenigwe, C.C.; Howard, A. Meat proteome as source of functional biopeptides. *Food Res. Int.* **2013**, *54*, 1021–1032. [CrossRef]
105. Shahidi, G.; Zhong, J. Bioactive peptides. *J. AOAC Int.* **2008**, *91*, 914–931. [PubMed]
106. Ahmed, A.M.; Mugurama, M. A review of meat protein hydrolysates and hypertension. *Meat Sci.* **2010**, *86*, 110–118. [CrossRef] [PubMed]
107. Fu, Y.; Young, J.F.; Dalsgaard, T.K.; Therkildsen, M. Separation of angiotensin I- converting enzyme inhibitory peptides from bovine connective tissue and their sta- bility towards temperature, pH and digestive enzymes. *Int. J. Food Sci. Technol.* **2015**, *50*, 1234–1243. [CrossRef]
108. Gómez-Guillén, M.C.; Giménez, B.; López-Caballero, M.E.; Montero, M.P. Functional and bioactive properties of collagen and gelatin from alternative sources: A review. *Food Hydrocoll.* **2011**, *25*, 1813–1827. [CrossRef]
109. Young, J.F.; Therkildsen, M.; Ekstrand, B.; Che, B.N.; Larsen, M.K.; Oksbjerg, N.; Stagsted, J. Novel aspects of health promoting compounds in meat. *Meat Sci.* **2013**, *95*, 904–911. [CrossRef] [PubMed]
110. Milan, B.Z.; Marija, B.; Jelena, I.; Jelena, J.; Marija, D.; Radmila, M.; Baltic, T. Bioactive peptides from meat and their influence on human health. *Tehnol. Mesa* **2014**, *55*, 8–21.

111. Udenigwe, C.C.; Aluko, R.E. Food protein-derived bioactive peptides: Production, processing and potential health benefits. *J. Food Sci.* **2012**, *77*, 11–24. [CrossRef] [PubMed]
112. Ryan, J.T.; Ross, R.P.; Bolton, D.; Fitzgerald, G.F.; Stanton, C. Bioactive peptides from muscle sources: Meat and fish. *Nutrients* **2011**, *3*, 765–791. [CrossRef] [PubMed]
113. Segura-Campos, M.; Chel-Guerrero, L.; Batancur-Ancona, D.; Hernandez-Escalante, V.M. Bioavailability of bioactive peptides. *Foods Rev. Int.* **2011**, *27*, 213–226. [CrossRef]
114. Ohara, H.; Matsumoto, H.; Itoh, K.; Iwai, K.; Sato, K. Comparison of quantity and structures of hydroxyproline-containing peptides in human blood after oral ingestion of gelatin hydrolysates from different sources. *J. Agric. Food Chem.* **2007**, *55*, 1532–1535. [CrossRef] [PubMed]
115. Shimizu, K.; Sato, M.; Zhang, Y.; Kouguchi, T.; Takahata, Y.; Morimatsu, F.; Shimizu, M. The bioavailable octapeptide Gly-Ala-Hyp-Gly-Leu-Hyp-Gly-Pro- stimulates nitric oxide synthesis in vascular endothelial cells. *J. Agric. Food Chem.* **2010**, *58*, 6960–6965. [CrossRef] [PubMed]
116. Hartmann, R.; Meisel, H. Food-derived peptides with biological activity: From research to food applications. *Current Opin. Biotechnol.* **2007**, *18*, 163–169. [CrossRef] [PubMed]
117. Kovacs-Nolan, J.; Phillips, M.; Mine, Y. Advances in the Value of Eggs and Egg Components for Human Health. *J. Agric. Food Chem.*, **2005**, *53*, 8421–8431. [CrossRef] [PubMed]
118. Pellegrini, A.; Thomas, U.; Wild, P.; Schraner, E.; von Fellenberg, R. Effect of lysozyme or modified lysozyme fragments on DNA and RNA synthesis and membrane permeability of *Escherichia coli*. *Microbiol. Res.* **2000**, *155*, 69–77. [CrossRef]
119. Mine, Y.; Ma, F.; Lauriau, S. Antimicrobial peptides released by enzymatic hydrolysis of hen egg white lysozyme. *J. Agric. Food Chem.* **2004**, *52*, 1088–1094. [CrossRef] [PubMed]
120. Pellegrini, A.; Hulsmeier, A.J.; Hunziker, P.; Thomas, U. Proteolytic fragments of ovalbumin display antimicrobial activity. *Biochim. Biophys. Acta* **2004**, *1672*, 76–85. [CrossRef] [PubMed]
121. Tezuka, H.; Yoshikawa, M. Presented at the Annual Meeting of the Japan Society for Bioscience, Biotechnology, and Agrochemistry, Tokyo, Japan, 1995; p. 163.
122. Davalos, A.; Miguel, M.; Bartolome, B.; Lopez-Fandino, R. Antioxidant activity of peptides derived from egg white proteins by enzymatic hydrolysis. *J. Food Prot.* **2004**, *67*, 1939–1944. [CrossRef] [PubMed]
123. Yamada, Y.; Matoba, N.; Usui, H.; Onishi, K.; Yoshikawa, M. Design of a highly potent anti-hypertensive peptide based on ovokinin(2-7). *Biosci. Biotechnol. Biochem.* **2002**, *66*, 1213–1217. [CrossRef] [PubMed]
124. Miguel, M.; Recio, I.; Gomez-Ruiz, J.A.; Ramos, M.; Lopez- Fandino, R. Angiotensin I-converting enzyme inhibitory activity of peptides derived from egg white proteins by enzymatic hydrolysis. *J. Food Prot.* **2004**, *67*, 1914–1920. [CrossRef] [PubMed]
125. Yu, Z.; Liu, B.; Zhao, W.; Yin, Y.; Liu, J.; Chen, F. Primary and secondary structure of novel ACE-inhibitory peptides from egg white protein. *Food Chem.* **2012**, *133*, 315–322. [CrossRef] [PubMed]
126. You, S.J.; Udenigwe, C.C.; Aluko, R.E.; Wua, J. Multifunctional peptides from egg white lysozyme. *Food Res. Int.* **2010**, *43*, 848–855. [CrossRef]

foods

MDPI

Article

Glycomacropeptide Reduces Intestinal Epithelial Cell Barrier Dysfunction and Adhesion of Entero-Hemorrhagic and Entero-Pathogenic *Escherichia coli* in Vitro

Shane Feeney [1,2], Joseph Thomas Ryan [1], Michelle Kilcoyne [2], Lokesh Joshi [2] and Rita Hickey [1,*]

1 Teagasc Food Research Centre, Moorepark, Fermoy, P61C996 Co. Cork, Ireland; shanefeeney1518@gmail.com (S.F.); josephthomas.ryan@gmail.com (J.T.R.)

2 Advanced Glycoscience Research Cluster, National Centre for Biomedical Engineering Science, National University of Ireland Galway, H91TK33 Galway, Ireland; michelle.kilcoyne@nuigalway.ie (M.K.); lokesh.joshi@nuigalway.ie (L.J.)

* Correspondence: rita.hickey@teagasc.ie; Tel.: +353-254-2227

Received: 4 October 2017; Accepted: 25 October 2017; Published: 27 October 2017

Abstract: In recent years, the potential of glycosylated food components to positively influence health has received considerable attention. Milk is a rich source of biologically active glycoconjugates which are associated with antimicrobial, immunomodulatory, anti-adhesion, anti-inflammatory and prebiotic properties. Glycomacropeptide (GMP) is the C-terminal portion of kappa-casein that is released from whey during cheese-making by the action of chymosin. Many of the biological properties associated with GMP, such as anti-adhesion, have been linked with the carbohydrate portion of the protein. In this study, we investigated the ability of GMP to inhibit the adhesion of a variety of pathogenic *Escherichia coli* strains to HT-29 and Caco-2 intestinal cell lines, given the importance of *E. coli* in causing bacterial gastroenteritis. GMP significantly reduced pathogen adhesion, albeit with a high degree of species specificity toward enteropathogenic *E. coli* (EPEC) strains O125:H32 and O111:H2 and enterohemorrhagic *E. coli* (EHEC) strain 12900 O157:H7. The anti-adhesive effect resulted from the interaction of GMP with the *E. coli* cells and was also dependent on GMP concentration. Pre-incubation of intestinal Caco-2 cells with GMP reduced pathogen translocation as represented by a decrease in transepithelial electrical resistance (TEER). Thus, GMP is an effective in-vitro inhibitor of adhesion and epithelial injury caused by *E. coli* and may have potential as a biofunctional ingredient in foods to improve gastrointestinal health.

Keywords: *Escherichia coli*; adherence; enterohemorrhagic; enteropathogenic; glycomacropeptide; milk

1. Introduction

Prevention and treatment of infectious diseases requires a thorough understanding of the complex interactions between pathogenic bacteria and the human host. It is estimated that at least 90% of all bacteria in the environment survive attached to or in close association with a surface, where they can thrive [1]. Bacterial colonisation and infection of the gastrointestinal tract involves the binding of bacterial adhesins to specific ligands present on the intestinal epithelium [2]. Bacterial survival then increases after this attachment is made, as bacteria are more resistant to cleansing mechanisms, immune factors, bacteriolytic enzymes, antibiotics and physical removal by hydrodynamic forces [3]. Hence, early prevention of bacterial adherence to the host epithelium should reduce the incidence of disease.

Pathogenic *Escherichia coli* is one of the leading causes of intestinal (enteritis, diarrhea, or dysentery) disease. When *E. coli* adheres to cells lining the intestine, disruption of normal intestinal barrier function occurs. This can result in the leakage of water and plasma proteins into the lumen and translocation of intestinal bacteria into the systemic circulation, contributing to the development of systemic septicaemia [4,5]. After adhesion, enteropathogenic *E. coli* (EPEC) and enterohemorrhagic *E. coli* (EHEC) insert bacterial effector proteins and translocated intimin receptor (TIR) into the host cell via the type III secretion system (TTSS), which are known to cause gut barrier dysfunction leading to increased cell permeability through recruitment of pro-inflammatory cytokines such as interleukin 8 (IL-8) and tumor necrosis factor alpha (TNF-α) [6–12]. These inflammatory cytokines modulate intracellular signalling pathways within the host that promote and redistribute tight junction (TJ) proteins such as zonula occludens-1 and claudin, which increases membrane permeability and paracellular movement of bacteria [13].

Previous studies have suggested that treatment with certain glycoproteins could help prevent considerable structural and functional damage caused by inflammation [12,14,15]. Milk glycoproteins have also been shown to obstruct specific host–pathogen interactions including bacterial adhesion to the host ligands [16–18]. Glycomacropeptide (GMP) is a casein-derived whey protein found in "sweet" whey, and is formed when kappa-casein is hydrolysed by chymosin during cheese production [19]. Previously, it has been reported that GMP can inhibit viral or bacterial adhesion to cells [20,21], promote proliferation of beneficial bacteria [22,23], neutralize enterotoxin, inhibit gastrointestinal secretions and exert immune regulation [24]. These bioactivities are mainly attributed to the O-linked glycosylation [25] associated with GMP and particularly the sialic acid (*N*-acetylneuraminic acid) component (reviewed by [26]). Sialic acid is also present on surface receptors of intestinal cells and has been identified as a ligand component for bacterial adhesion [27–29].

In previous studies, GMP has been shown to reduce the adherence of pathogens such as *Salmonella enteritidis, S. fyris, S. typhimurium, Vibro cholera, Helico pylori, Shigella flexneri* and *E. coli* to certain intestinal cell lines [2,16,20,21,30,31]. Nakajima et al. [30] found that GMP could inhibit the association of EHEC O157 with Caco-2 cells and it has also been shown to inhibit the association of EPEC with Caco-2 cells based on pathogen-binding to its sialic acid component [20]. The glycopeptide was also found to inhibit the adhesion of certain strains of EPEC to human HT29 cells [2] and the ETEC strain K88 to porcine intestinal cells and porcine mucus [32,33].

In this study, we aimed to directly compare the ability of GMP to inhibit adhesion of EPEC and EHEC strains to two intestinal cell lines to determine any bias of GMP towards inhibition of certain pathotypes or when using different cell lines. The study also aimed to establish the mechanism of anti-adhesion, whether it be through either direct (bacterial binding) or indirect (cell-line binding) inhibition, and to assess the ability of GMP to suppress pathogen-induced tight junction (TJ) barrier function impairment.

2. Materials and Methods

2.1. Materials

GMP containing approximately 8.5% sialic acid on a GMP basis was kindly provided by Agropur Ingredients (Eden Prairie, MN, USA). The human colonic adenocarcinoma cell lines, HT-29 and Caco-2, were purchased from the American Type culture collection (ATCC). Cell culture reagents were purchased from Sigma-Aldrich (Wicklow, Ireland).

2.2. Bacteria and Culture Conditions

The EPEC strains O111:H19 (NCTC 8007) and O125:H2 (NCTC8623) and EHEC strains NCTC12900, DAF 454 and DPC 6055 were obtained from the Leibniz Institute DSMZ-German Collection of Microorganisms and Cell Cultures (DSM; Braunschweig, Germany), the National Collection of Type Cultures (NCTC; London, UK) and the Dairy Products Research Centre culture collection (DPC; Teagasc Food Research Centre, Cork, Ireland). Strains were stocked in brain heart infusion (BHI) broth (Oxoid® Ltd., Basingstoke, Hampshire, England) containing 50% glycerol (*v/v*) and stored at −20 °C. All strains were cultured directly from storage into BHI broth and incubated under aerobic conditions at 37 °C.

2.3. Mammalian Cell Culture

HT-29 and Caco-2 cells were routinely grown in McCoy's 5A modified medium and Dulbecco's modified eagle medium (DMEM), respectively, both supplemented with 10% fetal bovine serum (FBS). All cells were maintained in 75 cm^2 tissue-culture flasks and incubated at 37 °C in 5% (v/v) CO_2 in a humidified atmosphere. Cells were passaged when the confluency of the flask was approximately 90% as previously described [34]. Cells were trypsinized and seeded onto 12-well PVDF (Polyvinylidene fluoride) membrane plates (Corning) at a density of 1×10^5 cells/well as previously described [34]. For translocation studies, Caco-2 cells were seeded onto transwell inserts (0.4 μm pore size, 12 mm diameter) (Corning), where they formed a tight monolayer. The media was changed every other day and the cells were grown until they reached confluence (1000 Ω) after 17–20 days.

2.4. Adhesion Assay

A series of adhesion assays, adapted from [35,36], were performed with HT-29 and Caco-2 cells and *E. coli* in the absence (control) and presence of GMP resuspended in either McCoy's 5A modified medium or DMEM, respectively. Cells were cultured as above for 48 h and the media was supplemented with 2% FBS 24 h prior to inhibition studies. *E. coli* were harvested from the BHI, washed three times in phosphate-buffered saline, pH 7.2 (PBS), and diluted to an OD_{600nm} of 0.2 (approximately 1×10^8 colony forming units (CFU)/mL) in either McCoy's 5A modified medium or DMEM. Prior to infecting the cell line, *E. coli* was pre-incubated with filter-sterilised GMP (5 mg/mL), for 1 h at 37 °C. The adhesion assays were conducted on three separate occasions in triplicate.

2.4.1. Standard Inhibition Assay

Confluent monolayers of HT-29 cells were washed twice in PBS and infected with *E. coli* (1×10^8 CFU/mL) which had been pre-incubated with 5 mg/mL of GMP or the respective buffer (control) for 1 h at 37 °C (5% CO_2). The cells were then incubated for 1 h at 37 °C and 5% (v/v) CO_2. To determine the number of cell-associated bacteria, each well was washed five times with PBS and the cells were lysed with 500 μL of 0.1% Trition X-100 in PBS for 30 min. Serial dilutions of the cell lysates were then plated onto BHI agar and incubated for 12 h at 37 °C, after which CFU/mL was calculated.

2.4.2. Concentration Dependency Assay

Prior to infecting the HT-29 cell line, *E. coli* NCTC12900 (1×10^8 CFU/mL) was pre-incubated for 1 h at 37 °C (5% CO_2) with various concentrations of GMP (1, 2, 3, 4, 5 and 6 mg/mL) in McCoy's 5A media (2% FBS). After this step, the inhibition assay was performed as described above.

2.4.3. Bacterial Interaction with GMP

Prior to infecting the HT-29 or Caco-2 cell lines, *E. coli* NCTC12900 (1×10^8 CFU/mL) was pre-incubated for 1 h at 37 °C (5% CO_2) with GMP (5 mg/mL) in McCoy's 5A media with 2% FBS. The samples were centrifuged at 4700 rpm for 7 min to pellet the bacterial cells. Media containing unbound GMP was removed and the bacterial pellet was then re-suspended in McCoy's 5A media with 2% FBS. After this step, the inhibition assay was performed as described above.

2.4.4. Cell-Line Interaction with GMP

The confluent monolayer was washed twice in PBS and 500 μL of the GMP (5 mg/mL) in McCoy's 5A media with 2% FBS was added to the wells. Unbound GMP was removed by washing the mammalian cells five times in PBS prior to infection with non-pre-incubated bacteria and incubated for 1 h at 37 °C (5% CO_2). The inhibition assay was then continued as described above.

2.4.5. Instantaneous Effect of GMP

The confluent monolayer was washed twice in PBS and was simultaneously exposed to 500 μL of GMP (5 mg/mL) and non-pre-incubated bacteria and incubated for 1 h at 37 °C (5% CO_2). The inhibition assay was then continued as described above.

2.5. Translocation and Transepithelial Electrical Resistance (TEER) Analysis

To investigate the effect of GMP on *E. coli* translocation, Caco-2 cells were grown on transwell inserts for 21 days in DMEM containing 10% (*v*/*v*) FBS, 1% (*v*/*v*) non-essential amino acids and 0.5% penicillin–streptomycin (5000 U/mL). Prior to infection, *E. coli* NCTC12900 (1×10^8 CFU/mL) was pre-incubated with GMP (5 mg/mL) for 1 h at 37 °C before being applied to the apical side of the transwell plate. Serum-free media (1 mL) were added to the basolateral chamber and the cells were incubated for 14 h at 37 °C (5% CO_2). Transwell inserts containing cells under the same conditions were also established for the introduction of non-treated bacteria in serum-free media (negative control). The number of translocating bacteria was determined by plating the basolateral medium onto BHI agar at 1, 2, 3, 4, 6, 8, 12 h post-apical infection as above. To confirm the formation of a tight cell monolayer, TEER measurements were carried out at 7, 14 and 21 days using an EVOM X meter coupled to an STX2 manual electrode (World Precision Instruments, Sarasota, FL, USA). To control for any disruption in the integrity of the cell monolayer during bacterial infection, TEER reads were taken at 1, 2, 3, 4, 6, 8, 12 h post-apical infection.

2.6. Statistical Analysis

Results from the inhibition and translocation studies are presented as mean ± standard deviations of replicate experiments. Statistical significance was determined using the unpaired student *t*-test and $p < 0.05$ was considered significant.

3. Results and Discussion

3.1. Anti-Adhesive Activity of GMP

The ability of GMP to prevent pathogen adhesion was assessed using both HT-29 and Caco-2 cell lines (Table 1). Concentration dependency assays were performed to determine if inhibition was dependent on GMP concentration (Figure 1). EHEC 12900 adhesion was selected for this purpose and a concentration of 5 mg/mL was chosen for all studies thereafter. The range of concentrations used in this assay were selected given that 1.2–1.5 mg/mL of GMP is found in whey from cheese manufacturing [37]. Prior to mammalian cell infection, *E. coli* strains were pre-incubated with GMP. Subsequently, GMP was shown not to kill the bacteria, did not influence bacterial growth over the course of the assay and did not affect the viability of the HT-29 cells as confirmed by real-time analysis of cell viability using an xCELLigence system (Roche).

Table 1. The percentage inhibition of *E. coli* adherence to HT-29 and Caco-2 cells relative to the control.

E. coli strain	% Inhibition	
	HT-29	Caco-2
EHEC 12900 O157:H7	70 ***	62 ***
EPEC O111:H2	26 *	25 *
EPEC O125:H32	24 *	25 *
EHEC DAF 454	21	N/T
EHEC DPC 6055	15	N/T

EHEC = enterohemorrhagic *E. coli*; EPEC = enteropathogenic *E. coli*, N/T = not tested (*** $p < 0.01$ and * $p < 0.05$).

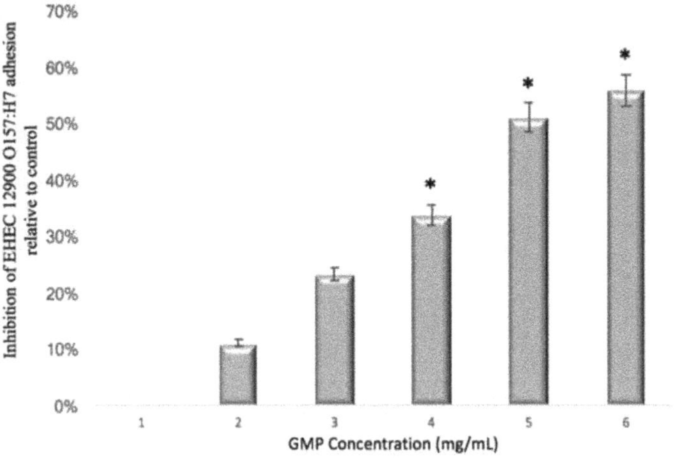

Figure 1. The effect of GMP concentration on inhibition of adhesion of EHEC 12900 O157:H7 to HT-29 cells (* $p < 0.05$). EHEC = enterohemorrhagic *E. coli*; GMP = Glycomacropeptide.

The greatest anti-adhesive activity in response to GMP was observed against EHEC 12900 ($p < 0.001$) with 70% and 62% inhibition of adhesion to HT-29 and Caco-2 cells, respectively, relative to the control (Table 1). Similarly, EPEC 0111:H2 ($p < 0.05$) and EPEC 0125:H32 ($p < 0.05$) adhesion to HT-29 and Caco-2 cells was also reduced in the presence of GMP by 26% and 24%, and 25% and 25%, respectively, relative to the control (Table 1). Inhibition of adhesion to HT-29 cells was observed for the other EHEC strains tested, but was not statistically significant. These results suggest that the effect of GMP on *E. coli* adhesion was not associated with a particular pathotype but instead appears to be strain specific. Interestingly, similar inhibition of the same strains by GMP was observed for both the Caco-2 and HT-29 cell lines, which indicates that the effect is not cell-line associated but instead may be strain specific (Table 1).

Previous studies [2,20,21,30] indicated that GMP inhibited adhesion of *E. coli* to mammalian cells by interacting with the bacteria, thereby preventing human-cell association. However, GMP may also interact with the cell line to prevent bacterial interaction. Therefore, three different possibilities were assessed:

(1) That GMP interacts with bacterial binding sites, thereby preventing the bacteria from binding to HT29 cell receptors. This was investigated by removing unbound GMP from bacterial GMP mixture using centrifugation prior to infection. Anti-adhesive activity was still observed against EHEC 12900 adhesion to HT29 cells ($p < 0.05$). However, no significant anti-adhesive activity was evident against EPEC O111 and O125 (Figure 2A).

(2) That GMP binds to epithelial cell receptors, thereby preventing bacteria from interacting with the host cell surface. This was investigated by pre-incubating GMP with the cell line and washing off unbound GMP prior to bacterial challenge. No inhibition was observed against either EHEC or EPEC strains. These results suggest that GMP interacts with the bacteria but not the mammalian cells. It is interesting to note that removing unbound GMP from the bacteria reduced the anti-adhesive activity compared to co-incubating GMP with the bacterial and mammalian cells (Figure 2B).

(3) That pre-incubation of the bacteria with GMP is not required and that inhibition of adhesion can occur with simultaneous exposure of GMP and bacteria to the cell lines as would be more realistic of the in-vivo situation. No anti-infective activity was observed against the EPEC strains; however, an instantaneous anti-infective activity against EHEC 12900 ($p < 0.05$) was evident (Figure 2C) at

levels comparable to using a pre-incubation step (Figure 2A). These results suggest that GMP does not require pre-incubation with the bacteria to exert its maximal inhibitory effect on *E. coli* cellular association, and a reduction in binding was evident instantaneously. No significant inhibition of adhesion of any of the strains was observed when using the Caco-2 cells in the three experiments outlined above.

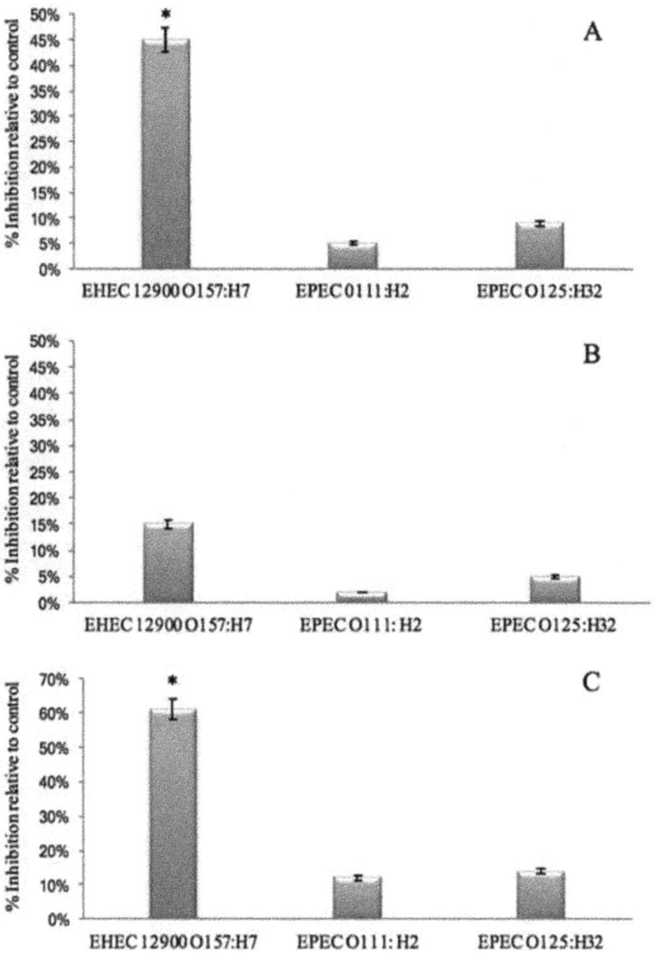

Figure 2. The effect of GMP on the association of *E. coli* strains with HT-29 cells. (**A**) Effect of pre-incubation of bacteria with the GMP for 1 h. (**B**) Effect of pre-incubation of HT29 cells with GMP prior to bacterial infection. (**C**) Effect of no pre-incubation step on anti-adherence effect of GMP (* $p < 0.05$). EHEC = enterohemorrhagic *E. coli*; EPEC = enteropathogenic *E. coli*.

It is known that *E. coli* adhere to epithelial cells of the intestine using adhesins such as pili and fimbriae, as well as capsular material found on the bacterial cells' surfaces, and that these receptors adhere to specific glycosylated ligands found on intestinal epithelial cell surfaces [38,39]. Sialic acids and mannose have been found to be important specific host-cell receptors for the initial phase of infection (adhesion of lectins on the surface of bacteria to specific receptors on host cell [3,39–42]). Therefore, the current study and previous studies [30,43] further confirm that the bovine milk glycopeptide GMP, which contains multiple sialic acid residues, may inhibit the colonisation of intestinal cell lines by mimicking receptor sites on eukaryotic cell membranes, thereby blocking bacterial infection. Rhoades et al. [2] demonstrated that GMP did not significantly inhibit the adhesion of EPEC O111:H27 to ileal mucosa tissues obtained from piglets. However, in the present study, EPEC O111:H2 adhesion was significantly decreased. Therefore, the difference in anti-adhesive activity of GMP amongst *E. coli* strains could in part be flagella mediated. The results from the current study and previous studies suggest that GMP interacts with the bacterial surface and not the mammalian cell surface. Further studies are required in order to determine if GMP is blocking the bacteria and stopping cellular interactions solely as a mode of action, or if bacterial cell surface changes are occurring also in its presence (e.g., upregulation or downregulation of adhesin expression).

3.2. Effect of GMP on Caco-2 Tight Junction Integrity

In order to monitor the ability of GMP to prevent barrier dysfunction, the TEER of Caco-2 cells infected with EHEC or EPEC was measured in the absence and presence of GMP. The ability of the *E. coli* strains to translocate through the cell monolayer was also measured at various time points. At 3 h post-infection with EPEC and EHEC strains, the TEER measurement of Caco-2 cells was reduced, which indicated an alteration of membrane integrity (Figure 3). Simultaneously, each of the bacteria tested—EPEC 0111:H2 and 0125:H32—penetrated the Caco-2 cell monolayer and were found in the basolateral medium 4 hours after inoculation, and bacterial numbers continued to increase over time (Figure 4). EHEC 12900 failed to penetrate the Caco-2 polarized monolayer and translocate into the basolateral media in both treated and control transwells. After 14 h, GMP's inhibitory activity ceased, and at this point both the test and control membranes had a TEER of less than 145 Ω, which is the approximate TEER of an un-seeded well (no cell line). A decrease in TEER suggested that each strain of *E. coli* disrupted the tight junctions (TJs) of Caco-2 monolayers. GMP significantly delayed both the EHEC- and EPEC-induced decrease in TEER for 3 h ($p < 0.01$ and $p < 0.05$, respectively) (Figure 3). GMP also significantly inhibited translocation of both EPEC strains through Caco-2 monolayers into the basolateral medium for a period of 2 h ($p < 0.01$) (Figure 4).

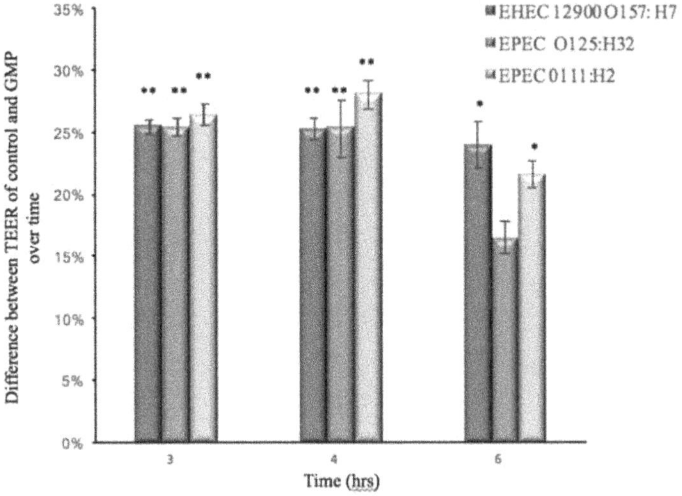

Figure 3. Percentage of prevention of reduction in Transepithelial Electrical Resistance (TEER) values of Caco-2 monolayers after infection with *E. coli* strains pretreated with GMP in comparison to untreated control (** $p < 0.01$ and * $p < 0.05$).

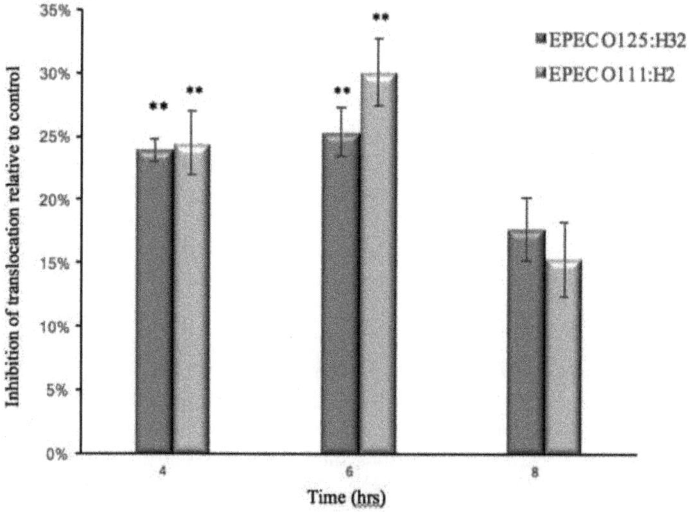

Figure 4. Percentage reduction of *E. coli* translocation across Caco-2 monolayers when pretreated with GMP in comparison to untreated control (** $p < 0.01$).

In order for EHEC and EPEC to cause infection, the bacteria must first adhere to epithelial cells where their TTSSs, which inject TIR, are activated. After an intercellular phosphorylation process, these TIRs penetrate the host membrane from the inside-out and bind to intimin on the bacterial capsule, making a secure bond to the host cell [39]. From this position, enteropathogenic bacteria, such as EPEC and EHEC, cause attaching and effacing (A/E) lesions in the small or large intestine, which in turn causes diarrhoea [44]. Overall, the mechanism in which EHEC generate A/E lesions is similar to EPEC, but a few important differences have been identified. Unlike EPEC, EHEC do not have the plasmid that facilitates initial local adherence to epithelial cells by bundle-forming pili (BFP) that facilitate bacterial autoaggregation [45]. Using the TTSS, EHEC and EPEC insert the locus of enterocyte effacement (LEE) effector molecules into host cells. EPEC effector molecules such as EspG and EspG 2 are known to modulate paracellular permeability and tight junction disruption during infection [46,47]. The absence of such molecules from EHEC's arsenal may explain why only EPEC translocated through the Caco-2 monolayers (Figure 4). Both EPEC and EHEC share the ability to initiate host-cell inflammation by inducing the expression of cytokines IL-8, TNF-α and IL-1β, which are pro-inflammatory cytokines that elicit the loosening and eventual destruction of TJs in human intestinal epithelial cells [10,48–51].

In agreement with the present study, it was previously shown that EHEC do not translocate through polarized Caco-2 monolayers. However, their LEE do, resulting in TNF-α and IL-8 secretion, which could cause intestinal blood-vessel destruction in humans. A number of studies [11,30,52,53] have found that GMP suppressed IL-8 production in Caco-2 cells that were challenged with other pathogens, which regulate IL-8 in a similar manner, to *E. coli*. EHEC and EPEC have been shown to activate the nuclear factor-kappa B (NF-κB) pathway, which triggers IL-8 secretion in intestinal epithelial cell lines [54]. Furthermore, IL-8 production in intestinal cells increases in the presence of TNF-α, which extenuates the degradation of NF-κB inhibitor Iκ-Bα [54–56]. NF-κB is considered the most important nuclear transcription factor for IL-8 and TNF-α-induced tight junction degradation in human cells [57–60].

TNF-α down-regulates production of the TJ protein ZO-1, and alters the junctional distribution of ZO-1 proteins via the NF-κB pathway, which increases permeability of Caco-2 tight junctions [59]. Gong et al. [61] demonstrated that GMP inhibits the NF-κB signalling pathway in HT-29 cells when challenged with EPEC lipopolysaccharide (LPS). Therefore, GMP may promote TJ integrity by suppressing NF-κB pathway activation, which would reduce IL-8 and TNF-α-associated permeability. This could explain the delay in the *E. coli*-induced TEER decrease and EPEC translocation in GMP-treated Caco-2 monolayers, as observed in the present study.

Interestingly, Rong et al. [62] demonstrated the ability of GMP in significantly alleviating the increase of pathogenic bacteria counts in intestinal contents, intestinal morphology and acute inflammatory responses induced by *E. coli* K88 infection. Importantly, similar to the current study, the researchers found that GMP prevents intestinal barrier permeability damage induced by *E. coli* K88 infection. It was concluded that GMP supplementation in the diet protects the weaning piglets against *E. coli* infection.

The myosin light-chain kinase (MLCK) pathway is another important TJ-associated intercellular pathway triggered by EHEC and EPEC exposure [63,64]. The interaction of IL-1β with the NF-kB pathway stimulates MLCK mRNA transcription, which upregulates MLCK protein expression leading to an increase of TJ permeability in Caco-2 cell monolayers [65]. It has been demonstrated that GMP treatment suppressed IL-1β mRNA levels in rats with colitis [66]. Similarly, Monnai and Otani [67] found that GMP stimulates the release of the IL-1β antagonist known as IL-1ra in monocytes. In the current study, GMP may prevent the activation of other pathways such as MLCK, which are associated with rearrangement of TJ-associated proteins ZO-1 and occludin after infection, leading to a decrease in TEER for both *E. coli* types and subsequent paracellular movement of EPEC (Figures 3 and 4). An alternative mechanism for EHEC-induced TEER reduction and inflammation, not involving LEE effector proteins, has been proposed. Ma et al [59] suggested that initial attachment of the H7 flagella of EHEC to Caco-2 cells triggers mitogen-activated protein kinase signalling pathways and the transcription factor NF-κB. This leads to the chemotaxis of IL-8 from the basolateral surface of colonic epithelial cells. It is possible that during the incubation period in the present study, GMP could be obstructing the initial attachment of EHEC's H7 flagella to Caco-2 cell monolayers via presentation of decoy ligands, thereby lowering the number of *E. coli* available to infect, which delays the decrease in TEER.

4. Conclusions

GMP prevents the adhesion of various strains of EHEC and EPEC to Caco-2 and HT-29 cells, and this adhesive ability was not linked to either pathotype or cell line but appeared to be strain-specific. To the best of our knowledge, this is the first study that details GMP's ability to maintain the structural integrity of Caco-2 TJs. Furthermore, GMP delays the paracellular movement of EPEC through the TJs of Caco-2 monolayers. Future work will focus on further establishing the strain-specificity of GMP's anti-adhesive effects, understanding this phenomenon and investigating the mechanism of action of GMP-induced reduction of cell permeability. This study indicates that GMP has potential as a bioactive ingredient which could assist in improving the gastrointestinal health of individuals. Indeed, recently GMP has shown promise as a nutritional therapy in patients with active distal ulcerative colitis [68]. Moreover, methods for producing GMP based on its glycosylation and the effect of its glycosylation on the peptide's techno-functional properties have been documented [69–74]. The fact that some dairy-ingredient companies have already realized the potential of GMP and are promoting this glycopeptide as a premium ingredient isolated from whey makes these findings particularly relevant.

Acknowledgments: Shane Feeney is in receipt of a Teagasc Walsh Fellowship. This work was funded by the Department of Agriculture and Food, Ireland, under the Food Institutional Research Measure, project reference number 10/RD/NUIG/707.

Author Contributions: Rita Hickey and Lokesh Joshi conceived the study design; Shane Feeney performed the experiments and analyzed the data; Joseph Thomas Ryan and Michelle Kilcoyne contributed to data analysis. All authors contributed to writing the paper.

Conflicts of Interest: The authors declare no conflict of interest.

References

1. Klemm, P.; Vejborg, R.M.; Hancock, V. Prevention of bacterial adhesion. *Appl. Microbiol. Biotechnol.* **2010**, *88*, 451–459. [CrossRef] [PubMed]
2. Rhoades, J.R.; Gibson, G.R.; Formentin, K.; Beer, M.; Greenberg, N.; Rastall, R.A. Caseinoglycomacropeptide inhibits adhesion of pathogenic *Escherichia coli* strains to human cells in culture. *J. Dairy Sci.* **2005**, *88*, 3455–3459. [CrossRef]
3. Ofek, I.; Hasty, D.L.; Sharon, N. Anti-adhesion therapy of bacterial diseases: Prospects and problems. *FEMS Immunol. Med. Microbiol.* **2003**, *38*, 181–191. [CrossRef]
4. Wheeler, A.P.; Bernard, G.R. Treating patients with severe sepsis. *N. Engl. J. Med.* **1999**, *340*, 207–214. [CrossRef] [PubMed]
5. Ammori, B.J.; Fitzgerald, P.; Hawkey, P.; McMahon, M.J. The early increase in intestinal permeability and systemic endotoxin exposure in patients with severe acute pancreatitis is not associated with systemic bacterial translocation: Molecular investigation of microbial DNA in the blood. *Pancreas* **2003**, *26*, 18–22. [CrossRef] [PubMed]
6. Deitch, E.A.; Xu, D.Z.; Qi, L.; Berg, R.D. Bacterial translocation from the gut impairs systemic immunity. *Surgery* **1991**, *109*, 269–276. [PubMed]
7. Ciancio, M.J.; Chang, E.B. Epithelial secretory response to inflammation. *Ann. N. Y. Acad. Sci.* **1992**, *664*, 210–221. [CrossRef] [PubMed]
8. Ciancio, M.J.; Vitiritti, L.; Dhar, A.; Chang, E.B. Endotoxin-induced alterations in rat colonic water and electrolyte transport. *Gastroenterol* **1992**, *103*, 1437–1443. [CrossRef]
9. Deitch, E.A.; Specian, R.D.; Berg, R.D. Endotoxin-induced bacterial translocation and mucosal permeability: Role of xanthine oxidase, complement activation, and macrophage products. *Crit. Care Med.* **1991**, *19*, 785–791. [CrossRef] [PubMed]
10. Bernet-Camard, M.F.; Coconnier, M.H.; Hudault, S.; Servin, A.L. Differentiation-associated antimicrobial functions in human colon adenocarcinoma cell lines. *Exp. Cell Res.* **1996**, *226*, 80–89. [CrossRef] [PubMed]
11. Izumikawa, K.; Hirakata, Y.; Yamaguchi, T.; Takemura, H.; Maesaki, S.; Tomono, K.; Igimi, S.; Kaku, M.; Yamada, Y.; Kohno, S.; et al. *Escherichia coli* O157 interactions with human intestinal Caco-2 cells and the influence of fosfomycin. *J. Antimicrob. Chemother.* **1998**, *42*, 341–347. [CrossRef] [PubMed]
12. Hirotani, Y.; Ikeda, K.; Kato, R.; Myotoku, M.; Umeda, T.; Ijiri, Y.; Tanaka, K. Protective effects of lactoferrin against intestinal mucosal damage induced by lipopolysaccharide in human intestinal Caco-2 cells. *Yakugaku Zasshi* **2008**, *128*, 1363–1368. [CrossRef] [PubMed]
13. Howe, K.L.; Reardon, C.; Wang, A.; Nazli, A.; McKay, D.M. Transforming growth factor-beta regulation of epithelial tight junction proteins enhances barrier function and blocks enterohemorrhagic *Escherichia coli* O157:H7-induced increased permeability. *Am. J. Pathol.* **2005**, *167*, 1587–1597. [CrossRef]
14. Requena, P.; Gonzalez, R.; Lopez-Posadas, R.; Abadia-Molina, A.; Suarez, M.D.; Zarzuelo, A.; de Medina, F.S.; Martinez-Augustin, O. The intestinal antiinflammatory agent Glycomacropeptide has immunomodulatory actions on rat splenocytes. *Biochem. Pharmacol.* **2010**, *79*, 1797–1804. [CrossRef] [PubMed]
15. Posadas, S.J.; Caz, V.; Caballero, I.; Cendejas, E.; Quilez, I.; Largo, C.; Elvira, M.; De Miguel, E. Effects of mannoprotein E1 in liquid diet on inflammatory response and TLR5 expression in the gut of rats infected by *Salmonella typhimurium*. *BMC Gastroenterol.* **2010**, *10*, 58. [CrossRef] [PubMed]
16. Coppa, G.V.; Zampini, L.; Galeazzi, T.; Facinelli, B.; Ferrante, L.; Capretti, R.; Orazio, G. Human milk oligosaccharides inhibit the adhesion to Caco-2 cells of diarrheal pathogens: *Escherichia coli*, *Vibrio cholerae*, and *Salmonella fyris*. *Pediatr. Res.* **2006**, *59*, 377–382. [CrossRef] [PubMed]
17. Cravioto, A.; Tello, A.; Villafan, H.; Ruiz, J.; del Vedovo, S.; Neeser, J.R. Inhibition of localized adhesion of enteropathogenic *Escherichia coli* to Hep-2 cells by immunoglobulin and oligosaccharide fractions of human colostrum and breast milk. *J. Infect. Dis.* **1991**, *163*, 1247–1255. [CrossRef] [PubMed]
18. Simon, P.M.; Goode, P.L.; Mobasseri, A.; Zopf, D. Inhibition of *Helicobacter pylori* binding to gastrointestinal epithelial cells by sialic acid-containing oligosaccharides. *Infect. Immun.* **1997**, *65*, 750–757. [PubMed]
19. Delfour, A.; Jolles, J.; Alais, C.; Jolles, P. Casein-glycopeptides: Characterization of a methionine residue and of the N-terminal sequence. *Biochem. Biophys. Res. Commun.* **1965**, *19*, 425. [CrossRef]

20. Bruck, W.M.; Kelleher, S.L.; Gibson, G.R.; Graverholt, G.; Lonnerdal, B.L. The effects of alpha-lactalbumin and Glycomacropeptide on the association of Caco-2 cells by enteropathogenic *Escherichia coli*, *Salmonella typhimurium* and *Shigella flexneri*. *FEMS Microbiol. Lett.* **2006**, *259*, 158–162. [CrossRef] [PubMed]
21. Bruck, W.M.; Redgrave, M.; Tuohy, K.M.; Lonnerdal, B.; Graverholt, G.; Hernell, O.; Gibson, G.R. Effects of bovine alpha-lactalbumin and casein Glycomacropeptide-enriched infant formulae on faecal microbiota in healthy term infants. *J. Pediatr. Gastroenterol. Nutr.* **2006**, *43*, 673–679. [CrossRef] [PubMed]
22. Ntemiri, A.; Ní Chonchúir, F.; O'Callaghan, T.F.; Stanton, C.; Ross, R.P.; O'Toole, P.W. Glycomacropeptide sustains microbiota diversity and promotes specific taxa in an artificial colon model of elderly gut microbiota. *J. Agric. Food Chem.* **2017**, *65*, 1836–1846. [CrossRef] [PubMed]
23. Sawin, E.A.; De Wolfe, T.J.; Aktas, B.; Stroup, B.M.; Murali, S.G.; Steele, J.L.; Ney, D.M. Glycomacropeptide is a prebiotic that reduces *Desulfovibrio* bacteria, increases cecal short-chain fatty acids, and is anti-inflammatory in mice. *Am. J. Physiol. Gastrointest Liver Physiol.* **2015**, *309*, G590–G601. [CrossRef] [PubMed]
24. Brody, E.P. Biological activities of bovine Glycomacropeptide. *Br. J. Nutr.* **2000**, *84* (Suppl. 1), S39–S46. [CrossRef] [PubMed]
25. Saito, T.; Itoh, T. Variations and distributions of O-glycosidically linked sugar chains in bovine kappa-casein. *J. Dairy Sci.* **1992**, *75*, 1768–1774. [CrossRef]
26. O'Riordan, N.; Kane, M.; Joshi, L.; Hickey, R.M. Structural and functional characteristics of bovine milk protein glycosylation. *Glycobiology* **2014**, *24*, 220–236. [CrossRef] [PubMed]
27. Severi, E.; Hood, D.W.; Thomas, G.H. Sialic acid utilization by bacterial pathogens. *Microbiology* **2007**, *153*, 2817–2822. [CrossRef] [PubMed]
28. Varki, N.M.; Varki, A. Diversity in cell surface sialic acid presentations: Implications for biology and disease. *Lab. Investig.* **2007**, *87*, 851–857. [CrossRef] [PubMed]
29. Wilbrink, M.H.; ten Kate, G.A.; van Leeuwen, S.S.; Sanders, P.; Sallomons, E.; Hage, J.A.; Dijkhuizen, L.; Kamerling, J.P. Galactosyl-lactose sialylation using *Trypanosoma cruzi* trans-sialidase as the biocatalyst and bovine kappa-casein-derived Glycomacropeptide as the donor substrate. *Appl. Environ. Microbiol.* **2014**, *80*, 5984–5991. [CrossRef] [PubMed]
30. Nakajima, K.; Tamura, N.; Kobayashi-Hattori, K.; Yoshida, T.; Hara-Kudo, Y.; Ikedo, M.; Sugita-Konishi, Y.; Hattori, M. Prevention of intestinal infection by Glycomacropeptide. *Biosci. Biotechnol. Biochem.* **2005**, *69*, 2294–2301. [CrossRef] [PubMed]
31. Stromqvist, M.; Falk, P.; Bergstrom, S.; Hansson, L.; Lonnerdal, B.; Normark, S.; Hernell, O. Human milk kappa-casein and inhibition of *Helicobacter pylori* adhesion to human gastric mucosa. *J. Pediatr. Gastroenterol. Nutr.* **1995**, *21*, 288–296. [CrossRef] [PubMed]
32. Gonzalez-Ortiz, G.; Hermes, R.G.; Jimenez-Diaz, R.; Perez, J.F.; Martin-Orue, S.M. Screening of extracts from natural feed ingredients for their ability to reduce enterotoxigenic *Escherichia coli* (ETEC) K88 adhesion to porcine intestinal epithelial cell-line IPEC-j2. *Vet. Microbiol.* **2013**, *167*, 494–499. [CrossRef] [PubMed]
33. Gonzalez-Ortiz, G.; Perez, J.F.; Hermes, R.G.; Molist, F.; Jimenez-Diaz, R.; Martin-Orue, S.M. Screening the ability of natural feed ingredients to interfere with the adherence of enterotoxigenic *Escherichia coli* (ETEC) K88 to the porcine intestinal mucus. *Br. J. Nutr.* **2014**, *111*, 633–642. [CrossRef] [PubMed]
34. Lane, J.A.; Marino, K.; Naughton, J.; Kavanaugh, D.; Clyne, M.; Carrington, S.D.; Hickey, R.M. Anti-infective bovine colostrum oligosaccharides: *Campylobacter jejuni* as a case study. *Int. J. Food Microbiol.* **2012**, *157*, 182–188. [CrossRef] [PubMed]
35. Horemans, T.; Kerstens, M.; Clais, S.; Struijs, K.; van den Abbeele, P.; Van Assche, T.; Maes, L.; Cos, P. Evaluation of the anti-adhesive effect of milk fat globule membrane glycoproteins on *Helicobacter pylori* in the human NCI-N87 cell line and C57BL/6 mouse model. *Helicobacter* **2012**, *17*, 312–318. [CrossRef] [PubMed]
36. Salcedo, J.; Barbera, R.; Matencio, E.; Alegria, A.; Lagarda, M.J. Gangliosides and sialic acid effects upon newborn pathogenic bacteria adhesion: An in vitro study. *Food Chem.* **2013**, *136*, 726–734. [CrossRef] [PubMed]
37. Thoma-Worringer, C.S.J.; Lopez-Findino, R. Health effects and technological features of caseinomacropeptide. *Int. Dairy J.* **2006**, *16*, 1324–1333. [CrossRef]
38. Fairbrother, J.M.; Nadeau, E.; Gyles, C.L. *Escherichia coli* in postweaning diarrhea in pigs: An update on bacterial types, pathogenesis, and prevention strategies. *Anim. Health Res. Rev.* **2005**, *6*, 17–39. [CrossRef] [PubMed]

39. Croxen, M.A.; Law, R.J.; Scholz, R.; Keeney, K.M.; Wlodarska, M.; Finlay, B.B. Recent advances in understanding enteric pathogenic *Escherichia coli*. *Clin. Microbiol. Rev.* **2013**, *26*, 822–880. [CrossRef] [PubMed]
40. Sharon, N.; Ofek, I. Safe as mother's milk: Carbohydrates as future anti-adhesion drugs for bacterial diseases. *Glycoconj. J.* **2000**, *17*, 659–664. [CrossRef] [PubMed]
41. Kaper, J.B.; Nataro, J.P.; Mobley, H.L. Pathogenic *Escherichia coli*. *Nat. Rev. Microbiol.* **2004**, *2*, 123–140. [CrossRef] [PubMed]
42. Spears, K.J.; Roe, A.J.; Gally, D.L. A comparison of enteropathogenic and enterohaemorrhagic *Escherichia coli* pathogenesis. *FEMS Microbiol. Lett.* **2006**, *255*, 187–202. [CrossRef] [PubMed]
43. Neelima; Sharma, R.; Rajput, Y.S.; Mann, B. Chemical and functional properties of Glycomacropeptide (GMP) and its role in the detection of cheese whey adulteration in milk: A review. *Dairy Sci. Technol.* **2013**, *93*, 21–43. [CrossRef] [PubMed]
44. Matthews, L.; Reeve, R.; Gally, D.L.; Low, J.C.; Woolhouse, M.E.; McAteer, S.P.; Locking, M.E.; Chase-Topping, M.E.; Haydon, D.T.; Allison, L.J.; et al. Predicting the public health benefit of vaccinating cattle against *Escherichia coli* O157. *Proc. Natl. Acad. Sci. USA* **2013**, *110*, 16265–16270. [CrossRef] [PubMed]
45. Zahavi, E.E.; Lieberman, J.A.; Donnenberg, M.S.; Nitzan, M.; Baruch, K.; Rosenshine, I.; Turner, J.R.; Melamed-Book, N.; Feinstein, N.; Zlotkin-Rivkin, E.; et al. Bundle-forming pilus retraction enhances enteropathogenic *Escherichia coli* infectivity. *Mol. Biol. Cell* **2011**, *22*, 2436–2447. [CrossRef] [PubMed]
46. Miyake, M.; Hanajima, M.; Matsuzawa, T.; Kobayashi, C.; Minami, M.; Abe, A.; Horiguchi, Y. Binding of intimin with TIR on the bacterial surface is prerequisite for the barrier disruption induced by enteropathogenic *Escherichia coli*. *Biochem. Biophys. Res. Commun.* **2005**, *337*, 922–927. [CrossRef] [PubMed]
47. Matsuzawa, T.; Kuwae, A.; Abe, A. Enteropathogenic *Escherichia coli* type III effectors EspG and EspG2 alter epithelial paracellular permeability. *Infect. Immun.* **2005**, *73*, 6283–6289. [CrossRef] [PubMed]
48. Bernet-Camard, M.F.; Coconnier, M.H.; Hudault, S.; Servin, A.L. Differential expression of complement proteins and regulatory decay accelerating factor in relation to differentiation of cultured human colon adenocarcinoma cell lines. *Gut* **1996**, *38*, 248–253. [CrossRef] [PubMed]
49. Welinder-Olsson, C.; Eriksson, E.; Kaijser, B. Virulence genes in verocytotoxigenic *Escherichia coli* strains isolated from humans and cattle. *APMIS* **2005**, *113*, 577–585. [CrossRef] [PubMed]
50. Welinder-Olsson, C.; Kaijser, B. Enterohemorrhagic *Escherichia coli* (EHEC). *Scand. J. Infect. Dis.* **2005**, *37*, 405–416. [CrossRef] [PubMed]
51. Glotfelty, L.G.; Hecht, G.A. Enteropathogenic *E. coli* effectors ESPG1/G2 disrupt tight junctions: New roles and mechanisms. *Ann. N. Y. Acad. Sci.* **2012**, *1258*, 149–158. [CrossRef] [PubMed]
52. Izumikawa, K.; Hirakata, Y.; Yamaguchi, T.; Yoshida, R.; Nakano, M.; Matsuda, J.; Mochida, C.; Maesaki, S.; Tomono, K.; Yamada, Y.; et al. Analysis of genetic relationships and antimicrobial susceptibility of verotoxin-producing *Escherichia coli* strains isolated in Nagasaki prefecture, Japan in 1996. *Microbiol. Immunol.* **1998**, *42*, 677–681. [CrossRef] [PubMed]
53. Luck, S.N.; Turner, S.A.; Rajakumar, K.; Adler, B.; Sakellaris, H. Excision of the *Shigella* resistance locus pathogenicity island in *Shigella flexneri* is stimulated by a member of a new subgroup of recombination directionality factors. *J. Bacteriol.* **2004**, *186*, 5551–5554. [CrossRef] [PubMed]
54. Savkovic, S.D.; Koutsouris, A.; Hecht, G. Activation of NF-kappaB in intestinal epithelial cells by enteropathogenic *Escherichia coli*. *Am. J. Physiol.* **1997**, *273*, C1160–C1167. [PubMed]
55. Savkovic, S.D.; Ramaswamy, A.; Koutsouris, A.; Hecht, G. EPEC-activated ERK1/2 participate in inflammatory response but not tight junction barrier disruption. *Am. J. Physiol. Gastrointest Liver Physiol.* **2001**, *281*, G890–G898. [PubMed]
56. Lang, A.; Lahav, M.; Sakhnini, E.; Barshack, I.; Fidder, H.H.; Avidan, B.; Bardan, E.; Hershkoviz, R.; Bar-Meir, S.; Chowers, Y. Allicin inhibits spontaneous and TNF-alpha induced secretion of proinflammatory cytokines and chemokines from intestinal epithelial cells. *Clin. Nutr.* **2004**, *23*, 1199–1208. [CrossRef] [PubMed]
57. Mukaida, N.; Okamoto, S.; Ishikawa, Y.; Matsushima, K. Molecular mechanism of Interleukin-8 gene expression. *J. Leukoc. Biol.* **1994**, *56*, 554–558. [PubMed]
58. Al-Sadi, R.; Ye, D.; Dokladny, K.; Ma, T.Y. Mechanism of IL-1beta-induced increase in intestinal epithelial tight junction permeability. *J. Immunol.* **2008**, *180*, 5653–5661. [CrossRef] [PubMed]

59. Ma, T.Y.; Iwamoto, G.K.; Hoa, N.T.; Akotia, V.; Pedram, A.; Boivin, M.A.; Said, H.M. TNF-alpha-induced increase in intestinal epithelial tight junction permeability requires NF-kappaB activation. *Am. J. Physiol. Gastrointest Liver Physiol.* **2004**, *286*, G367–G376. [CrossRef] [PubMed]
60. Ye, D.; Ma, I.; Ma, T.Y. Molecular mechanism of tumor necrosis factor-alpha modulation of intestinal epithelial tight junction barrier. *Am. J. Physiol. Gastrointest Liver Physiol.* **2006**, *290*, G496–G504. [CrossRef] [PubMed]
61. Gong, J.; Chen, Q.; Yan, Y.; Pang, G. Effect of casein Glycomacropeptide on subunit p65 of nuclear transcription factor-κB in lipopolysaccharide-stimulated human colorectal tumor HT-29 cells. *Food Sci. Hum. Wellness* **2014**, *3*, 51–55. [CrossRef]
62. Rong, Y.; Lu, Z.; Zhang, H.; Zhang, L.; Song, D.; Wang, Y. Effects of casein Glycomacropeptide supplementation on growth performance, intestinal morphology, intestinal barrier permeability and inflammatory responses in *Escherichia coli* K88 challenged piglets. *Anim. Nutr.* **2015**, *1*, 54–59. [CrossRef]
63. Yuhan, R.; Koutsouris, A.; Savkovic, S.D.; Hecht, G. Enteropathogenic *Escherichia coli*-induced myosin light chain phosphorylation alters intestinal epithelial permeability. *Gastroenterology* **1997**, *113*, 1873–1882. [CrossRef]
64. Bellmeyer, A.; Cotton, C.; Kanteti, R.; Koutsouris, A.; Viswanathan, V.K.; Hecht, G. Enterohemorrhagic *Escherichia coli* suppresses inflammatory response to cytokines and its own toxin. *Am. J. Physiol. Gastrointest Liver Physiol.* **2009**, *297*, G576–G581. [CrossRef] [PubMed]
65. Al-Sadi, R.M.; Ma, T.Y. IL-1beta causes an increase in intestinal epithelial tight junction permeability. *J. Immunol.* **2007**, *178*, 4641–4649. [CrossRef] [PubMed]
66. Daddaoua, A.; Puerta, V.; Zarzuelo, A.; Suarez, M.D.; Sanchez de Medina, F.; Martinez-Augustin, O. Bovine Glycomacropeptide is anti-inflammatory in rats with hapten-induced colitis. *J. Nutr.* **2005**, *135*, 1164–1170. [PubMed]
67. Monnai, M.O.H. Effect of bovine k-caseinoglycopeptide on secretion of interleukin-1 family cytokines by P388D1 cells, a line derived from mouse monocyte/macrophage. *Milchwissenschaft* **1997**, *52*, 192–196.
68. Hvas, C.L.; Dige, A.; Bendix, M.; Wernlund, P.G.; Christensen, L.A.; Dahlerup, J.F.; Agnholt, J. Casein Glycomacropeptide for active distal ulcerative colitis: A randomized pilot study. *Eur. J. Clin. Investig.* **2016**, *46*, 555–563. [CrossRef] [PubMed]
69. Kreuß, M.; Krause, I.; Kulozik, U. Separation of a glycosylated and non-glycosylated fraction of caseinomacropeptide using different anion-exchange stationary phases. *J. Chromatogr. A* **2008**, *1208*, 126–132. [CrossRef] [PubMed]
70. Kreuß, M.; Krause, I.; Kulozik, U. Influence of glycosylation on foaming properties of bovine caseinomacropeptide. *Int. Dairy J.* **2009**, *19*, 715–720. [CrossRef]
71. Kreuß, M.; Strixner, T.; Kulozik, U. The effect of glycosylation in the interfacial properties of bovine caseinomacropeptide. *Food Hydrocoll.* **2009**, *23*, 1818–1826. [CrossRef]
72. Kreuß, M.; Kulozik, U. Separation of glycosylated caseinomacropeptide at pilot scale using membrane adsorption in direct-capture mode. *J. Chromatogr. A* **2009**, *1216*, 8771–8777. [CrossRef] [PubMed]
73. Koliwer-Brandl, H.; Siegert, N.; Umus, K.; Kelm, A.; Tolkach, A.; Kulozik, U.; Kuballa, J.; Cartellieri, S.; Kelm, S. Lectin inhibition assays for the analysis of bioactive milk sialoglycoconjugates. *Int. Dairy J.* **2011**, *21*, 413–420. [CrossRef]
74. Siegert, N.; Tolkach, A.; Kulozik, U. The pH-dependent thermal and storage stability of glycosylated caseinomacropeptide. *LWT-Food Sci. Technol.* **2012**, *47*, 407–412. [CrossRef]

foods

MDPI

Article

Immunodetection of Porcine Red Blood Cell Containing Food Ingredients Using a Porcine-Hemoglobin-Specific Monoclonal Antibody

Jack A. Ofori and Yun-Hwa P. Hsieh *

Department of Nutrition, Food and Exercise Sciences, 420 Sandels Building, Florida State University, Tallahassee, FL 32306-1493, USA; papajackx@yahoo.co.uk

* Correspondence: yhsieh@fsu.edu; Tel.: +1-850-644-1744; Fax: +1-850-645-5000

Received: 14 October 2017; Accepted: 13 November 2017; Published: 20 November 2017

Abstract: Monoclonal antibody (mAb) 24C12-E7 has been found to bind to a 12 kDa antigenic protein in the red blood cell (RBC) of porcine blood. The purpose of this study was to determine the identity of this 12 kDa protein and consequently examine its potential as a marker for monitoring porcine RBC-containing ingredients (PRBCIs) in foods. Proteomic techniques identified the 12 kDa antigenic protein to be a monomer of the tetrameric hemoglobin molecule. Further heat-processing of spray-dried PRBCIs diminishes its detectability. Whereas mAb 24C12-E7-based indirect enzyme-linked immunosorbent assay (iELISA) could detect 1% (v/v) or less of PRBCIs in raw and cooked ground meats (beef, pork and chicken), the detection limits were 3 to 30 times higher for spiked cooked beef and pork. The assay is effective for monitoring the presence of PRBCIs in foods to protect the billions of people that avoid consuming blood. In situations where these PRBCIs are present as ingredients in foods that have undergone further heat processing, the assay, however, may not be as sensitive depending on the types of sample matrix, types of PRBCIs and the level of inclusion.

Keywords: monoclonal antibody; porcine blood; ingredients; red blood cells; hemoglobin; ELISA

1. Introduction

The use of blood-derived proteins mainly produced from porcine or bovine origin as food ingredients is increasing in both conventional and unconventional food products and is a source of worry for those that avoid consuming blood for various reasons. Consumer concerns are heightened by the fact that these protein ingredients are often declared by their brand names on the food labels and hence unknown to the consumer that they are blood-derived. To protect the interest of the particular sector of consumers, efforts have been made towards developing analytical methods for monitoring the presence of these blood-derived ingredients in dietary products. However, the task is arduous as these proteins are diverse in nature in the sense that they could be produced from whole blood, the plasma or red blood cell (RBC) fraction of the blood, or hydrolyzed isolated blood proteins.

From a panel of previously developed monoclonal antibodies (mAbs) that bind to bovine blood, either raw or heat-treated (100 °C for 15 min or 121 °C at 1.2 bar for 15 min) [1], Subsequently, a sandwich enzyme-linked immunorsobent assay (sELISA) was developed for detecting bovine plasma-derived food ingredients [2] and a competitive ELISA (cELISA) for detecting bovine RBC-derived food ingredients [3]. However, complementary methods to detect porcine-derived blood ingredients are still lacking. The only work performed apart from our laboratory was the work by Raja Nhari and others [4] who reported of mAbs that they have developed against heat-treated porcine blood. They mentioned that two mAbs found to be bind to a 60 kDa protein in porcine plasma could be useful for the detection of porcine plasma in processed food. However, judging from the very low absorbance readings (<0.06) obtained for the mAbs against blood samples, the usefulness of this assay in detecting blood in processed food is dubious. In addition, given the diverse nature of these blood proteins, actual samples would have to be tested to guarantee the usefulness of the assay. In the case of food ingredients derived from porcine red blood cells, there is no study reported in the literature on the subject.

Recently, we have raised porcine-blood-specific mAbs with the intent to use them as probes in immunoassays either individually or in various combinations for monitoring porcine blood-derived food ingredients. Two of these porcine blood antibodies that bind cooperatively to a 90 kDa antigenic protein in the plasma fraction of porcine blood, when used in sELISA, have been found effective in detecting porcine plasma-derived ingredients in foods and dietary supplements [5] Among the same panel of developed mAbs that bind to porcine blood, mAb 24C12-E7 was found to be porcine blood-specific and to react with a 12 kDa antigenic protein in the RBC fraction of porcine blood. In a previous study where we raised bovine-blood-specific antibodies for the purpose of probing the presence of bovine blood material in both feed and food using immunoassay [1], further studies revealed that the 12 kDa antigenic protein recognized by one of the antibodies, mAb Bb1H9, was a monomer of the tetrameric bovine hemoglobin molecule [6]. Based on this hind sight information, we anticipated this 12 kDa protein recognized by the porcine-blood-specific mAb 24C12-E7 likely to be a peptide of porcine hemoglobin. Thus, the objectives of this study were to (1) establish the identity of this 12 kDa antigenic protein recognized by mAb 24C12-E7; and (2) consequently determine its potential as a suitable marker for the detection of porcine RBC-derived food ingredients using indirect non-competitive ELISA (iELISA) probed by mAb 24C12-E7.

2. Materials and Methods

Hydrogen peroxide, β-mercaptoethanol, isotyping kit (ISO-2 1 Kit), ABTS (2,2′-azino-bis 3 ethylbenzthiazoline-6-sulfonic acid), EZblue™ gel staining reagent, and porcine hemoglobin were purchased from Sigma-Aldrich Co. (St. Louis, MO, USA). Alkaline phosphates (AP) conjugate substrate kit, Protein Assay Kit, goat anti-mouse IgG (H + L) AP (anti-Ig-AP) conjugate, Tris-buffered saline (TBS), 0.5 M Tris-HCl buffer (pH 6.8), 1.5 M Tris-HCl (pH 8.8), TEMED (*N,N,N′,N′*-tetra-methyl-ethylenediamine), Precision Plus Protein Kaleidoscope Standards, 30% acrylamide/bis solution, Tris/glycine buffer, Tris/glycine/SDS buffer, supported nitrocellulose membrane (0.2 μm), 7 cm 7 to 10 immobilized pH gradient (IPG) strip, rehydration/sample buffer, equilibration buffer I and II, and thick blot paper were purchased from Bio-Rad Laboratories Inc., Hercules, CA, USA. Egg albumin, bovine serum albumin (BSA) and all other chemicals were purchased from Fisher Scientific (Fair Lawn, NJ, USA). All chemicals and reagents were analytical grade and solutions were prepared using distilled deionized pure water (DD water) from a NANOpure DIamond ultrapure water system (Barnstead International, Dubuque, IA, USA).

Liquid samples of whole blood from pigs, cattle, donkey, horse, goat, sheep, rabbit, turkey, and chicken, porcine plasma, porcine serum and porcine red blood cells were purchased from Lampire Biological Laboratories, Pipersville, PA. Porcine gelatin and soy protein powders were obtained from GELITA USA Inc. (Sioux City, IA, USA) and SoyLink (Oskaloosa, IA, USA), respectively. Commercially produced porcine (porcine plasma powder (PPP), porcine Fibrimex® powder (PFP), porcine hemoglobin powder (PHP), porcine hydrolyzed globin (PHG), Aprosan (APS), Aprothem (APT) Apropork (APP), and Aprored (APR)) and bovine (bovine plasma powder (BPP), bovine Fibrimex® powder (BFP), bovine hemoglobin powder (BHP), bovine fibrinogen powder (BFGP) and Immunolin® (ILN)) blood-derived food ingredients were obtained from Sonac BV, Eindhoven, Netherlands (PPP, PFP, PHP, PHG, BPP, BFP, BHP, and ILN) or from Proliant Inc., Barcelona, Spain (APP, APR, APS and APT) or from Proliant Inc., Ankeny, IA, USA. Nonfat dry milk was purchased from a local grocery store. Meat samples, including beef eye of round roast, pork loin, lamb shoulder, whole chicken, whole duck, whole goose, turkey breast, bison, and frozen dressed rabbit were purchased from a local supermarket. Horse meat was obtained from the College of Veterinary Medicine, Auburn University (Auburn, AL, USA). Deer, elk and African buffalo steak meats were provided by the Fats and Proteins Research Foundation (Bloomington, IL, USA).

2.1. *Sample Preparation*

2.1.1. Extraction of Soluble Proteins from Animal Blood and Non-Blood Materials

Soluble proteins were extracted from cooked (100 °C, 15 min) blood samples (whole blood, plasma, serum and RBCs), common food protein ingredients (soy powder, bovine gelatin, porcine gelatin, egg albumin and BSA) and meat samples (pork, beef, horse, elk, donkey, lamb, Bison, African buffalo, rabbit, turkey, chicken, goose, and duck) as previously described [1]. Extraction of soluble proteins from commercially produced porcine and bovine blood ingredients was performed as previously reported for commercially feedstuffs [1]. Raw blood samples were used as is. The protein extracts were stored at −20 °C until use.

2.1.2. Spiked Sample Extracts

To study the effect of matrix on the detectability of this 12 kDa antigenic protein, spiked samples were prepared as follows and probed with 24C12-E7 using iELISA.

To 9.5 g of ground chicken in a sampling bag was added 0.5 g of porcine RBC-containing product (PHP, APS, APT or APR), and the mixture stirred thoroughly with a glass rod to obtain 5% w/w spiked ground chicken. Then, 50 mL of 10 mM phosphate buffered saline (PBS, pH 7.2) was added to the mixture in the sampling bag and the mixture was homogenized first by hand and then for another 1 min using a stomacher (Model Number STO 400, Tekmar Company, Cincinnati, OH, USA). The homogenized samples were then stored overnight at 4 °C after which it was centrifuged at 3220 g for 1 h at 4 °C (Eppendorf 5810R centrifuge, Brinkman Instruments Inc., Westbury, NY, USA) and then passed through Whatman # 1 filter paper to obtain 5% w/w spiked raw ground chicken extracts. Non-spiked raw chicken (0%) containing no added target analyte was similarly prepared. Lower levels of spiking (0.01%, 0.03%, 0.05%, 0.1%, 0.3%, 0.5%, 1% and 3% v/v) were obtained by diluting 5% v/w spiked samples with the appropriate amount of cooked chicken meat (0%) to ensure homogeneity.

Another set of 5% w/w mixture of spiked ground chicken in a beaker as prepared as described above was cooked by immersing the beakers (covered with aluminum foil) in boiling water for 15 min. The cooked samples were broken down into finer particles, 20 mL of 10 mM PBS was added, and the mixture was homogenized for 2 min at 11,000 rpm using the ULTRA-TURRAX T25 basic homogenizer (IKA Works Inc., Wilmington, NC, USA). The homogenized samples were then stored, centrifuged, and passed through filter as described above to obtain 5% v/w spiked cooked chicken meat extracts. Lower levels of spiking (0.01%, 0.03%, 0.05%, 0.1%, 0.3%, 0.5%, 1% and 3% v/v) were obtained by diluting 5% w/w spiked samples with the appropriate amount of non-spiked cooked chicken meat (0%) that had been similarly prepared. All Spiked sample extracts were prepared on the same day and tested immediately after preparation. Ground beef and pork spiked with different amounts of RBC-containing product (PHP, APS, APT or APR) were prepared for studies on matrix effect as described above for spiked ground chicken.

2.2. *Non-Competitive Indirect Enzyme-Linked Immunosorbent Assay (iELISA)*

The selectivity of mAb 24C12-E7 was determined using antigen-coated indirect non-competitive iELISA as previously described [1] with modifications as follows. Plates were incubated for 1 h at 37 °C; 0.2% fish gelatin in 10 mM PBS and PBST (0.05% v/v Tween-20 in 10 mM PBS) used as blocking and antibody buffer respectively; and absorbance read at 415 nm using the PowerWave XS microplate reader (Bio-Tek Instruments, Winooski, VT, USA). Sample diluent (0.06 carbonate buffer) was run alongside test samples as blanks and the average absorbance subtracted from readings obtained for test samples.

2.3. *Sodium Dodecyl Sulfate–Polyacrylamide Gel Electrophoresis (SDS-PAGE) and Western Blot*

SDS-PAGE followed by western blot was performed to reveal the presence of the 12 kDa antigenic protein in various samples. Briefly, soluble proteins (loaded at 2 to 20 μg of protein in 10 μL sample buffer per lane depending on the sample) from the samples were loaded onto 5% stacking gels and separated on 15% polyacrylamide separating gels at 200 V with the aid of the Mini-Protein 3 Electrophoresis Cell (Bio-Rad) as per the method of Laemmli [7]. The separated proteins were transferred electrophoretically (1 h at 100 V, using the Mini Trans-Blot Electrophoretic Transfer Cell, Bio-Rad) according to the protocol by Towbin and others [8] and the membrane blocked with 0.2% fish gelatin in tris buffered saline (TBS). The blotted membrane was then incubated with selected mAb 24C12-E7 followed by incubation with secondary antibody (goat anti-mouse IgG (H + L)-AP conjugate) and subsequent color development as previously described [1]. Precision Plus Protein Kaleidoscope standards were used for the molecular-weight estimations on gels and blot.

2.4. Two-Dimensional Gel Electrophoresis and Western Blot

Two-dimensional electrophoresis (TDGE) was carried out on purified porcine hemoglobin (25 μg per 125 μL of rehydrating buffer) with isoelectric focusing (using the PROTEAN IEF Cell, Bio-Rad) as the first dimension and SDS-PAGE as the second dimension as previously described [6]. The gel was subsequently subjected to western blot as described above to determine the isoelectric point of the 12 kDa protein that is recognized by mAb 24C12-E7.

2.5. N-Terminal Sequencing

Purified porcine hemoglobin that has been subjected to TDGE was first transferred unto a Westran S PVDF (polyvinylidene fluoride) (0.2 μm) membrane as described above. Subsequently, the transferred proteins were stained with EZ-Blue and spot on the PVDF membrane that corresponds to spots on the nitrocellulose membrane that reacted with mAb 24C12-E7, was excised and subjected to *N*-terminal amino acid sequence using an ABI 477A sequencer with an online 120A HPLC system (Applied Biosystems Inc., Foster City, CA, USA). The sequence data was collected using the ABI 610 software (Applied Biosystems Inc.) and analyzed with FASTA programming (European Bioinformatics, http://www2.ebi.ac.uk/fasta3/)

3. Results and Discussion

3.1. Identity of the 12 kDa Antigenic Protein

In this experiment, extracts of porcine hemoglobin were probed with mAb 24C12-E7 using both iELISA and western blot. Porcine blood was run alongside as the control for comparison. From the results of iELISA (Figure 1a), mAb 24C12-E7 reacted very strongly (OD > 2) with both non-heated porcine blood and porcine hemoglobin extracts. Western blot results (Figure 1b) also corroborates the iELISA results as mAb 24C12-E7 reacted with a 12 kDa in both porcine blood and porcine hemoglobin samples suggesting that this 12 kDa antigenic protein is likely to be a monomer of porcine hemoglobin as anticipated. Both samples also revealed a band at around 26 kDa which could be a dimer of this 12 kDa protein as will be discussed below.

The hemoglobin tetramer is a weak complex of 2 alpha-beta dimers held together by a bond between an alpha and a beta subunit. The beta and alpha subunits were reported to have pIs of 7.1 and 8.76, respectively [9]. Thus, as further proof of the 12 kDa antigenic protein being a monomer of porcine hemoglobin, extract of porcine hemoglobin was subjected to TDGE followed by blotting with mAb 24C12-E7, in order to determine the pI of this porcine 12 kDa protein. As shown in Figure 2a, mAb 24C12-E7 reacted with several poorly separated spots around the molecular weight of 12 kDa and between the pIs of 7.5 and 9.5, instead of two distinct spots with pIs of 7.1 and 8.76. Several such poorly separated spots were also visible at a molecular weight of around 26 kDa. A similar observation was reported previously where TDGE of bovine hemoglobin produced several spots with molecular weight around 12 kDa [6]. Other researchers have also reported the presence of multiple spots upon subjecting hemoglobin from other species to TDGE [10], an observation that has been ascribed to the presence of impurities, dimerization, or variable phosphorylation [11]. Finally, a spot, circled and labeled "a" (Figure 2b) on PVDF, corresponding to the spot circled and labeled "1" (Figure 2a) on the nitrocellulose membrane that reacted with mAb 24C12-E7, was excised and subjected to N-terminal sequencing as a conclusive proof if this 12 kDa antigenic protein is indeed a monomer of the tetrameric hemoglobin molecule. Results of the sequencing revealed that the spot was a mixed sequence of two overlapping spots with the first 12 amino acids of both sequence as shown in Table 1. Sequences 1 and 2 showed 100% homology with the alpha and beta subunit of porcine hemoglobin, respectively, when the sequences obtained was blasted using the NCBI Blast: Protein Sequence software (http://blast.ncbi.nlm.nih.gov/Blast.cgi?PAGE=Proteins). These results confirm this 12 kDa protein in porcine blood to be indeed a monomer of the tetrameric hemoglobin molecule.

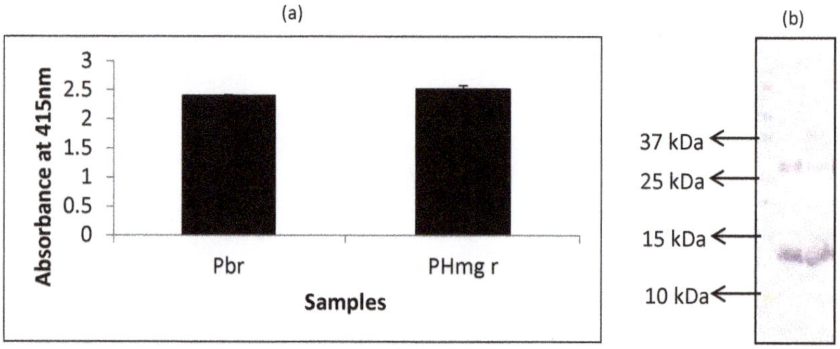

Figure 1. Analysis of raw porcine blood (Pbr) and porcine hemoglobin extracts (PHmg) using (**a**) iELISA and (**b**) western blot using mAb 24C12-E7 as probe. Samples were coated at 0.5 µg per 100 µL of carbonate buffer in iELISA. mAb 24C12-E7 supernatant was diluted 1:50 in 0.2% fish gelatin in PBST and iELISA results are expressed as $A_{415} \pm$ SD, n = 3 (SD = standard deviation). In the case of western blot samples were loaded at 10 µg per 10 µL of sample buffer alongside the Precision Plus Protein Kaleidoscope pre-stained standards.

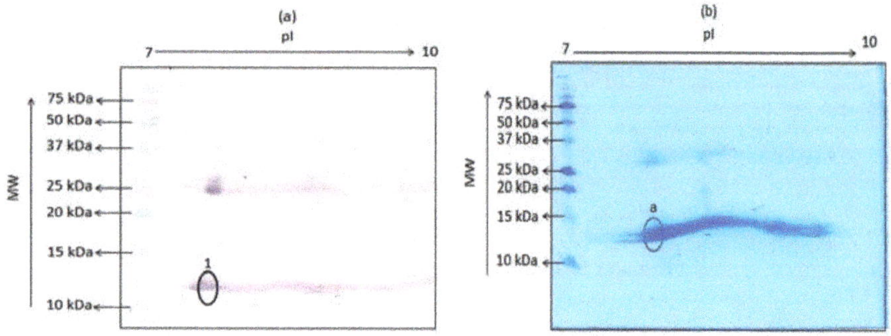

Figure 2. Two-dimensional gel electrophoresis of porcine hemoglobin extract containing 25 µg total protein followed by transfer unto (**a**) nitrocellulose membrane and detection with mAb 24C12-E7 and (**b**) PVDF membrane and subsequent staining with EZBlueTM. Precision Plus Protein Kaleidoscope pre-stained standards were run alongside for estimation of molecular weight of proteins. Circled spot "a" on the PVDF membrane was subjected to *N*-terminal sequencing.

Table 1. *N*-terminal sequence data for the excised spot labeled "a" on PVDF (polyvinylidene fluoride).

Sequence	First 12 Amino Acids
Sequence 1	V L S A A D K A N V K A
Sequence 2	V H L S A E E K E A V L

3.2. Selectivity of mAb 24C12-E7 Based iELISA

Having demonstrated that the 12 kDa antigenic protein recognized by mAb 24C12-E7 is a monomer of the tetrameric hemoglobin molecule, a simple iELISA was optimized and developed to demonstrate the utility of mAb 24C12-E7 for monitoring the presence of porcine blood or porcine RBC-derived proteins in foods based on its reaction with the 12 kDa protein. The first in the series of experiments was to determine the ability of the iELISA to discriminate (1) between porcine blood and blood from other species; (2) porcine RBCs from porcine plasma; and (3) porcine blood from non-blood proteins that are likely to be present in a food matrix.

The ability of the assay to distinguish porcine blood from other species was determined by testing the assay against porcine blood and blood from other species, including bovine, horse, donkey, goat, sheep, rabbit, chicken and turkey. Since commercially produced blood proteins typically undergo heat-treatment by way of spray-drying, soluble proteins extracted from heated blood from the various species were used as samples for testing. As shown in Figure 3a, the assay reacted strongly with heated porcine blood (A_{415} = 2.0) but did not react with heated blood from the other eight species tested (A_{415} < 0.1) indicating the porcine selectivity of this mAb 24C12-E7 and the heat-stability of its epitope. Subsequently, analysis of extracts of heated porcine blood and porcine blood fractions (porcine serum, porcine plasma and porcine RBCs) revealed the assay to be specific for porcine RBC as the assay reacted with porcine blood (A_{415} = 1.95) and porcine RBCs (A_{415} = 0.78), but not with porcine plasma (A_{415} < 0.1) or porcine serum (A_{415} < 0.1). The much lower reading for porcine RBCs in comparison with porcine blood is attributed to the much lower amounts of the 12 kDa antigenic protein (hemoglobin) present in the former which was purchased as a 10% diluted solution.

The mAb 24C12-E7 based iELISA was further tested against extracts prepared from a number of common food proteins (whey, soy, egg albumin, non-fat dry milk, bovine serum albumin, and porcine gelatin) and the meat of thirteen species (pig, cow, deer, elk, horse, African buffalo, rabbit, lamb, bison, chicken, turkey, goose and duck) for examining the assay's cross-reactivity. Results clearly showed that the assay did not cross-react (A_{415} < 0.1) with any of these non-blood proteins tested. The results demonstrated the excellent selectivity of this mAb24C12-E7 based assay.

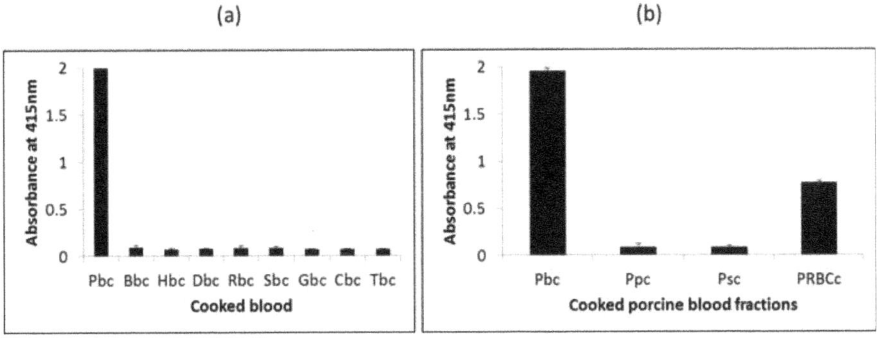

Figure 3. Reactivity of mAb 24C12-E7 with (**a**) cooked blood extracts from nine species and (**b**) cooked porcine blood and blood fractions using iELISA. Soluble proteins extracted from cooked whole blood, plasma, serum or RBCs (red blood cells) were coated at 0.5 µg per 100 µL. mAb 24C12-E7 supernatant was diluted 1:50 in 0.2% fish gelatin in PBST . Results are expressed as $A_{415} \pm$ SD, n = 3. Pbc: cooked porcine blood; Bbc: cooked bovine blood; Hbc: cooked horse blood; Dbc: cooked donkey blood; Rbc: cooked rabbit blood; Sbc: cooked sheep blood; Gbc: cooked goat blood; Cbc: cooked chicken blood; Tbc: cooked turkey blood; Ppc: cooked porcine plasma; Psc: cooked porcine serum; and PRBCc: cooked porcine RBCs.

3.3. *Reactivity of the iELISA with Commercial Blood Proteins*

From the results above, this iELISA based on the detection of the 12 kDa antigenic protein in porcine RBCs has potential for accurately detecting the presence of raw and processed porcine blood and porcine RBC-derived proteins in foods. Therefore, the assay was tested against commercially produced spray-dried porcine and bovine blood proteins that are obtained either from whole blood, or from the plasma or RBC fraction. Non-heated and heated (further heated by immersing in boiling water for 15 min) versions of these spray-dried proteins were analyzed. Heated versions of these proteins were included to examine how the assay will perform in situations where they are used as ingredients in foods that may undergo further heat-treatment in their production. As shown in Figure 4, for the non-heated spray-dried products, the assay did not react with any of the bovine blood proteins (bovine plasma powder (BPP), bovine hemoglobin powder (BHP), bovine Fibrimex® powder (BFP), Immunolin® (ILN), and bovine fibrinogen powder (BFGP)) ($A_{415} < 0.1$) tested as expected (Figure 4). In the case of the porcine proteins, the assay reacted strongly with the porcine RBC-derived proteins, namely, porcine hemoglobin powder (PHP) ($A_{415} = 2.07$), Aprothem (APT) ($A_{415} = 1.97$), and Aprored (APR) ($A_{415} = 1.33$) but weakly with porcine hydrolyzed globin (PHG) ($A_{415} = 0.286$). Apparently, the weak reaction with PHG is due to destruction of the epitope by the hydrolysis involved in the production of this product. The assay also exhibited a strong reaction to the whole porcine blood product, Aprosan (APS) ($A_{415} = 1.84$). However, the assay also unexpectedly reacted moderately with the two porcine plasma-derived proteins, porcine plasma powder (PPP) ($A_{415} = 0.94$) and porcine Fibrimex® powder (PFP) ($A_{415} = 0.97$). Based on the excellent porcine hemoglobin selectivity of this assay demonstrated in the above section, this positive reaction obviously can be attributed to the poor quality control practice of the commercial manufacturing by introducing contamination with hemoglobin either from the poor separation of the plasma from the red blood cells, or as a result of cross contamination with porcine RBC-derived proteins that are also produced by the same company.

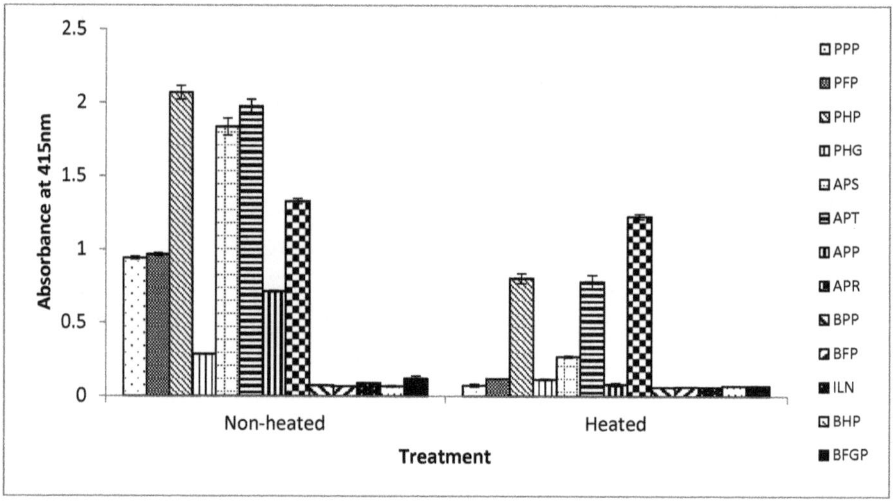

Figure 4. Reactivity of mAb 24C12-E7 with non-heated and heated bovine and porcine blood-derived food ingredients using iELISA. Soluble proteins extracted from these blood-derived proteins were coated at 0.5 μg/100 μL. mAb 24C12-E7 supernatant was diluted 1:50 in 0.2% fish gelatin in PBST. Results are expressed as $A_{415} \pm$ SD, $n = 3$. PPP: porcine plasma powder; PFP: porcine Fibrimex® powder; PHP: porcine hemoglobin powder; PHG: porcine hydrolyzed globin; APS: Aprosan; APT: Aprothem; APP: Apropork; APR: Aprored; BPP: bovine plasma powder; BFP: bovine Fibrimex® powder; ILN: Immunolin®; BHP: bovine hemoglobin powder; and BFGP: bovine fibrinogen powder.

For the further heated protein samples, the overall reaction signals were reduced compared to the non-heated spray-dried samples. The assay neither reacted with the bovine products (BPP, BFP, BHP, ILN, and BFGP) ($A_{415} < 0.1$) nor with any of the porcine plasma-derived proteins (PPP, PFP, and APP) ($A_{415} < 0.1$) but reacted only with the porcine RBC-containing proteins, namely, PHP ($A_{415} = 0.8$), APT ($A_{415} = 0.78$), APR ($A_{415} = 1.23$), and APS ($A_{415} = 0.27$), except in the case of the hydrolyzed product, PHG ($A_{415} < 0.1$) (Figure 4). According to the manufacturers, these spray-dried protein products were already heat-processed to internal temperatures of 70 °C or 80 °C. Thus, in the case of the heated samples, the further heat-treatment may have destroyed and insolubilized significant amount of the antigenic proteins, and hence the much lower reaction signals. For those proteins containing porcine RBCs, the reduction in signals (absorbance) for non-heated versions compared to heated versions were 61%, 60%, 85%, 60%, and 7% for PHP, PHG, APS, APT, and APR, respectively. This may explain the lack of binding with the hemoglobin contaminated porcine plasma-derived proteins in the case of the heated versions, as the further heat-treatment (boiling for 15 min) may have diminished any contaminated antigenic protein present to levels below the detection limit of the assay.

3.4. *Effect of Heat-Treatment on the 12 kDa Antigenic Protein*

It is apparent that the epitope of 12 kDa protein recognized by mAb 24C12-E7 may be affected when the protein is subjected to excessive heat processing which consequently may affect its detectability using ELISA. Accordingly, extracts of these porcine blood proteins that are derived from both plasma and RBCs were further analyzed with western blot to observe the changes of this antigenic protein. Porcine plasma-derived proteins were included as negative controls.

In the case of the non-heated proteins, SDS-PAGE revealed the presence of the 12 kDa proteins in the porcine RBC-derived proteins, PHP (lane 5) and APT (lane 7), and the whole porcine blood product, APS (lane 4). In the case of the other RBC-derived product, APR (lane 6), a slightly higher band at around 13 kDa was revealed (Figure 5a(i)). APR is a branded product with scarce information about its nature except for the disclosed fact that it is a pigment obtained from porcine RBCs. Perhaps modifications to the hemoglobin by the processing involved in its production may have caused it to move at a slightly slower rate and hence the 13 kDa band observed instead of the 12 kDa band. The 12 kDa band, however, was absent in the hydrolyzed product, PHG (lane 8). It appears that the hydrolysis action breaks down the protein into lower molecular weight peptides which can be seen as a smear below 10 kDa. A band at around 25 kDa was also present in the samples, PHP, APT, and APS, and also present faintly in PHG (Figure 5a(i)). A slightly higher band at around 27 kDa was present in APR. Other protein bands (>50 kDa) were present in all of these samples (PHP, APT, APS, APR, and PHG) (Figure 5a(i)). When these proteins were blotted and probed with mAb 24C12-E7, the antibody reacted with the 12 kDa protein and the 25 kDa protein bands in PHP, APT, and APS, and the corresponding slightly higher bands at 13 kDa and 27 kDa, in the case of APR (Figure 5a(ii)). It is reasonable to consider that the 25 or 27 kDa antigenic proteins are a dimer of the 12 and 13 kDa proteins, respectively. Lack of binding with the higher molecular weight bands (>50 kDa) revealed on SDS-PAGE may be because these proteins are non-specific contaminants. The control samples, porcine plasma-derived proteins PPP (lane 1), PFP (lane 2) and APP (lane 3) also revealed the 12 kDa band in increasing order of intensity, although much fainter in comparison with RBC-containing products (PHP, APS, APT and APR). This corroborates our earlier assertion that the observed reaction with non-heated porcine plasma proteins (PPP, PFP and APP) using iELISA was due to contamination with porcine hemoglobin. Interestingly, unlike in the case of iELISA, mAb 24C12-E7 did not react with the 12 kDa protein in these plasma-derived samples. Studies have shown that the mean total quantity of all proteins transferred onto the western blot membrane over a 2 h period using conventional tank transfer (which was employed in this study) is 57% [11]. Thus, the little amount of contaminant protein residues, judging from the faint to almost indiscernible bands, could have easily been lost during the transfer or diminished to levels below the Western blot detection limit (Figure 5a(ii)).

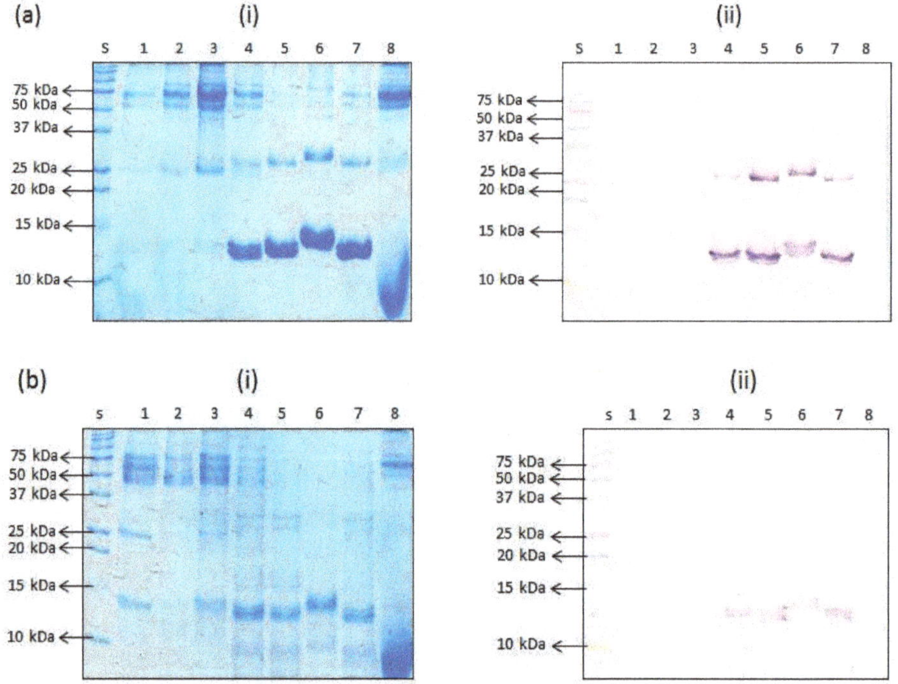

Figure 5. SDS PAGE (**i**) and western blot (**ii**) analysis of non-heated (**a**) and heated (**b**) extracts of porcine blood-derived proteins. Protein extracts were loaded at 10 μg per 10 μL of sample buffer per lane alongside Precision Plus Protein kaleidoscope pre-stained standards. Lane s: Precision Plus Protein Kaleidoscope standards; Lane 1: porcine plasma powder (PPP); Lane 2: porcine Fibrimex® powder (PFP); Lane 3: Apropork (APP); Lane 4: Aprosan (APS); Lane 5: porcine hemoglobin powder (PHP); Lane 6: Aprored (APR); Lane 7: Aprothem (APT); and Lane 8: porcine hydrolyzed globin (PHG).

From the SDS-PAGE results for the heat-treated counterparts, a similar observation as the iELISA was evident showing overall weaker intensity of the 12 kDa band in the RBC-containing products, namely, APS (lane 4), PHP (lane 5), and APT (lane 7) (Figure 5b(i)). Likewise, the slightly heavier band (13 kDa) in the product APR was also weaker in intensity compared to its non-heated counterparts. The band intensity for the dimers (25 or 27 kDa proteins) were even weaker or absent in the product APR (lane 6). Upon subsequent transfer and blotting with mAb 24C12-E7, the 12 kDa protein also appeared in the heated products, APS, PHP and APT, and the 13 kDa protein in APR, but with much weaker band intensities compared to the non-heated versions. The protein bands revealed on SDS-PAGE for the porcine plasma-derived protein samples were also not revealed by Western blot analysis (Figure 5b(ii)) due to overall heat-induced (from spray-drying, cooking for 15 min and sample heat-treatment for 5 min prior to SDS-PAGE) reduction in binding between mAb 24C12-E7 and its 12 kDa antigen.

In summary, from the submissions made above, it is apparent that excessive heat treatment as in spray-drying followed by subsequent heating for 15 min, affects the integrity of the antigenic protein in a manner that diminishes its detectability.

3.5. Detection Limit of mAb 24C12-E7 Based iELISA

The interference of using PRBCIs in various food matrices was investigated by testing both raw and cooked meat products spiked with different amounts of each PRBCI (PHP, APT, and APR) and porcine whole blood protein (APS). The detection limits of porcine materials in different food matrices were determined using the iELISA. The detection limit of the assay was defined as the minimum amount of spiked sample that was statistically different from un-spiked meat (0%) samples. As summarized in Table 2, the assay could sensitively detect 0.3%, 1%, 0.5% and 0.3% (*v*/*v*) of porcine hemoglobin powder (PHP), Aprored (APR), Aprothem (APT), and Aprosan (APS) in raw ground chicken, respectively (Figure 6a(i))). Surprisingly, for spiked cooked ground chicken samples, the detection limits were even lower in some cases as in 0.3% (*v*/*v*), 0.5% (*v*/*v*), 0.3% (*v*/*v*) and 0.3% (*v*/*v*) PHP, APR, APT, and APS in cooked ground chicken, respectively (Figure 6a(ii)). In the case of spiked ground beef, the detection limit for the assay was 0.1% (*v*/*v*), 1% (*v*/*v*), 0.3% (*v*/*v*) and 0.3% (*v*/*v*) PHP, APR, APT, and APS in raw spiked ground, respectively (Figure 6b(i)); and 1.0% (*v*/*v*), 5.0% (*v*/*v*), 3.0% (*v*/*v*), and 1.0% (*v*/*v*) PHP, APR, APT, and APS in cooked spiked counterparts, respectively (Figure 6b(ii)). The assay could also detect 0.1% (*v*/*v*), 1.0% (*v*/*v*), 0.1% (*v*/*v*), and 0.3% (*v*/*v*) of PHP, APR, APT, and APS in raw ground pork respectively (Figure 6c(i)). For spiked cooked ground pork, 3.0% (*v*/*v*) of PHP, APT, and APS could be detected (Figure 6b(ii)). However, in the case of cooked pork spiked with APR, the assay could not detect APR up to inclusion levels of 10% (*w*/*w*) (Figure 6b(ii)).

From the results, it is obvious that different matrices have different effects on the measurability of the 12 kDa antigen. Despite the differences, for raw samples the assay could sensitively detect 1.0% (*v*/*v*) or less of these porcine RBC-derived or porcine whole blood proteins in all spiked meat samples. In the case of cooked spiked samples, heating affects the antigen in a manner that reduces its immunoreactivity as seen with higher readings of non-heated proteins in comparison with heated versions. The detection limits were several times higher than those of their raw spiked sample counterparts. This observation, however, seems to apply to red meats (beef and pork) only, as the detection limits for cooked spiked samples in chicken meat were comparable to those of the raw samples. The higher detection limit for spiked cooked sample is understandable as these proteins had already undergone heat-treatment by way of spray-drying and the extra heat treatment (boiling in water for 15 min) can further destroy the epitope leading to a lower immunoreactivity (and hence higher detection limits) for cooked samples compared to spiked raw meat samples. It is unclear why cooking of spiked ground chicken did not affect detectability of the antigenic protein as was the case with cooked spiked ground beef and pork. Perhaps some components in the ground chicken sample may have protected the 12 kDa protein from the effects of heat.

Table 2. Summary of detection limit of spiked meat samples using 24C12-E7 based iELISA.

Spiked Samples	Detection Limit	
	Raw	Cooked
1. Porcine hemoglobin powder in ground chicken	0.3% (*v*/*v*)	0.3%
2. Aprored in ground chicken	1.0% (*v*/*v*)	0.5%
3. Aprothem in ground chicken	0.5% (*v*/*v*)	0.3%
4. Aprosan in ground chicken	0.3% (*v*/*v*)	0.3%
1. Porcine hemoglobin powder in ground beef	0.1% (*v*/*v*)	1% (*v*/*v*)
2. Aprored in ground beef	1.0% (*v*/*v*)	5.0% (*v*/*v*)
3. Aprothem in ground beef	0.3% (*v*/*v*)	3.0% (*v*/*v*)
4. Aprosan in ground beef	0.3% (*v*/*v*)	1.0% (*v*/*v*)
1. Porcine hemoglobin powder in ground pork	0.1% (*v*/*v*)	3.0% (*v*/*v*)
2. Aprored in ground pork	1.0% (*v*/*v*)	Undetectable up to 10.0% (*w*/*w*)
3. Aprothem in ground pork	0.1% (*v*/*v*)	3.0% (*v*/*v*)
4. Aprosan in ground pork	0.3% (*v*/*v*)	3.0% (*v*/*v*)

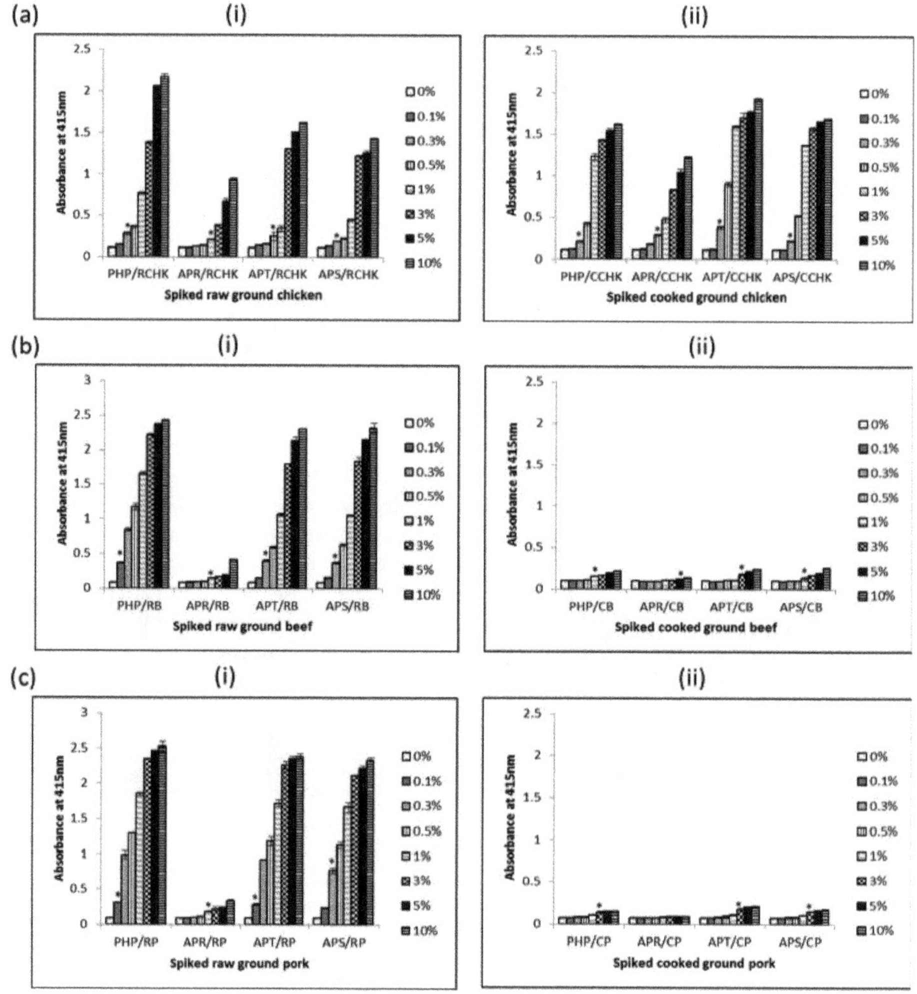

Figure 6. Detection limit of PHP, APR, APT, and APS in raw (**i**) and cooked (**ii**) ground chicken (**a**), ground beef (**b**) and ground pork (**c**) using 24C12-E7 based iELISA. Results are expressed as $A_{415} \pm$ SD, $n = 3$. An asterisk (*) indicates a significant difference ($p < 0.05$) from background (0%). RCHK: raw ground chicken; CCHK: cooked ground chicken; RB: raw beef; CB: cooked beef; RP: raw pork; and CP: cooked pork.

4. Conclusions

In conclusion, the 12 kDa antigenic protein recognized by mAb 24C12-E7 has been identified as a monomer of the tetrameric hemoglobin molecule based on the immunoreactivity, molecular size, relative heat-stability and amino acid analysis. Further studies have also shown that in situations where this antigenic protein has previously been subjected to severe processing as in spray-drying at internal temperatures of 70 °C or 80 °C, followed by cooking in boiling water for 15 min, its integrity is affected in a manner that makes it less detectable. Consequently, the developed iELISA using mAb 24C12-E7 could detect 1% (v/v) or less of diversely processed porcine hemoglobin-containing food ingredients in raw ground meats. However, for spiked cooked samples, except for cooked spiked chicken where the assay could detect 0.5% (v/v) or less of these porcine blood proteins, the detection limits were 3 to 10 times higher for cooked spiked beef, and even higher ($\geq$30 times) for cooked spiked pork. These results demonstrate that the mAb 24C12-E7 based iELISA has the potential for use as an analytical tool for monitoring the presence of spray-dried porcine hemoglobin-containing blood protein ingredients in foods for the protection of those that avoid consuming blood for various reasons. However, in situations where these spray-dried proteins have been used in formulations that have undergone further heat-processing, the assay may not be suitable when these proteins have been included at levels below 3% or even 10% for some of these proteins.

Acknowledgments: This project was funded by the Florida State University Research Foundation GAP Commercialization Grant Program.

Author Contributions: Yun-Hwa P. Hsieh conceived the research idea, and designed the experiments with Jack A. Ofori; Jack A. Ofori performed the experiment, analyzed the data and drafted the manuscript. Yun-Hwa P. Hsieh edited and finalized the paper.

Conflicts of Interest: The authors declare no conflict of interest.

References

1. Hsieh, Y.H.; Ofori, J.A.; Rao, Q.; Bridgeman, C.R. Monoclonal antibodies specific to thermostable proteins in animal blood. *J. Agric. Food Chem.* **2007**, *55*, 6720–6725. [CrossRef] [PubMed]
2. Ofori, J.A.; Hsieh, Y.H. Characterization of a 60-kDa Thermally Stable Antigenic Protein as a Marker for the Immunodetection of Bovine Plasma-Derived Food Ingredients. *J. Food Sci.* **2015**, *80*, C1654–C1660. [CrossRef] [PubMed]
3. Ofori, J.A.; Hsieh, Y.H. Immunodetection of bovine hemoglobin-containing food ingredients using monoclonal antibody Bb 1H9. *Food Control.* in press.
4. Raja Nhari, R.M.; Hamid, M.; Rasli, N.M.; Omar, A.R.; El Sheikha, A.F.; Mustafa, S. Monoclonal antibodies specific to heat-treated porcine blood. *J. Sci. Food Agric.* **2016**, *96*, 2524–2531. [CrossRef] [PubMed]
5. Ofori, J.A.; Hsieh, Y.H. Monoclonal Antibodies as Probes for the Detection of Porcine Blood-Derived Food Ingredients. *J. Agric. Food Chem.* **2016**, *64*, 3705–3711. [CrossRef] [PubMed]
6. Ofori, J.A.; Hsieh, Y.H. Characterization of a 12 kDa thermal-stable antigenic protein in bovine blood. *J. Food Sci.* **2011**, *76*, C1250–C1256. [CrossRef] [PubMed]
7. Laemmli, U.K. Cleavage of structural proteins during the assembly of the head of bacteriophage T4. *Nature* **1970**, *227*, 680–685. [CrossRef] [PubMed]
8. Towbin, H.; Staehelin, T.; Gordon, J. Electrophoretic transfer of proteins from polyacrylamide gels to nitrocellulose sheets: Procedure and some applications. *Proc. Natl. Acad. Sci. USA* **1979**, *76*, 4350–4354. [CrossRef] [PubMed]
9. Zhang, K.; Liu, Y.; Shang, Y.; Zheng, H.; Guo, J.; Tian, H.; Jin, Y.; He, J.; Liu, X. Analysis of pig serum proteins based on shotgun liquid chromatography-tandem mass spectrometry. *Afr. J. Biotechnol.* **2012**, *11*, 12118–12127.
10. Xu, L.; Glatz, C.E. Predicting protein retention time in ion-exchange chromatography based on three-dimensional protein characterization. *J. Chromatogr. A* **2009**, *1216*, 274–280. [CrossRef] [PubMed]
11. Tovey, E.R.; Baldo, B.A. Comparison of semi-dry and conventional tank-buffer electrophoresis of proteins from polyacrylamide gels to nitrocellulose membranes. *Electrophoresis* **1987**, *8*, 384–387. [CrossRef]

foods

MDPI

Article

Predicted Release and Analysis of Novel ACE-I, Renin, and DPP-IV Inhibitory Peptides from Common Oat (*Avena sativa*) Protein Hydrolysates Using in Silico Analysis

Stephen Bleakley [1,2], Maria Hayes [1,*], Nora O' Shea [3], Eimear Gallagher [4] and Tomas Lafarga [5]

1 Food Biosciences Department, Teagasc Food Research Centre, Ashtown, D15 Dublin, Ireland; Stephen.Bleakley@teagasc.ie
2 School of Biological Sciences, College of Sciences and Health and Environment, Sustainability and Health Institute, DIT Kevin Street, D08 NF82 Dublin, Ireland
3 Food Chemistry and Technology Department, Teagasc Food Research Centre, Moorepark, Fermoy, Co. Cork P61 C996, Ireland; Norah.O'Shea@teagasc.ie
4 Food Quality and Sensory Science Department, Teagasc Food Research Centre, Ashtown, Dublin 15, Ireland; Eimear.Gallagher@teagasc.ie
5 Parc Científic I Tecnològic Agroalimentari de Lleida, Parc de Gardeny, Edifici Fruit Centre, Institut de Recerca, Tecnològia Agroalimentàries (IRTA), 25003 Lleida, Spain; tomas.lafarga@irta.cat
* Correspondence: Maria.Hayes@teagasc.ie; Tel.: +353-1-8059957

Received: 26 October 2017; Accepted: 27 November 2017; Published: 4 December 2017

Abstract: The renin-angiotensin-aldosterone system (RAAS) plays an important role in regulating hypertension by controlling vasoconstriction and intravascular fluid volume. RAAS itself is largely regulated by the actions of renin (EC 3.4.23.15) and the angiotensin-I-converting enzyme (ACE-I; EC 3.4.15.1). The enzyme dipeptidyl peptidase-IV (DPP-IV; EC 3.4.14.5) also plays a role in the development of type-2 diabetes. The inhibition of the renin, ACE-I, and DPP-IV enzymes has therefore become a key therapeutic target for the treatment of hypertension and diabetes. The aim of this study was to assess the bioactivity of different oat (*Avena sativa*) protein isolates and their ability to inhibit the renin, ACE-I, and DPP-IV enzymes. In silico analysis was carried out to predictthe likelihood of bioactive inhibitory peptides occurring from oat protein hydrolysates following in silico hydrolysis with the proteases papain and ficin. Nine peptides, including FFG, IFFFL, PFL, WWK, WCY, FPIL, CPA, FLLA, and FEPL were subsequently chemically synthesised, and their bioactivities were confirmed using in vitro bioassays. The isolated oat proteins derived from seven different oat varieties were found to inhibit the ACE-I enzyme by between 86.5 $\pm$ 10.7% and 96.5 $\pm$ 25.8%, renin by between 40.5 $\pm$ 21.5% and 70.9 $\pm$ 7.6%, and DPP-IV by between 3.7 $\pm$ 3.9% and 46.2 $\pm$ 28.8%. The activity of the synthesised peptides was also determined.

Keywords: oats; *Avena sativa*; bioactive peptides; ACE-I; renin; DPP-IV; renin-angiotensin-aldosterone system

1. Introduction

High blood pressure is the single largest risk factor attributed to deaths worldwide. It is responsible for 12.8% of deaths, and affects all countries and income groups [1]. Furthermore, high systolic blood pressure is globally attributable to 51% of strokes, 45% of ischaemic heart disease, and between 37% (Southeast Asia) and 54% (European countries) of cardiovascular deaths [1]. Hypertension is therefore a considerable problem in our society, not only placing a great burden on our healthcare system, but also having a substantial impact on the economy, with direct medical costs of cardiovascular disease (CVD) estimated to increase three-fold from $273 billion in 2010 to $818 billion in 2030 in the United States alone [2]. The renin-angiotensin-aldosterone system (RAAS) plays an important role in regulating blood pressure by controlling arteriolar vasoconstriction and intravascular fluid volume. RAAS itself is largely regulated by the actions of the enzyme angiotensin-I-converting enzyme (ACE-I; EC 3.4.15.1), which increases blood pressure [3].

ACE-I inhibitors have therefore become one of the most commonly studied drugs, with global annual sales exceeding $6 billion USD. ACE-I inhibitory drugs can be considered one of the major protease inhibitor success stories [4]. Synthetic ACE-I inhibitor drugs such as captopril, enalapril, and alacepril, often come with several side effects, including hypotension, dry cough, and impaired renal function [5]. Functional foods with antihypertensive bioactivities have therefore become a popular alternative to synthetic drugs, especially for individuals who are borderline hypertensive and do not warrant the prescription of pharmaceutical drugs [6].

Gliptins or dipeptidyl peptidase IV (DPP-IV) inhibitors can block the action of the enzyme DPP-IV, and may be used to treat diabetes mellitus type 2. DPP-IV inhibitors increase incretin levels, inhibit glucagon release, increase insulin secretions, and decrease gastric emptying and blood glucose levels.

In addition to providing energy and amino acids essential for growth, it has become increasingly recognised that some dietary proteins contain biologically active peptides that can impart a beneficial physiological effect [7]. Such bioactive peptides are typically 2–30 amino acids in length, and can exert their physiological response as a result of their hormone-like properties or by acting as antagonistic receptor inhibitors [8]. Bioactive peptides are inactive within the precursor protein, and can be released through hydrolysis by digestive enzymes or fermentation. The function of these peptides covers a variety of activities, including antihypertensive [9], hypoglycemic [10], antiamnestic [11], antimicrobial [12], antithrombotic [13], antioxidative [14], hypocholesterolemic [15], gastrointestinal absorption modulation [16], appetite suppression [17], opioids [18], immunomodulation [19], and cytomodulation [20].

Milk proteins remain the most common food resource for bioactive peptide generation [7,21]. The most common ACE-I inhibitory peptides, Ile-Pro-Pro and Val-Pro-Pro, are derived from casein in milk [9]. However, bioactive peptides have also been identified in a number of plant and animal protein sources, including rice [22], soybean [23], wheat [24], seaweed [25], pea [26], broccoli [27], garlic [28], egg [29], meat [30], blood [31] and fish [32].

The common oat (*Avena sativa*) is a promising source of bioactive peptides, with several peptides already identified, but it has not been fully explored yet [33]. This study isolated proteins from seven different oat varieties, and identified the bioactivities of these extracts. In addition, it identified ACE-I, renin (EC 3.4.23.15), and DPP-IV inhibitory peptides following in silico digestion of common oat proteins. Selected peptides were synthesised and tested for their ability to inhibit ACE-I, renin, and DPP-IV.

2. Materials and Methods

2.1. Materials

Seven strains of de-hulled and milled oats were used for protein extraction, including Barra, Husky, Maesbro, Mascan, Rhapsody, Selwyn, and Vodka. The oat samples were harvested in 2014/2015 by John Finnan, Teagasc Oakpark, and were also supplied by the Tillage manager at Glanbia PLC (Glanbia PLC, Kilkenny, Ireland). The naked, de-husked grains were used to obtain a higher yield, as they are significantly higher in protein and lower in crude fiber [34]. Ammonium sulfate was obtained from Sigma Aldrich (Wicklow, Ireland).

2.2. Oat Protein Extraction

A 2% (*w*/*v*) solution of milled oats in distilled deionised water (ddH_2O) was sonicated using an Ultrasonic Bath Branson 3510 at 42 kHz for 1 h. This was followed by stirring overnight at 4 °C. The oats were then centrifuged at 10,000× *g* for 1 h at 4 °C using a Sorvall™ Lynx 6000 centrifuge (Thermo

Scientific™, Dublin, Ireland). The supernatant was then collected and stored at 4 °C, while the pellet was re-suspended in 4% (*w*/*v*) ddH_2O. The resuspended oats were sonicated for 1 h, stirred overnight at 4 °C, and centrifuged at 10,000× *g* for 1 h at 4 °C. The supernatant was then pooled with the first supernatant for further processing. An 80% ammonium sulfate-saturated solution was made with the pooled oat supernatant, and stirred for 1 h at 4 °C, followed by centrifugation at 17,000× *g* for 1 h at 4 °C. The pellet was resuspended and dialysed overnight at 4 °C. Samples were stored at −20 °C and subsequently freeze-dried using an industrial-scale freeze drier, FD 80 model (Cuddon Engineering, Marlborough, New Zealand).

2.3. *In Silico Digestion*

The primary proteins found in oats were identified from the literature, with protein sequences obtained from the UniProt database, available at http://www.uniprot.org (Table 1). Each protein sequence was digested in silico with papain (EC 3.4.22.2) or ficin (EC 3.4.22.3) using BIOPEP, which is available at http://www.uwm.edu.pl/biochemia/index.php/en/biopep [35], and the method shown in Figure 1.

Table 1. The main storage proteins found in oats.

Protein	UniProt ID	Sequence **	Amino Acid Length	Molecular Mass (Da)
11S globulin	Q38780	MATTSFPSMLFYFCIFLLFHGSMAQLFGQSSTP WQSSRQGGLRGCRFDRLQAFEPLRQVRSQAGI TEYFDEQNEQFRCTGVSVIRRVIEPQGLVLPQY HNAPALVYILQGRGFTGLTFPGCPATFQQQFQ PFDQSQFAQGQRQSQTIKDEHQRVQRFKQGD VVALPAGIVHWCYNDGDAPIVAIYVFDVNNN ANQLEPRQKEFLLAGNNKREQQSGNNIFSGLS VQLLSEALGISQQAAQRIQSQNDQRGEIIRVSQ GLQFLKPIVSQQVPGEQQVYQPIQTQEGQATQ YQVGQSTQYQVGKSTPYQGGQSSQYQAGQSW DQSFNGLEENFCSLEARKNIENPQHADTYNPR AGRITRLNSKNFPILNIVQMSATRVNLYQNAIL SPFWNINAHSVIYMIQGHARVQVVNNNGQTV FNDILRRGQLLIVPQHFVVLKKAEREGCQYISF KTNPNSMVSHIAGKSSILRALPIDVLANAYRISR QEARNLKNNRGEEFGAFTPKLTQKGFQSYQDI EEGSSSPVRASE	527	59,406
12S globulin	P12615	MATTRFPSLLFYSCIFLLCNGSMAQLFGQSFTP WQSSRQGGLRGCKFDRLQAFEPLRQVRSQAGI TEYFDEQNEQFRCAGVSVIRRVIEPQGLLLPQY HNAPGLVYILQGRGFTGLTFPGCPATFQQQFQ QFDQARFAQGQSKSQNLKDEHQRVHHIKQGD VVALPAGIVHWCYNDGDAPIVAVYVFDVNNN ANQLEPRQKEFLLAGNNKREQQFGQNIFSGFS VQLLSEALGISQQAAQKIQSQNDQRGEIIRVSQ GLQFLKPFVSQQGPVEHQAYQPIQSQQEQSTQ YQVGQSPQYQEGQSTQYQSGQSWDQSFNGLE ENFCSLEARQNIENPKRADTYNPRAGRITHLNS KNFPTLNLVQMSATRVNLYQNAILSPYWNINA HSVMHMIQGRARVQVVNNHGQTVFNDILRR GQLLIIPQHYVVLKKAEREGCQYISFKTTPNSM VSYIAGKTSILRALPVDVLANAYRISRQESQNLK NNRGEEFGAFTPKFAQTGSQSYQDEGESSSTEKASE	518	58,545
Avenin-3	P80356	MKTFLIFALLAMAATMATAQFDPSEQYQPYPE QQQPILQQQQMLLQQQQQMLLQQQPLLQVL QQQLNPCRQFLVQQCSPVAVVPFLRSQILQQSS CQVMRQQCCRQLEQIPEQLRCPAIHSVVQAII MQQQQFFQPQMQQQFFQPQMQQVTQGIFQP QMQQVTQGIFQPQLQQVTQGIFQPQMQGQIE GMRAFALQALPAMCDVYVPPHCPVATAPLGGF	220	25,275
Avenin-E	Q09114	TTTVQYNPSEQYQPYPEQQEPFVQQQPFVQQQ QQPFVQQQQMFLQPLLQQQLNPCKQFLVQQC SPVAVVPFLRSQILRQAICQVARQQCCRQLAQI PEQLRCPAIHSVVQAIILQQQQQQQFFQPQLQ QQVFQPQLQQVFNQPQQQAQFEGMRAFALQ ALPAMCDVYVPPQCPVATAPLGGF	182	21,036
Avenin-F	Q09097	TTTVQYDPSEQYQPYPEQQEPFVQQQPPFVQQ QQPFVQQQEPF	43	5214
Avenin-A	Q09095	PSEQYQPYPEQQQPFLQQQPLELQQQQXXLVLFLQK	36	4393

Table 1. *Cont.*

Protein	UniProt ID	Sequence **	Amino Acid Length	Molecular Mass (Da)
Avenin	P27919	MKIFFFLALLALVVSATFAQYAESDGSYEEVEG SHDRCQQHQMKLDSCREYVAERCTTMRDFPIT WPWKWWKGGCEELRNECCQLLGQMPSECRC DAIWRSIQRELGGFFGTQQGLIGKRLKIAKSLPT QSTWALSAISPNSMVSHIAGKSSILRALPVDVL ANAYRISRQEARNLKNNRGQESGVFTPKFTQT SFQPYPEGEDESSLINKASE	214	24,230
Tryptophanin/2S albumin	A7U440	MKALFLLAFLALAASAAFAQQYADTGVGGW DGCMPEKARLNSCKDYVVERCLTLKDIPITWP WKWWKGGCESEVRSQCCMELNQIAPHCRCK AIWRAVQGELGGFLGFQQSEIMKQVHVAQSLP SRCNMGPNCNFPTNLGYY	142	15,901

** Amino acid nomenclature: A, ala, alanine; C, cys, cysteine; D, asp, aspartic acid; E, glu, glutamic acid; F, phe, phenylalanine; G, gly, glycine; H, his, histidine; I, Ile, isoleucine; K, lys, lysine; L, leu, leucine; M, met, methionine; N, asn, asparagine; P, pro, proline; Q, gln, glutamine; R, arg, arginine; S, ser, serine; T, thr, threonine; V, val, valine; W, trp, tryptophan; Y, tyr; tyrosine; X, undetermined amino acid. Protein sequences were obtained from the UniProt database, which is available at http://www.uniprot.org/.

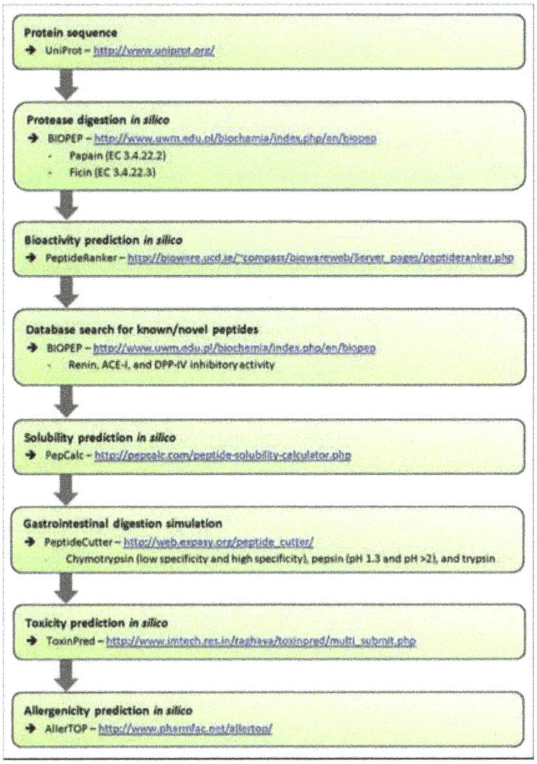

Figure 1. Methodology for in silico digestion and bioactivity prediction of oat protein hydrolysates.

2.4. *Bioactivity Prediction In Silico*

The peptides that resulted from oat protein hydrolysates were ranked for bioactivity according to their PeptideRanker score and known inhibitory peptide characteristics (Table 2), as previously described [36]. PeptideRanker, available at http://bioware.ucd.ie/~compass/biowareweb/Server_pages/peptideranker.php [37], is a server that predicts how likely a peptide is to be bioactive based on an N-to-1 neural network algorithm [37]. PeptideRanker predicts how likely peptides are to be bioactive, but does not indicate the targets for which they are most suitable. A literature search was therefore carried out to identify the characteristics of peptides that have been shown to increase the likelihood of inhibition with the enzyme targets in this study (Table 2).

Table 2. Characteristic criteria used to identify tripeptides with predicted renin, angiotensin-I-converting enzyme (ACE-I), and dipeptidyl peptidase-IV (DPP-IV) inhibition activity.

	Location	Group	Amino Acids	Reference
Renin inhibitory peptide characteristics	N1/*N*-terminus	Hydrophobic	Ala, Gly, Val, Leu, Ile, Pro, Phe, Met, Trp	[38]
	N2	N/A		
	N3/C-terminus	Bulky	Trp, Val, Ile, Leu, Tyr, Met, Phe,	[38]
ACE-I inhibitory peptide characteristics	N1/*N*-terminus	Hydrophobic (Small with low lipophilicity)	Val, Ile, Leu	[39]
	N2	Positively charged (Large with high lipophilicity & low electronic properties)	Leu, Arg	[39]
	N3/C-terminus	Aromatic acids (Small with low lipophilicity & high electronic properties)	Pro, Phe, Trp	[39]
		-	Leu	[40]
		Positively charged	Lys, Arg	[41]
		-	Pro	[41]
DPP-IV inhibitory peptide characteristics	N1/*N*-terminus	Hydrophobic or aromatic	Leu, Ile, Val, Phe, Trp, Try	[42]
	N2	N/A		
	N3/C-terminus	-	Pro, Ala	

Additional in silico analysis was carried out to predict water solubility, resistance to gastrointestinal digestion, toxicity, and allergenicity (Figure 1). Solubility in water was predicted using PepCalc, which is available at http://pepcalc.com. Resistance to digestion was predicted using PeptideCutter, which is available at http://web.expasy.org/peptide_cutter/ [36] with the enzymes' chymotrypsin-low specificity, chymotrypsin-high specificity, pepsin (pH 1.3), pepsin (pH > 2), and trypsin. Toxicity was scanned with default settings using ToxinPred, which is available at http://www.imtech.res.in/raghava/toxinpred/multi_submit.php [36]. Allergenicity was predicted using AllerTOP, which is available at http://www.pharmfac.net/allertop/ [36] (Figure 1).

2.5. *Chemical Synthesis of Peptides*

The selected peptides (FFG, IFFFL, PFL, WWK, WCY, FPIL, CPA, FLLA, FEPL) were synthesised by microwave-assisted solid phase peptide synthesis (MW-SPPS) performed on a Liberty microwave peptide synthesiser (Mathews, NC, USA). Peptides were synthesised on H-Ala-HMPB-ChemMatrix and H-Ile-HMPB-ChemMatrix resins (PCAS Biomatrix Inc., Quebec, QC, Canada) and purified using RP-HPLC on a Semi Preparative Jupiter Proteo (4u, 90A) column (Phenomenex, Cheshire, UK). Fractions containing the desired molecular mass were identified using matrix-assisted laser desorption ionisation time-of-flight mass spectrometry (MALDI-TOFMS) and were pooled and freeze-dried on a Genevac HT 4X lyophilizer (Genevac Ltd., Ipswich, UK).

2.6. *Renin Inhibition Assay*

Protein isolates from all seven oat varieties and selected synthesised peptides were tested in vitro for renin inhibition activity. The Renin Inhibition Screening Assay (Cambridge BioSciences, Cambridge, UK) was carried out as per the manufacturer's instructions. Briefly, 10 μL of each sample inhibitor at a concentration of 1 mg/mL dimethyl sulfoxide (DMSO) was added to 20 μL renin substrate, 150 μL assay buffer, and 10 μL renin, in triplicate. The samples were incubated at 37 °C for 15 min, and read with excitation wavelengths of 340 nm and emission wavelengths of 500 nm. Fluorescence was read using a FLUOstar Omega microplate reader (BMG LABTECH GmbH, Offenburg, Germany). The percentage inhibition was calculated using the following equation:

$$\%\ \text{Renin inhibition} = 100\%\ \text{Initial activity} - \text{Inhibitor} \times 100/100\%\ \text{Initial activity} \quad (1)$$

2.7. ACE-I Inhibition Assay

Protein isolates from all seven oat varieties and selected synthesised peptides were tested in vitro for ACE-I inhibition. The bioassay (ACE Kit—WST, Dojindo Laboratories, Kumamoto, Japan) was carried out according to the manufacturer's instructions. First, 20 μL of each sample inhibitor at a concentration of 1 mg/mL ddH_2O was added to 20 μL substrate and 20 μL enzyme working solution in triplicate. Samples were incubated at 37 °C for 1 h. Each well then had 200 μL indicator working solution added, followed by a further incubation at room temperature for 10 min. Absorbance at 450 nm was read using a FLUOstar Omega microplate reader (BMG LABTECH GmbH, Offenburg, Germany). The percentage inhibition was calculated using the following equation:

$$\text{\% ACE-I inhibition} = 100\% \text{ Initial activity} - \text{Inhibitor} \times 100/100\% \text{ Initial activity} \quad (2)$$

2.8. DPP-IV Inhibition Assay

Protein isolates from all seven oat varieties and selected synthesised peptides were tested in vitro for DPP-IV inhibition. The bioassay (DPP-IV Inhibitor Screening Assay Kit, Cayman Chemical, Ann Arbor, MI, USA) was carried out as per the manufacturer's instructions. First, 10 μL of each sample inhibitor at a concentration of 1 mg/mL assay buffer was added to 30 μL diluted assay buffer, 10 μL diluted DPP-IV, and 50 μL substrate solution, in triplicate. Samples were incubated at 37 °C for 30 min. Fluorescence was read with excitation wavelengths of 355 nm and emission wavelengths of 460 nm using a FLUOstar Omega microplate reader (BMG LABTECH GmbH, Offenburg, Germany). The percentage inhibition was calculated using the following equation:

$$\text{\% DPP-IV inhibition} = 100\% \text{ Initial activity} - \text{Inhibitor} \times 100/100\% \text{ Initial activity} \quad (3)$$

3. Results

3.1. In Silico Bioactivity Prediction

In silico analysis of oat protein isolates identified a number of bioactive peptides that had previously been reported in the BIOPEP database (Table 3). These peptides were not selected for chemical synthesis.

Table 3. Previously identified peptides from oat (*Avena sativa*) generated using in silico hydrolysis with papain or ficin.

Peptide	Peptide Ranker Score	BIOPEP ID	Activity Description	IC_{50}	Mass (Da)	Reference
FG	0.99	7605	ACE-I inhibitor	3700.0 μM	222.229	[43–46]
PF	0.99	8854	DPP-IV inhibitor	N/A	262.294	[47]
FL	0.99	8555	DPP-IV inhibitor	399.58 μM	278.337	[43]
FY	0.98	3556	ACE-I inhibitor	25 μM	328.347	[48]
FA	0.96	3176	DPP-IV inhibitor	N/A	236.256	[49]
FN	0.95	8778	DPP-IV inhibitor	N/A	279.281	[47]
MG	0.94	7609	ACE-I inhibitor	4800 μM	206.25	[46]
PG	0.88	7625	ACE-I inhibitor	17,000 μM	172.169	[46]
PG	0.88	8855	DPP-IV inhibitor	N/A	172.169	[47]
PG	0.88	3285	N/A	N/A	172.169	[50]
MR	0.85	8836	DPP-IV inhibitor	N/A	305.386	[47]
PL	0.81	7513	ACE-I inhibitor	337.32 μM	228.277	[51]
PL	0.81	8638	DPP-IV inhibitor	N/A	228.277	[52]
PLG	0.8	7510	ACE-I inhibitor	4.74 μM	285.329	[51]
PR	0.79	3537	ACE-I inhibitor	4.10 μM	271.305	[53]
RG	0.74	8882	DPP-IV inhibitor	N/A	231.24	[47]
LG	0.72	7619	ACE-I inhibitor	8800 μM	188.212	[46]
MA	0.69	3173	DPP-IV inhibitor	N/A	220.277	[49]
RL	0.63	8886	DPP-IV inhibitor	N/A	287.348	[47]

Based on the known characteristics of renin, ACE-I and DPP-IV inhibitory peptides (Table 2), novel peptides were identified for in vitro analysis in this study. The peptides chosen had the amino acid sequences FFG, IFFFL, PFL, WWK, WCY, FPIL, CPA, FLLA, and FEPL (Table 4). The tables predicted to have the greatest bioactivities are shown in Table A1 (Appendix A).

Table 4. Selected peptides and predicted solubility, resistance to digestion, toxicity, and allergenicity for chosen oat peptides.

Peptide	Solubility in Water	Resistance to Digestion	Toxicity	Allergenicity Probability
FFG	Poor	No	Non-toxin	Non-allergen
IFFFL	Poor	No	Non-toxin	Non-allergen
PFL	Poor	No	Non-toxin	Non-allergen
WWK	Good	No	Non-toxin	33.3%
WCY	Poor	No	Non-toxin	33.3%
FPIL	Poor	No	Non-toxin	Non-allergen
CPA	Poor	Yes	Non-toxin	Non-allergen
FLLA	Poor	No	Non-toxin	Non-allergen
FEPL	Good	No	Non-toxin	Non-allergen

Bioactive peptides need to survive degradation by gastrointestinal digestion, reach their target intact, and maintain bioavailability in order to exert a beneficial physiological effect [44]. The novel peptides were therefore analysed in silico for predicted solubility, resistance to digestion, toxicity, and allergenicity (Table 4). Most of the peptides were expected to be poorly soluble due to their high hydrophobic residue content. Most of the peptides were also expected to be broken down by gastrointestinal digestive enzymes, although this could be overcome by methods such as encapsulation [45]. The peptides were also expected to be non-toxic and non-allergenic (Table 4).

3.2. Renin Inhibition

Oat protein isolates displayed renin inhibition values ranging between 40.5% (±2.16%) and 70.9% (±7.7%) (Figure 2). Barra oat had the highest levels of renin inhibition, with 70.9% (±7.7%) inhibition, followed by Vodka oat, which had 66.1% (±22.1%) inhibition. There was significantly lower renin inhibition activity from synthesised peptides compared with that of the oat protein isolates.

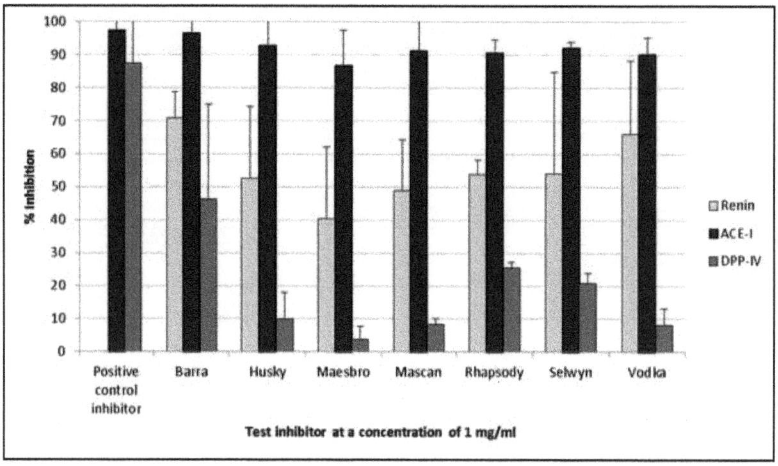

Figure 2. Inhibition bioactivity for the targets renin, angiotensin-I-converting enzyme (ACE-I), and dipeptidyl peptidase-IV (DPP-IV) using protein isolates from seven oat varieties at a test concentration of 1 mg of extract (dry weight)/mL. Values are the mean of triplicate samples. Captopril was used as the positive control for ACE-I inhibition, while sitagliptin was used as the positive control for DPP-IV inhibition.

However, renin inhibition was not observed with the synthesised peptides. Only the peptide IFFFL was found to inhibit renin by 17.1% (±2.6%) when tested at a concentration of 1 mg/mL compared with the control.

3.3. ACE-I Inhibition

The crude oat protein extracts inhibited ACE-I by between 86.6% (±10.7%) and 96.5% (±25.8%). These results were therefore comparable with that of the positive control captopril, which was found to inhibit ACE-I by 97.7% (±23.2%) when tested at a concentration of 1 mg/mL (Figure 2). Protein extracted from Barra oat varieties displayed the highest ACE-I inhibition values, and inhibited the enzyme by 96.5% (±25.8%) when assayed at a concentration of 1 mg/mL compared with the positive control.

The highest levels of ACE-I inhibition with the synthesised peptides were observed for the peptides WCY (97.8 ± 21.7%), FLLA (97 ± 16.2%), and WWK (95.3 ± 14.2%) (Figure 3). The lowest levels of inhibition were seen with IFFFL (53 ± 41.2%) and FEPL (48.9 ± 7.8%), contradicting previous findings of the beneficial effect of leucine at the C-terminus of an ACE-I inhibiting bioactive peptide [40]. However, the poor activity from IFFFL and FEPL could also be due to the larger peptide size as is the case with the peptide PFL, which also has a C-terminal leucine, and displayed higher levels of ACE-I inhibition (81.4 ± 33.8%). Alternatively, the lower ACE-I inhibitory values observed for FEPL could also be due to the presence of proline at the penultimate position within the peptide, which has been suggested to reduce the binding affinity of peptides with the ACE-I enzyme [46].

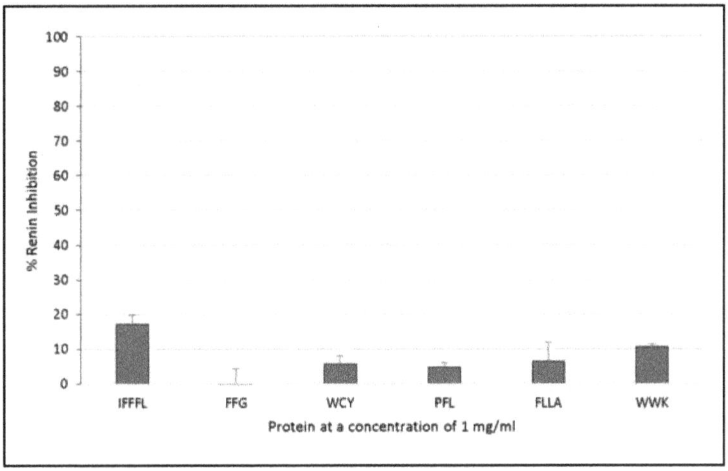

Figure 3. Renin inhibition assay of several chemically synthesised peptides from predicted oat protein hydrolysates following in silico digestion with the proteases papain and ficin. Peptides were assayed at a concentration of 1 mg protein/mL. Values are the mean of triplicate samples.

3.4. DPP-IV Inhibition

The protein isolates of the seven oat varieties did not display significant DPP-IV inhibitory activities when assayed at a concentration of 1 mg/mL, and values ranged from 3.7% (±3.9%) to 46.3% (±8.8%) compared with the positive control sitagliptin (87.4 ± 4%) (Figure 2). The highest DPP-IV inhibition activity was seen with the protein extracts of Barra oat (46.3 ± 8.8%), Rhapsody oat (25.4 ± 1.8%), and Selwyn oat (20.7 ± 3.1%).

Peptides containing the sequence Xaa-Pro (where Xaa represents any amino acid, and proline is present at the second residue from the N-terminus) have been found to be effective DPP-IV inhibitors [43]. The peptides CPA (22.2 ± 4.8%) and FPIL (13.1 ± 3.2%) were the only chemically synthesised peptides that were found to inhibit DPP-IV at a concentration of 1 mg/mL.

4. Discussion

Functional foods in the form of bioactive peptides offer additional physiological benefits beyond their basic nutritional value. Bioactive peptides that inhibit the enzymes within RAAS, such as renin and ACE-I, are used as alternatives to antihypertensive pharmaceutical drugs [7]. Similarly, DPP-IV inhibitory bioactive peptides have also been shown to effectively prevent the onset of type-2 diabetes by preventing the cleavage of the glucagon-like peptide 1 (GLP-1) and glucose-dependent insulinotropic peptide (GIP) incretins [43]. Bioactive peptides have been identified from a variety of food sources, including milk proteins, seaweed, and meat [21,28,30], as well as a number of grains, including rice, soybean, wheat, and barley [22–24,54]. This study determined the ACE-I, renin, and DPP-IV inhibitory activities of oat protein isolates and peptides synthesised in vitro.

While there are pharmaceutical therapies for hypertension and type-2 diabetes, they are often accompanied by adverse side effects, such as a dry cough, anaphylaxis, renal impairment, hyperkalaemia, and inflammation-related pain [55–57]. Pharmaceuticals have several reported side effects, but are active at lower concentrations than food-derived bioactive peptides. However, peptides consumed with IC_{50} values of 100–500 μM have been shown to be active in vivo and inhibit ACE-I, renin, and DPP-IV enzymes [58,59]. Furthermore, functional foods with bioactivities can be a beneficial alternative to synthetic drugs for individuals who have borderline disease states and do not warrant the prescription of pharmaceutical drugs [6].

Oat protein isolates were found to inhibit renin by between 40.5% (±2.16%) and 70.9% (±7.7%) (Figure 2) in vitro when assayed at a concentration of 1 mg/mL protein, but the selected synthesised peptides did not inhibit renin to the same degree (Figure 3). This demonstrates that it is necessary to carry out in vitro work and characterise all of the peptides present in a hydrolysate. The observed activities could also be due to other compounds present in the protein isolate, such as phenolic compounds, and phenolic compounds may still be present despite the use of dialysis to concentrate the protein fraction. The active peptide(s) could have been previously identified (Table 4) or were not chosen for synthesis. The characteristics of renin inhibitory peptides are not as well defined as other bioactive peptide targets (such as ACE-I inhibitory peptides, for example) due to the notably poor potency that has been observed with renin peptide inhibitors [60]. The peptides IFFFL, FLLA, and WWK (Figure 3) were chosen for chemical synthesis based on previously published literature by Udenigwe and colleagues (2012), who observed that the bulky amino acids at the *N*-terminus aided in renin inhibition (Table 3).

The oat protein isolates generated in this work were all found to inhibit the enzyme ACE-I by between 86.6% (±10.7%) and 96.5% (±25.8%) (Figure 2). This was significantly greater than a similar study assessing crude protein extracts derived from barley [54]. Unlike the renin assay, there were comparable levels of activity in the selected synthesised peptides and the oat protein extracts (Figure 4). This suggests that the correct peptides were selected for chemical synthesis, which is largely because the mechanism of the action of ACE-I inhibitory peptides is better understood.

The Barra oat protein isolate had the highest activity with 46.3% (±28.8%) inhibition, followed by Rhapsody oat (25.4 ± 1.8%) and Selwyn oat (20.7 ± 3.1%). Similarly, the selected synthesised peptides also had relatively poor DPP-IV inhibition (Figure 5). DPP-IV is a proline-specific endopeptidase that cleaves dipeptides from the *N*-terminus [61]. Of the nine peptides synthesised (Table 4), those containing a proline residue were therefore selected for testing DPP-IV inhibition.

In silico methods were used heavily in this study to evaluate the potential of oat protein to generate bioactive peptides, as well as predict generated bioactive peptides following hydrolysis with the food-grade proteases papain and ficin. Similar bioinformatic techniques have already been described in other studies [36], which highlighted their value in reducing time and expense for the preliminary screening of novel sources of bioactive peptides. The preparatory in silico screening of potential bioactive peptides for characteristics such as allergenicity, toxicity, solubility, and degradation are additional important uses for such online software and tools, especially when presenting these peptides for possible human consumption.

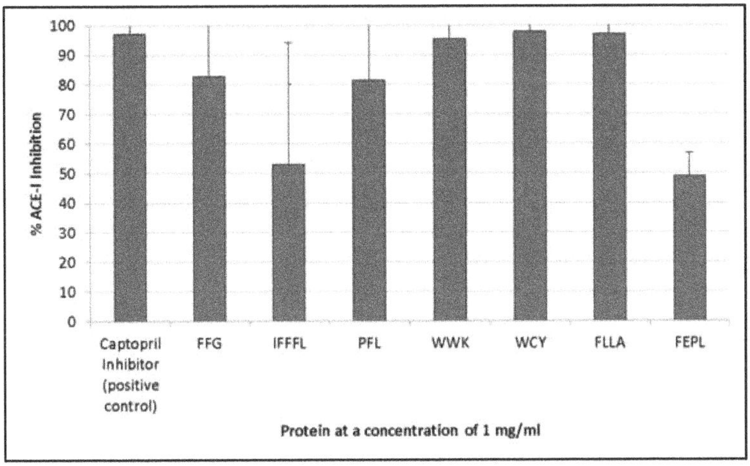

Figure 4. ACE-I inhibition assay of several synthesised peptides that were identified in silico from the digestion of oat proteins with papain or ficin, which are both proteases. The activity of peptides was compared with that of a captopril inhibitor (positive control). Peptides were assayed at a concentration of 1 mg protein of extract/mL. Values are the mean of triplicate samples.

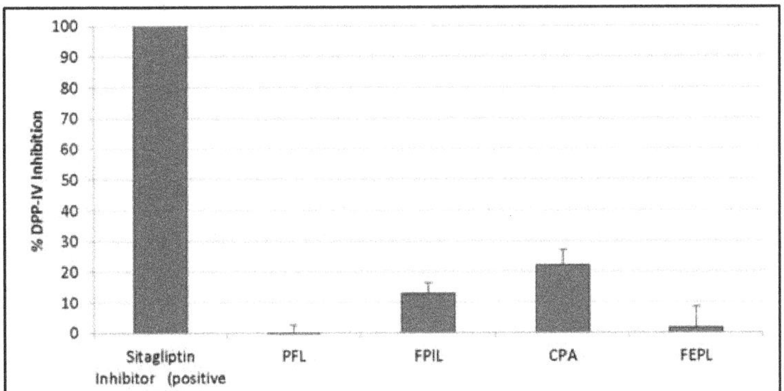

Figure 5. DPP-IV inhibition of several synthesised peptides that were identified in silico from the hydrolysis of oat proteins with papain or ficin, which are both proteases. The activity of peptides was compared with that of a sitagliptin inhibitor (positive control). Peptides were assayed at a concentration of 1 mg/mL. Values are the mean of triplicate samples.

5. Conclusions

There has been little work carried out to evaluate the potential of the common oat (*Avena sativa*) as a potential source of bioactive peptides. Bioinformatics techniques were used to predict whether oats were indeed a rich source of bioactive peptides following in silico hydrolysis with the proteases papain and ficin on the main storage proteins. Oat protein isolates displayed the highest inhibition bioactivity against the target ACE-I (86.6–96.5%), with lower inhibition levels observed with renin (40.5–70.9%) and DPP-IV (3.7–46.3%). Following the chemical synthesis of nine novel peptides, in vitro bioassays gave mixed results as to their efficacy in inhibiting the injurious enzyme targets ACE-I (48.9–97.8%), renin (0–17.1%), and DPP-IV (0–22.2%). The in silico methods utilised in this study correctly identified ACE-I inhibitory peptides, beyond that of renin and DPP-IV inhibitory peptides. This could reflect a more specific use of these in silico methods until the characteristics of renin and DPP-IV inhibitory peptides are better understood.

Acknowledgments: Stephen Bleakley is in receipt of a Teagasc Walsh Fellowship (Grant No. 2016073). This work forms part of the NutriCerealsIreland project funded under the Food Institutional Research Measure (FIRM 11/SF/317) granted by the Irish Department of Agriculture, Food and Marine (DAFM). Tomas Lafarga is in receipt of a Juan de la Cierva contract awarded by the Spanish Ministry of Economy, Industry, and Competitiveness (FJCI-2016-29541).

Author Contributions: M.H. and T.L. conceived and designed the experiments; S.B. and M.H. performed the experiments; M.H. and T.L. analysed the data; N.O.S. and E.G. contributed reagents/materials/analysis tools; S.B. and M.H. wrote the paper.

Conflicts of Interest: The authors declare no conflict of interest.

Appendix A

Table A1. Top 15 novel peptide hydrolysates identified following in silico hydrolysis of oat proteins with proteases papain and ficin that were predicted to have high bioactivity for the inhibition of ACE-I and DPP-IV.

Peptide	Predicted Activity	Peptide Ranker Score	Protein	Enzyme	Location	Mass (Da)
FFG *	ACE-I	1	Avenin	Papain, Ficin	f(109–111)	369.45
IFFFL *	ACE-I	0.99	Avenin	Ficin	f(3–7)	685.93
WCY *	ACE-I	0.98	11S globulin/12S globulin	Papain, Ficin/Papain	f(172–174)/f(172–174)	470.57
PFL *	ACE-I	0.98	Avenin-3/Avenin-E	Ficin/Ficin	f(84–86)/f(70–72)	375.5
WWK *	ACE-I	0.98	Avenin/Tryptophanins	Papain, Ficin/Papain, Ficin	f(70–72)/f(65–67)	518.65
IFFFLA	ACE-I	0.95	Avenin	Papain	f(3–8)	757.02
FPIL *	DPP-IV, ACE-I	0.92	11S globulin	Ficin	f(364–367)	488.68
IWR	ACE-I	0.91	Avenin/Tryptophanins	Papain/Papain	f(98–100)/f(94–96)	473.61
FPTL	DPP-IV,	0.84	12S globulin	Ficin	f(356–359)	476.62
PFV	DPP	0.82	12S globulin	Ficin	f(264–266)	361.47
CPA *	DPP-IV	0.79	11S globulin/12S globulin/Avenin-3/Avenin-E	Papain, Ficin/Papain, Ficin/Papain/Papain	f(121–123)/f(121–123)/f(116–118)/f(102–104)	289.37
FLLA *	DPP	0.79	11S globulin/12S globulin	Papain/Papain	f(203–206)/f(203–206)	462.62
FEPL *	ACE-I	0.75	12S globulin	Papain	f(53–56)	504.63
PLR	ACE-I	0.74	11S globulin/12S globulin	Papain/Papain	f(55–57)	384.51

* Peptides that were chosen for chemical synthesis and tested with in vitro assays.

References

1. WHO. *Global Health Risks: Mortality and Burden of Disease Attributable to Selected Major Risks*; World Health Organization: Geneva, Switzerland, 2009.
2. Heidenreich, P.A.; Trogdon, J.G.; Khavjou, O.A.; Butler, J.; Dracup, K.; Ezekowitz, M.D.; Finkelstein, E.A.; Hong, Y.; Johnston, S.C.; Khera, A.; et al. Forecasting the future of cardiovascular disease in the united states: A policy statement from the American heart association. *Circulation* **2011**, *123*, 933–944. [CrossRef] [PubMed]
3. Ondetti, M.A.; Cushman, D.W. Enzymes of the renin-angiotensin system and their inhibitors. *Annu. Rev. Biochem.* **1982**, *51*, 283–308. [CrossRef] [PubMed]
4. Turk, B. Targeting proteases: Successes, failures and future prospects. *Nat. Rev. Drug Discov.* **2006**, *5*, 785–799. [CrossRef] [PubMed]

5. Fitzgerald, C.; Gallagher, E.; Tasdemir, D.; Hayes, M. Heart health peptides from macroalgae and their potential use in functional foods. *J. Agric. Food Chem.* **2011**, *59*, 6829–6836. [CrossRef] [PubMed]
6. Chen, Z.-Y.; Peng, C.; Jiao, R.; Wong, Y.M.; Yang, N.; Huang, Y. Anti-hypertensive nutraceuticals and functional foods. *J. Agric. Food Chem.* **2009**, *57*, 4485–4499. [CrossRef] [PubMed]
7. Korhonen, H.; Pihlanto, A. Bioactive peptides: Production and functionality. *Int. Dairy J.* **2006**, *16*, 945–960. [CrossRef]
8. Clare, D.A.; Swaisgood, H.E. Bioactive milk peptides: A prospectus. *J. Dairy Sci.* **2000**, *83*, 1187–1195. [CrossRef]
9. Nakamura, Y.; Yamamoto, N.; Sakai, K.; Okubo, A.; Yamazaki, S.; Takano, T. Purification and characterization of angiotensin I-converting enzyme inhibitors from sour milk. *J. Dairy Sci.* **1995**, *78*, 777–783. [CrossRef]
10. Hatanaka, T.; Inoue, Y.; Arima, J.; Kumagai, Y.; Usuki, H.; Kawakami, K.; Kimura, M.; Mukaihara, T. Production of dipeptidyl peptidase iv inhibitory peptides from defatted rice bran. *Food Chem.* **2012**, *134*, 797–802. [CrossRef] [PubMed]
11. Wilson, J.; Hayes, M.; Carney, B. Angiotensin-I-converting enzyme and prolyl endopeptidase inhibitory peptides from natural sources with a focus on marine processing by-products. *Food Chem.* **2011**, *129*, 235–244. [CrossRef]
12. Pellegrini, A. Antimicrobial peptides from food proteins. *Curr. Pharm. Des.* **2003**, *9*, 1225–1238. [CrossRef] [PubMed]
13. Chabance, B.; Marteau, P.; Rambaud, J.C.; Migliore-Samour, D.; Boynard, M.; Perrotin, P.; Guillet, R.; Jollès, P.; Fiat, A.M. Casein peptide release and passage to the blood in humans during digestion of milk or yogurt. *Biochimie* **1998**, *80*, 155–165. [CrossRef]
14. Suetsuna, K.; Ukeda, H.; Ochi, H. Isolation and characterization of free radical scavenging activities peptides derived from casein. *J. Nutr. Biochem.* **2000**, *11*, 128–131. [CrossRef]
15. Nagaoka, S.; Futamura, Y.; Miwa, K.; Awano, T.; Yamauchi, K.; Kanamaru, Y.; Tadashi, K.; Kuwata, T. Identification of novel hypocholesterolemic peptides derived from bovine milk beta-lactoglobulin. *Biochem. Biophys. Res. Commun.* **2001**, *281*, 11–17. [CrossRef] [PubMed]
16. Shimizu, M. Food-derived peptides and intestinal functions. *Biofactors* **2004**, *21*, 43–47. [CrossRef] [PubMed]
17. Yvon, M.; Beucher, S.; Guilloteau, P.; Le Huerou-Luron, I.; Corring, T. Effects of caseinomacropeptide (cmp) on digestion regulation. *Reprod. Nutr. Dev.* **1994**, *34*, 527–537. [CrossRef] [PubMed]
18. Teschemacher, H. Opioid receptor ligands derived from food proteins. *Curr. Pharm. Des.* **2003**, *9*, 1331–1344. [CrossRef] [PubMed]
19. Gill, H.S.; Doull, F.; Rutherfurd, K.J.; Cross, M.L. Immunoregulatory peptides in bovine milk. *Br. J. Nutr.* **2000**, *84* (Suppl. S1), S111–S117. [CrossRef]
20. Meisel, H.; FitzGerald, R.J. Biofunctional peptides from milk proteins: Mineral binding and cytomodulatory effects. *Curr. Pharm. Des.* **2003**, *9*, 1289–1295. [PubMed]
21. Saito, T. Antihypertensive peptides derived from bovine casein and whey proteins. In *Bioactive Components of Milk*; Bösze, Z., Ed.; Springer: New York, NY, USA, 2008; pp. 295–317.
22. Li, G.H.; Qu, M.R.; Wan, J.Z.; You, J.M. Antihypertensive effect of rice protein hydrolysate with in vitro Angiotensin-I-converting enzyme inhibitory activity in spontaneously hypertensive rats. *Asia Pac. J. Clin. Nutr.* **2007**, *16* (Suppl. S1), 275–280.
23. Rho, S.J.; Lee, J.S.; Chung, Y.I.; Kim, Y.W.; Lee, H.G. Purification and identification of an Angiotensin I-converting enzyme inhibitory peptide from fermented soybean extract. *Process Biochem.* **2009**, *44*, 490. [CrossRef]
24. Motoi, H.; Kodama, T. Isolation and characterization of angiotensin I-converting enzyme inhibitory peptides from wheat gliadin hydrolysate. *Nahrung* **2003**, *47*, 354–358. [CrossRef] [PubMed]
25. Suetsuna, K.; Nakano, T. Identification of an antihypertensive peptide from peptic digest of wakame (*Undaria pinnatifida*). *J. Nutr. Biochem.* **2000**, *11*, 450–454. [CrossRef]
26. Aluko, R.E. Determination of nutritional and bioactive properties of peptides in enzymatic pea, chickpea, and mung bean protein hydrolysates. *J. AOAC Int.* **2008**, *91*, 947–956. [PubMed]
27. Lee, J.-E.; Bae, I.Y.; Lee, H.G.; Yang, C.-B. Tyr-Pro-Lys, an Angiotensin-I-converting enzyme inhibitory peptide derived from Broccoli (*Brassica oleracea italica*). *Food Chem.* **2006**, *99*, 143–148. [CrossRef]

28. Suetsuna, K. Isolation and characterization of Angiotensin-I-converting enzyme inhibitor dipeptides derived from *Allium sativum* L (garlic) isolation and characterization of angiotensin I-converting enzyme inhibitor dipeptides derived from. *J. Nutr. Biochem.* **1998**, *9*, 415. [CrossRef]
29. Miguel, M.; Aleixandre, A. Antihypertensive peptides derived from egg proteins. *J. Nutr.* **2006**, *136*, 1457–1460. [PubMed]
30. Vercruysse, L.; Van Camp, J.; Smagghe, G. Ace inhibitory peptides derived from enzymatic hydrolysates of animal muscle protein: A review. *J. Agric. Food Chem.* **2005**, *53*, 8106–8115. [CrossRef] [PubMed]
31. Yu, Y.; Hu, J.; Miyaguchi, Y.; Bai, X.; Du, Y.; Lin, B. Isolation and characterization of angiotensin i-converting enzyme inhibitory peptides derived from porcine hemoglobin. *Peptides* **2006**, *27*, 2950–2956. [CrossRef] [PubMed]
32. Jung, W.K.; Mendis, E.; Je, J.Y.; Park, P.J.; Son, B.W.; Kim, H.C. Angiotensin-I-converting enzyme inhibitory peptide from yellowfin sole (*Limanda aspera*) frame protein and its antihypertensive effect in spontaneously hypertensive rats Angiotensin I-converting enzyme inhibitory peptide from yellowfin sole (*Limanda aspera*) frame protein and its antihypertensive effect in spontaneously hypertensive rats. *Food Chem.* **2006**, *94*, 26.
33. Cavazos, A.; Gonzalez de Mejia, E. Identification of bioactive peptides from cereal storage proteins and their potential role in prevention of chronic diseases. *Compr. Rev. Food Sci. Food Saf.* **2013**, *12*, 364–380. [CrossRef]
34. Biel, W.; Bobko, K.; Maciorowski, R. Chemical composition and nutritive value of husked and naked oats grain. *J. Cereal Sci.* **2009**, *49*, 413–418. [CrossRef]
35. Minkiewicz, P.; Dziuba, J.; Iwaniak, A.; Dziuba, M.; Darewicz, M. Biopep database and other programs for processing bioactive peptide sequences. *J. AOAC Int.* **2008**, *91*, 965–980. [PubMed]
36. Lafarga, T.; O'Connor, P.; Hayes, M. Identification of novel dipeptidyl peptidase-IV and angiotensin-I-converting enzyme inhibitory peptides from meat proteins using in silico analysis. *Peptides* **2014**, *59*, 53–62. [CrossRef] [PubMed]
37. Mooney, C.; Haslam, N.J.; Pollastri, G.; Shields, D.C. Towards the improved discovery and design of functional peptides: Common features of diverse classes permit generalized prediction of bioactivity. *PLoS ONE* **2012**, *7*, e45012. [CrossRef] [PubMed]
38. Li, H.; Aluko, R.E. Identification and inhibitory properties of multifunctional peptides from pea protein hydrolysate. *J. Agric. Food Chem.* **2010**, *58*, 11471–11476. [CrossRef] [PubMed]
39. Wu, J.; Aluko, R.E.; Nakai, S. Structural requirements of angiotensin-I-converting enzyme inhibitory peptides: Quantitative structure-activity relationship study of di- and tripeptides. *J. Agric. Food Chem.* **2006**, *54*, 732–738. [CrossRef] [PubMed]
40. Ruiz, J.Á.G.; Ramos, M.; Recio, I. Angiotensin converting enzyme-inhibitory activity of peptides isolated from Manchego cheese. Stability under simulated gastrointestinal digestion. *Int. Dairy J.* **2004**, *14*, 1075–1080.
41. Ondetti, M.A.; Rubin, B.; Cushman, D.W. Design of specific inhibitors of Angiotensin-converting enzyme: New class of orally active antihypertensive agents. *Science* **1977**, *196*, 441–444. [CrossRef] [PubMed]
42. Nongonierma, A.B.; Mooney, C.; Shields, D.C.; FitzGerald, R.J. In silico approaches to predict the potential of milk protein-derived peptides as dipeptidyl peptidase-IV (DPP-IV) inhibitors. *Peptides* **2014**, *57*, 43–51. [CrossRef] [PubMed]
43. Nongonierma, A.B.; FitzGerald, R.J. Inhibition of dipeptidyl peptidase IV (DPP-IV) by proline containing casein-derived peptides. *J. Funct. Foods* **2013**, *5*, 1909–1917. [CrossRef]
44. Picariello, G.; Ferranti, P.; Fierro, O.; Mamone, G.; Caira, S.; Di Luccia, A.; Monica, S.; Addeo, F. Peptides surviving the simulated gastrointestinal digestion of milk proteins: Biological and toxicological implications. *J. Chromatogr. B* **2010**, *878*, 295–308. [CrossRef] [PubMed]
45. Homayouni, A.; Azizi, A.; Ehsani, M.; Yarmand, M.; Razavi, S. Effect of microencapsulation and resistant starch on the probiotic survival and sensory properties of synbiotic ice cream. *Food Chem.* **2008**, *111*, 50–55. [CrossRef]
46. Cheung, H.S.; Wang, F.L.; Ondetti, M.A.; Sabo, E.F.; Cushman, D.W. Binding of peptide substrates and inhibitors of angiotensin-converting enzyme. Importance of the COOH-terminal dipeptide sequence. *J. Biol. Chem.* **1980**, *255*, 401–407. [PubMed]
47. Lan, V.T.; Ito, K.; Ohno, M.; Motoyama, T.; Ito, S.; Kawarasaki, Y. Analyzing a dipeptide library to identify human dipeptidyl peptidase iv inhibitor. *Food Chem.* **2015**, *175*, 66–73. [CrossRef] [PubMed]
48. Yano, S.; Suzuki, K.; Funatsu, G. Isolation from alpha-zein of thermolysin peptides with Angiotensin-I-converting enzyme inhibitory activity. *Biosci. Biotechnol. Biochem.* **1996**, *60*, 661–663. [CrossRef] [PubMed]

49. Bella, A.M.; Erickson, R.H.; Kim, Y.S. Rat intestinal brush border membrane Dipeptidyl aminopeptidase IV: Kinetic properties and substrate specificities of the purified enzyme. *Arch. Biochem. Biophys.* **1982**, *218*, 156–162. [CrossRef]
50. Ashmarin, I.; Karazeeva, E.; Lyapina, L.; Samonina, G. The simplest proline-containing peptides PG, GP, PGP, and GPGG: Regulatory activity and possible sources of biosynthesis. *BioChemistry* **1998**, *63*, 119–124. [PubMed]
51. Byun, H.-G.; Kim, S.-K. Structure and activity of angiotensin I converting enzyme inhibitory peptides derived from Alaskan pollack skin. *J. Biochem. Mol. Biol.* **2002**, *35*, 239–243. [CrossRef] [PubMed]
52. Gallego, M.; Aristoy, M.-C.; Toldrá, F. Dipeptidyl peptidase IV inhibitory peptides generated in Spanish dry-cured ham. *Meat Sci.* **2014**, *96*, 757–761. [CrossRef] [PubMed]
53. Saito, Y.; Wanezaki, K.; Kawato, A.; Imayasu, S. Structure and activity of Angiotensin I converting enzyme inhibitory peptides from sake and sake lees. *Biosci. Biotechnol. Biochem.* **1994**, *58*, 1767–1771. [CrossRef] [PubMed]
54. Gangopadhyay, N.; Wynne, K.; O'Connor, P.; Gallagher, E.; Brunton, N.P.; Rai, D.K.; Hayes, M. In silico and in vitro analyses of the angiotensin-I-converting enzyme inhibitory activity of hydrolysates generated from crude barley (*Hordeum vulgare*) protein concentrates. *Food Chem.* **2016**, *203*, 367–374. [CrossRef] [PubMed]
55. Sica, D.A.; Gehr, T. Angiotensin-Converting Enzyme Inhibitors. In *Hypertension*; Elsevier-Saunders: Philadelphia, PA, USA, 2005; pp. 669–682.
56. Fein, A. ACE inhibitors worsen inflammatory pain. *Med. Hypotheses* **2009**, *72*, 757. [CrossRef] [PubMed]
57. Sidorenkov, G.; Navis, G. Safety of ace inhibitor therapies in patients with chronic kidney disease. *Expert Opin. Drug Saf.* **2014**, *13*, 1383–1395. [CrossRef] [PubMed]
58. Li, G.-H.; Le, G.-W.; Shi, Y.-H.; Shrestha, S. Angiotensin-I–converting enzyme inhibitory peptides derived from food proteins and their physiological and pharmacological effects. *Nutr. Res.* **2004**, *24*, 469–486. [CrossRef]
59. FitzGerald, R.J.; Meisel, H. Milk protein-derived peptide inhibitors of angiotensin-I-converting enzyme. *Br. J. Nutr.* **2000**, *84*, 33–37. [CrossRef]
60. Fisher, N.D.; Hollenberg, N.K. Renin inhibition: What are the therapeutic opportunities? *J. Am. Soc. Nephrol.* **2005**, *16*, 592–599. [CrossRef] [PubMed]
61. Lawandi, J.; Gerber-Lemaire, S.; Juillerat-Jeanneret, L.; Moitessier, N. Inhibitors of prolyl oligopeptidases for the therapy of human diseases: Defining diseases and inhibitors. *J. Med. Chem.* **2010**, *53*, 3423–3438. [CrossRef] [PubMed]

foods

MDPI

Review

Future Protein Supply and Demand: Strategies and Factors Influencing a Sustainable Equilibrium

Maeve Henchion [1,*], Maria Hayes [2], Anne Maria Mullen [3], Mark Fenelon [4] and Brijesh Tiwari [5]

1. Department Agri-Food Business and Spatial Analysis, Rural Economy and Development Programme, Teagasc Food Research Centre, Ashtown, Dublin D15 KN3K, Ireland
2. Food BioSciences Department, Teagasc Food Research Centre, Ashtown, Dublin D15 KN3K, Ireland; Maria.Hayes@teagasc.ie
3. Food Quality and Sensory Science, Teagasc Food Research Centre, Ashtown, Dublin D15 KN3K, Ireland; Anne.Mullen@teagasc.ie
4. Teagasc Food Research Programme, Teagasc Food Research Centres, Ashtown and Moorepark, Fermoy, Co. Cork P61 C996, Ireland; Mark.Fenelon@teagasc.ie
5. Food Chemistry and Technology Department, Teagasc Food Research Centre, Ashtown, Dublin D15 KN3K, Ireland; Brijesh.Tiwari@teagasc.ie

* Correspondence: Maeve.Henchion@teagasc.ie; Tel.: +353-1-805-9515

Received: 28 June 2017; Accepted: 13 July 2017; Published: 20 July 2017

Abstract: A growing global population, combined with factors such as changing socio-demographics, will place increased pressure on the world's resources to provide not only more but also different types of food. Increased demand for animal-based protein in particular is expected to have a negative environmental impact, generating greenhouse gas emissions, requiring more water and more land. Addressing this "perfect storm" will necessitate more sustainable production of existing sources of protein as well as alternative sources for direct human consumption. This paper outlines some potential demand scenarios and provides an overview of selected existing and novel protein sources in terms of their potential to sustainably deliver protein for the future, considering drivers and challenges relating to nutritional, environmental, and technological and market/consumer domains. It concludes that different factors influence the potential of existing and novel sources. Existing protein sources are primarily hindered by their negative environmental impacts with some concerns around health. However, they offer social and economic benefits, and have a high level of consumer acceptance. Furthermore, recent research emphasizes the role of livestock as part of the solution to greenhouse gas emissions, and indicates that animal-based protein has an important role as part of a sustainable diet and as a contributor to food security. Novel proteins require the development of new value chains, and attention to issues such as production costs, food safety, scalability and consumer acceptance. Furthermore, positive environmental impacts cannot be assumed with novel protein sources and care must be taken to ensure that comparisons between novel and existing protein sources are valid. Greater alignment of political forces, and the involvement of wider stakeholders in a governance role, as well as development/commercialization role, is required to address both sources of protein and ensure food security.

Keywords: protein; novel protein; protein demand; in vitro meat; algae; insect; dairy; meat; vegetal; consumer

1. Introduction

UN figures projecting global population growth of almost 50% since 2000 to 9.5 billion by 2050 [1] seem to be generally accepted. In addition to giving rise to an increased demand for food as a result of more mouths to feed, other changes, such as increased incomes and urbanization, will result in changes in consumption patterns. Thus, not only will the amount of food required change, the type of foods demanded, and their relative contribution to diets, will change. Projected demand for protein is of particular interest, with projections that the world demand for animal-derived protein will double by 2050 [2], resulting in concerns for sustainability and food security. In part, this is because it is generally accepted that animal-based foods produce higher levels of greenhouse gases (GHG) than plant-based foods and these are associated with climate change [3]. This is compounded by the fact that increased demand for animal-based protein is expected to intensify pressure on land due to the need to produce more animal feed. This in turn will increase the conversion of forests, wetlands and natural grasslands into agricultural lands, which in itself has negative consequences for GHG emissions, biodiversity and other important ecosystem services [4]. This paper looks at a number of possible scenarios relating to future protein demand for human consumption, discusses factors influencing this demand and outlines some strategies to respond to the expected increased demand. It considers implications and opportunities for current and novel/emerging animal and plant-based protein sources. It concludes by discussing implications for future research, the agri-food industry and related stakeholders.

2. Protein Demand

In addition to increased demand arising from population growth, increased demand for protein globally is driven by socio-economic changes such as rising incomes, increased urbanisation, and aging populations whereby the contribution of protein to healthy aging is increasingly recognised [5,6], and recognition of the role of protein in a healthy diet. Economic development and increased urbanization is leading to major transitions in population-level dietary patterns in low and middle income countries in particular, such that most of the global increases in demand for foods of animal origin are seen in developing countries [6]. Some forces, however, provide a countervailing force slowing demand in developed countries. Such factors include increased awareness of the impact of food production and consumption on the environment and on health. In the context of protein, the negative impact is mainly associated with animal-derived protein with reports that 12% of GHG emissions derive from livestock production and that 30% of human-induced terrestrial biodiversity loss can be attributed to animal production [2]. Land use is also a concern; for example, in the EU two thirds of total agricultural area is used for livestock production and around 75% of protein-rich animal feed is imported from South America using large tracts of land there also [2]. Health concerns arise with over-consumption of protein, particularly when linked with saturated fatty acids and over consumption of processed meats. Ethical issues about animal production could also stifle demand with a trend towards flexitarianism and initiatives aimed at reducing meat consumption evident in some markets.

Figure 1 presents trends in protein consumption per capita between 1961 and 2011 for developed and developing countries. Differential levels are evident with developed countries showing higher absolute levels of consumption but developing countries showing significant growth rates over the period. Current protein demand for the 7.3 billion inhabitants of the world is approximately 202 million tonnes globally. However, even accepting a 2.3 billion growth in population, vastly different outcomes in terms of demand for protein result depending on assumptions made about average consumption for the future. Table 1 shows protein demand requirements for a range of possible scenarios. The scenarios presented are somewhat simplistic in that they are based on current knowledge about consumption levels in developed and developing countries as well as prevailing knowledge regarding the amount of protein required to meet basic nutritional needs. Developing detailed projections about future protein demand is beyond the scope of this paper. Nonetheless, the scenarios presented serve to illustrate the impact of variation in per capita consumption on projected demand, independent of population growth, and thus suggest the opportunity for demand side responses as a complement to supply side initiatives. Scenarios 1 to 3 assume maintenance of current consumption patterns for the existing population of 7.3 billion but have different assumptions for the additional 2.3 billion people that the UN project will exist. Scenario 1 assumes that the additional population will come from developing countries and that these consumers will have a per capita consumption level that is the same as current per capita consumption in developing countries. However, given the upward trajectory of demand in such countries, Scenario 3 assumes that per capita consumption for the additional population will be the same as for developed countries. Scenario 2 assumes an intermediate position with the additional population consuming the same level as the current global average. These scenarios show significant variation in additional demand, between 32% and 43% in these examples, depending on the level of per capita demand assumed.

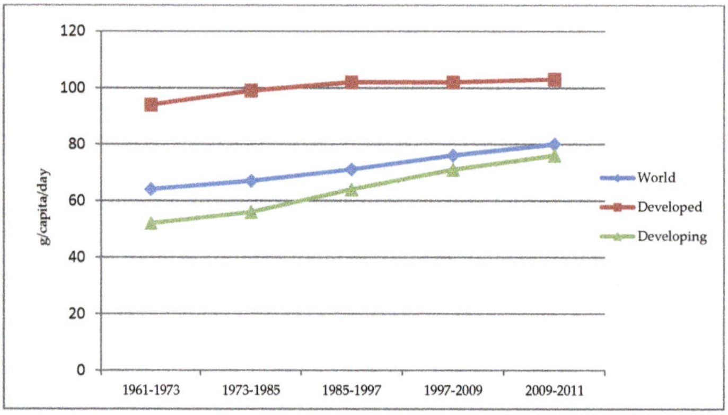

Figure 1. Evolution in protein consumption per capita (g/capita/day). Source: Author's analysis based on food balance and population data obtained from http://faostat3.fao.org.

Table 1. Impact of different consumption scenarios on annual protein demand [1].

Scenario	Pop. (000,000)	Consumption g/capita/day	Tonnes/Annum	% Change from 202.352 m Tonnes
1: Existing population at current consumption levels but increased population at average protein consumption for developing world for 2009–2011	9.6	76	263,802,000	+32%
2: Existing population at current consumption levels but increased population at average protein consumption for the world for 2009–2011	9.6	80	267,160,000	+33%
3: Existing population at current consumption levels but increased population at average protein consumption for the developed world for 2009–2011	9.6	103	286,468,500	+43%
4: Entire population at current max. consumption levels	9.6	103	360,912,000	+78%
5: Entire population at level required for sedentary adult	9.6	50 [2]	175,200,000	−13%

Source: Authors' calculations based on FAO/OECD data. [1] The data used measure "availability" as opposed to demand, i.e., demand is calculated as a residual based on "total available for human consumption = total food supply – feed – seed – industrial uses – waste". While subject to several limitations as a measure of demand and likely to overestimate food consumption, in the absence of data from household surveys, it is accepted as a good proxy for consumption levels of a population as a whole and therefore for developing projections of future food supply needs [7]; [2] based on 2000 calories per day, 10% calorie intake as protein and 4 calories per gram of protein.

At least two other scenarios could be proposed based on current knowledge about protein consumption and nutrition. The first assumes the entire global population will consume protein at the highest rate at which it is currently being consumed by a population (103 g/hd/day) and the other assumes the entire global population will consume protein at a level which is considered to be adequate for the average sedentary adult. These scenarios show significantly different outcomes in future demand ranging from a 78% increase to a reduction of more than 10%.

Which scenario is most likely is dependent on the balance between competing forces that promote and restrict protein consumption. Given that most population growth is occurring in developing countries and that socio-economic factors in such countries are driving a nutrition transition towards higher intakes of protein, combined with evidence that the market for protein in developed countries continues to be driven by demographic and lifestyle changes, Scenario 5 is highly unlikely. Indeed, over a third of 18–34 years old in North America claim to be trying to consume as much protein as possible, whilst FAO research [8] highlights the staggering growth expected in demand for poultry in South East Asia (725% increase between 2000 and 2030), which is primarily attributed to increasing per capita consumption rates rather than increasing population levels. Furthermore, new research on the benefits of protein beyond muscle development and maintenance, for example research on the impact of protein on satiety and weight management [9] and more recently on hunger stimulation in hunger-suppressed individuals such as the elderly, continues to emerge, further stimulating increased demand. Combining these factors with enduring protein undernourishment [10], one is led to the conclusion that increased protein demand in excess of one third is highly likely for the future. This figure is conservative in comparison with figures from other sources, e.g., Swiss company Bühler who argue that an additional 50% will be needed by 2050 [11].

However, all proteins are not the same; they vary in terms of nutritional profile, digestibility and bioavailability, environmental implications, consumer acceptance, and other factors. Thus, future protein supply cannot be merely a matter of producing more of the same in the same proportions. Factors that influence the potential of various current and future protein supply sources are considered below.

3. Existing Protein Sources

Currently vegetal sources of protein dominate protein supply globally (57%), with meat (18%), dairy (10%), fish and shellfish (6%) and other animal products (9%) making up the remainder [12].

3.1. Plant Based Protein (Cereals)

Cereal proteins account for the major portion of dietary protein intake globally [13,14] and are important for animals as well as humans. They are particularly important in the diet in developing nations and wheat accounts for largest group of plant protein sources in the Western diet [13]; in the form of bread, wheat is a key component of protein delivery in Europe with a typical loaf containing 8 g of protein per 100 g. Protein content ranges 10–15% with the largest protein content found in the storage proteins [14]. These storage proteins include the prolamins, globulins and germins [15]. Corn (maize), rice and wheat are the main staples consumed globally but, in some regions such as West Africa, millet is consumed extensively. In Southern India where protein malnutrition of infants is common, rice and millet are consumed regularly and, in Ethiopia, Teff, with an amino acid profile similar to egg protein, is preferred. Indeed, it has been calculated that a typical Ethiopian diet consists of 65 g of protein of which 41 g come from Teff while only 6 g come from animal protein consumption [16].

Maize (corn) is important for global food security. As a source of food (as opposed to feed), it accounts for 25% and 15% of total maize consumption in developing countries and globally, respectively. In addition, maize as a source of protein is very similar in terms of its contribution to calories intake globally. This ranges from 61% in Mesoamerica, 45% in Eastern and Southern Africa (ESA), 29% in the Andean region, 21% in West and Central Africa (WCA), to 4% in South Asia [17].

Compared to other cereals, the protein from oats is of high quality and the amino acid content and quality is comparable to soy protein [18]. Oat protein contains a higher content of the essential amino acid lysine compared to other cereals and has a lower proline and glutamic acid content. This attribute means that oat protein, once digested, can be tolerated by persons with gluten intolerance and allergies [19]. Rice does not contain large quantities of protein but rice protein flours have been prepared previously using enzymatic treatments with carbohydrate-hydrolysing enzymes to yield products with 91% protein [18,20].

While plant-based protein sources often lack one or more amino acids in sufficient quantity to meet human nutritional needs [21], combinations of different proteins, including cereal-pulse combinations, and supplementation, can help to overcome this in strict vegan or vegetarian diets. Cereals can also offer a significant health benefit as a rich source of bioactive peptides. A number of these are documented in the database BIOPEP [22]. Such food-derived bioactive peptides have gained increased interest as agents in the control of chronic diseases [22–24] and to reduce the risk of side effects arising from synthetic drugs usage. Bioactivities associated with cereal proteins include antioxidant, anti-inflammatory, cholesterol-lowering, satiety, anti-diabetic and others as reviewed recently [18]. They have a wide range of physiological effects that have been demonstrated in animal models previously. However, prolamins from some cereals, including wheat, barley, and rye, give rise to biologically-active anti-nutritional peptides following proteolysis that are able to adversely affect in vivo the intestinal mucosa of coeliac patients, whereas prolamins from other cereals such as maize and rice do not. Bioactivities associated with oat, barley and wheat protein derived peptides include opioid activity while rice protein contains the sequence RGD and lack celiac toxicity which could be beneficial in health and nutrition. Several groups have documented the use and potential of cereal derived bioactive peptides previously [18,25–28]. From an agronomical viewpoint, issues including cultivar use, environmental conditions and agricultural practices can affect the bioactive peptide content within cereal proteins. In terms of determining a physiological effect in humans, knowledge concerning the bioavailability of peptides and research on randomized clinical trials is necessary in order to understand the full potential of these proteins for health benefits and applications.

Notwithstanding the fact that plant-based protein is deemed to be preferable from a landuse and GHG emission perspective than animal-based proteins, plant protein is also subject to environmental criticisms. Such criticisms seem to be mainly related to the use of plant-based protein for animal feed and the increased environmental pressures associated with the development of industrialised farming systems to meet this demand. Soy, for example, is an important source of protein, however 85% of its production is used to feed animals and fish [29]. The drive to produce soy, to respond to the increased demand for animal protein, is associated with deforestation, and habitat loss, in South America in particular. There are ethical issues associated with this as around 12 million hectares of land outside Europe is required to produce protein-rich meal for European livestock production [10]. Other environmental issues such as water use, soil degradation, and pollution are also associated with high intensity plant protein production. The feeding of available edible plant protein to animals is itself a cause of controversy from an ethical perspective with previous authors arguing that if all such available edible plant protein were consumed by humans in an equitable manner then sufficient supplies of dietary protein would be available to feed the global population. The development of suitable vegetal protein products, which are accepted by consumers as preferred food items that compete with animal protein products rather than being positioned as imitations or extenders of animal protein products, is however required for a major shift to occur in the utilisation of plant-based protein.

3.2. Meat

Based on the FAO Food Balance Sheet data, it is clear that global meat consumption has increased significantly in recent decades. Analysis of such data by Henchion et al. [30] finds that overall meat consumption increased by almost 60% between 1990 and 2009. This trend is expected to continue, driven in particular by income growth in countries such as Asia, Latin America and the Middle East. Using 1997 as the base year, Rosegrant, Paiser, Meijer and Witcover [31] expect the quantity of meat demanded by consumers in developing countries to double by the year 2020. This growth is tempered somewhat by a suggestion that meat consumption per capita appears saturated in developed countries [32], compounded by aging populations, changing demographics and increased health and dietary awareness [30]. While the United Nations, some governments and several NGOs are implementing campaigns to reduce the amount of meat consumed [33,34], global meat consumption is expected to increase by 76% by 2050 [35]. This increased demand requires significant production growth. Given constraints on land and water availability and the impact of meat production on climate change, increased efficiencies in production practices and improvements throughout the wider food chain will be critical [36].

Meat is an important component of the human diet, and beef in particular has played a key role in food security in terms of providing energy, protein, and essential micronutrients. As highlighted by Gerber and colleagues [37], ruminants play a key role in converting fibrous material, which cannot be digested by humans, into protein which has a high nutritional value. Raw meat contains 20–25% protein depending on source and fat content, which, on loss of water due to cooking, can correspond to 28–36% in cooked meat. Protein from meat is an excellent source of essential amino acids and has high net protein utilization and digestibility [38]. In addition, meat is a key source of, often highly bioavailable, minerals (iron, zinc, and selenium) and vitamins (A, B9&12, D, and E), and has an ability, referred to as the "meat factor", to enhance iron availability from other sources [39,40].

While meat consumption carries clear health benefits, over-consumption can lead to negative health impacts. The International Agency for Research on Cancer (IARC) report [41] suggested processed meat should be classified as "carcinogenic to humans" (every 50 g portion of processed meat increases risk of colorectal cancer by 18%). Red meat, on the other hand, was classified as "probably carcinogenic to humans" (every 100 g of red meat consumption leads to a 17% increased risk). While causation has not been fully elucidated, heme iron is thought to play a critical role through catalysis of N-nitroso-compounds, generation of lipid oxidation products and a possible direct cytotoxic effect. However, as discussed by De Smet and Voosen [42] "the benefits and risks associated with red and processed meat consumption should not necessarily cause dilemmas, if these meats are consumed in moderate amounts as part of balanced diets". In relation to cardiovascular disease (CVD), however, research seems more inconsistent and there is debate regarding the paucity of data from randomized controlled trials (RCT). A recent meta-analysis of RCT supports the idea that consumption of under 0.5 servings, or 35 g, of total red meat per day does not negatively impact on CVD risk factors [43]. Public health concerns about livestock production are also relevant in relation to health, including zoonoses such as avian influenza, and concerns about emergence of novel diseases at animal-human-ecosystem interface.

From an environmental perspective, meat production, at a global level, contributes significantly to climate change and land use change [40]. High levels of greenhouse gases (GHG) are produced during meat production with ruminants contributing in a significant way. Land use, water, energy and chemical inputs (e.g., fertilizers) all reflect negatively on the environmental footprint from meat production. Of course, the relative level of impact varies depending on factors such as the species under consideration, production system (e.g., grass vs. concentrate production system), requirement for deforestation, and others. In particular, there is a case to be made for the promotion of meat produced from grass-fed ruminants [44].

In tandem to meat production, animals also produce a variety of other goods and services. Animals and cattle in particular, often serve as financial instruments and a route out of poverty in developing countries. Meat production is important for economic growth and poverty reduction in rural areas but needs to be managed carefully to ensure social effects are not negative.

Plant-derived alternatives to meat are being developed with many products already on the market. Quorn is the most well-known of these. It is derived from a fungus, using a natural fermentation process and wheat-derived glucose syrup. Quorn Foods, founded in 1985, currently sells about 22,000 tonnes of quorn in 16 countries, with investment underway to double production capacity in their UK plant. In vitro meat (see below) is presented as a longer-term alternative.

3.3. Dairy

Dairy based ingredients dominate the protein market, mainly because of a highly developed global milk industry and the fluid nature of milk that facilitates diversity through production of valuable by-product streams. They have functional properties and health benefits that are supported by scientific/medical studies, making them a suitable nutritional base for other foods. The global market for dairy protein is complex, multifaceted and driven by evolving markets and, more recently, trends in lifestyle nutrition. Protein ingredients are commonly manufactured as powder, but liquid-based convenience formats are growing rapidly. Developments in genetic selection coupled with feeding regimes of animals can be used to increase the protein content in milk, substantially increasing the yield of protein ingredients that can be manufactured. In Ireland for example, the protein content of milk has increased from an average of 3.2 to over 3.5% from 1996 to 2015. The development of membrane separation technologies such as micro- and ultra-filtration has created a diverse portfolio of protein ingredients, with tailored functional and/or nutritional attributes, generating multiple applications for a wide variety of foods across the globe. While global demand for protein is clear, predicting equilibrium between supply and demand for dairy protein is complicated by differences in rate at which intra- and inter-country milk production systems develop. India for example, with the highest milk production in the world, has a low input and low output system, comprising mainly small farms, and a large domestic market. A supply of high quality fluid milk, underpinned by quality control systems, is required to make viable any development in processing infrastructure (membrane, evaporation and drying equipment) for the manufacture of high value protein ingredients which could be traded globally.

While skim and whole milk products contain protein, more concentrated systems can be categorized into:

(1) total milk protein: milk protein concentrate (MPC); isolate (MPI); hydrolysate (MPH);
(2) casein-based: caseinates from acid or rennet source (as sodium or calcium salts) and micellar casein; and
(3) whey-based: whey powder; demineralized whey powder (DWP); whey protein concentrate (WPC); whey protein isolate (WPI); whey protein hydrolysate (WPH).

Many of these ingredients are further sub-divided based on their protein content, e.g., WPC35 and WPC80 have 35 and 80% protein respectively. Developments in the manufacture of MPC have led to its increased usage for protein fortification, partly due to a low lactose content and acceptable flavor profile [45]. During production of dairy-based ingredients, as the protein content is increased, lactose and mineral salts are removed generating a by-product permeate stream, i.e., ultrafiltration of skim milk generates MPC and milk permeate whereas filtration of whey generates WPC and a whey permeate. The market for permeate powder is a key factor underpinning the business model for manufacture of a high protein concentrate, and the ability to valorise the entire milk or whey stream is a key determinant influencing export volumes of the various dairy protein powders.

Many factors are driving the increased demand for dairy proteins, including the emergence of economies with increasing income. In countries like China, urbanization has provided new infrastructure capable of supporting distribution chains for dairy-based beverages. In Europe, the removal of quotas in 2015 resulted in a significant increase in milk production and protein ingredient manufacture has provided an avenue for utilization of this milk pool. The market for Greek-style yoghurt has expanded in the US, driven by the need for new consumer platforms for consumption of protein. In the US, liquid milk, ultra-filtrated to increase its protein content (50% more protein than regular milk), has been recently sold by Coca-Cola under the fairlife® brand. The hugely successful protein bar segment has recently been extended to popular chocolate based bars, offering high protein (e.g., 18 g of protein per bar) alternatives, based on skim milk, milk protein and/or whey protein concentrate. Incorporation of protein into mainstream confectionary like this, adds to the complexity of protein markets, as now, global distribution chains are not solely associated with the sports and lifestyle/nutritional beverage sector.

The global market for whey protein is continuing to grow based on increased demand for performance nutrition and beverages such as infant formula. Relatively new protein ingredients have been generated through microfiltration of milk to produce micellar casein protein and native whey.

The global whey market is even further developed with individual fractions such as α-lactalbumin, β-lactoglobulin, Immunoglobulins (IgG), Lactoferrin, Lactoperoxidase and Glycomacropeptide (GMP) having application in infant nutrition, medical and sports beverages or foods. New research demonstrating the role of whey proteins in maintaining muscle mass (delayed development of sarcopenia) provides further opportunities for expansion of whey protein markets and is of particular relevance to an ageing population; these applications and those for special medical purposes often require high protein WPI (>80% protein) alternatives. Scientific studies have shown the beneficial effect of whey protein on muscle development [46,47] creating applications for body composition, sports, exercise and/or sarcopenia and even for serotonin synthesis in the brain [48]. The quality of the protein is a key attribute for such application; the emphasis is on individual, digestible amino acid content, which is excellent for dairy proteins, and in particular the whey protein α-lactalbumin, which also has high lysine availability [49].

One of the most important determinants of international trade in protein ingredients is safety, mainly through the impact of contamination and/or spoilage, where the distribution chain can be disrupted virtually overnight. An example of a food safety scandal was the adulteration of milk in China (in the media in 2008) with the compound melamine. Addition of melamine increases nitrogen levels and thereby falsifies protein content. The infant formula sector has particularly stringent quality specifications, including microbial specifications and chemical residues (e.g., veterinary drug residues). In addition, given the globally traded nature of dairy products, regulatory and political restrictions associated with biosecurity are of significance.

Dairy production at farm level is subject to many of the negative environmental impacts associated with meat production due to its requirement for land and water and production of GHGs. However, grazing of dairy cows has been claimed to be preferable to grazing of suckler cows as they produce animal protein more efficiently [50]. Technological efficiency has already resulted in a significant reduction in the number of animals, and the quantity of food and water required to produce the same volume of milk. There has also been a significant reduction in manure, methane, N_2O and the overall carbon footprint associated with milk produced currently compared to 1944 [51]. However, there is potential for significant improvement based on existing technologies with the significance of different strategies varying regionally and according to farming system. For example, in South Asian mixed dairy farming systems, feasible improvements in feed quality, animal health and husbandry could reduce GHG emissions by 38% of the base line emissions while for dairy mixed systems in OECD countries adoption of improved manure management systems, feed supplementation and energy saving equipment could reduce emissions by 14–17% of the baseline [52]. At the processing level, there is recognition of a growing need to adopt sustainable manufacturing processes, including energy and water/wastewater conservation; this is currently contributing to the re-engineering of unit operations within dairy manufacturing plants [53]. Environmental impact and sustainability of production during powder manufacture are key factors for milk processors when determining the type of protein ingredient to produce. A recent study by Finnegan et al., 2017 [54] applied environmental lifecycle assessment to estimate global warming potential (GWP) of manufacture of a range of dairy products. The authors used a system boundary covering the dairy, the farm, and transportation through to processing. They identified that during processing operations for powder production (SMP, WMP, and protein powders) the highest contribution to GWP was due to the energy intensive evaporation and drying steps. From an environmental perspective, dairy production/processing systems have been reviewed and the carbon footprint determined [54,55].

Country of origin labelling has become another key marketing tool for many countries and providing proof of provenance has become a prerequisite for purchase in some markets. A claim such as "Pasture based", i.e., animals fed on grass outdoors, is an example of a trending marketable attribute associated with dairy products. Countries such as Ireland are leading the way in their approach to promoting a grass based feeding system, substantiating claims by scientific research, demonstrating the benefits of milk and its products produced by a grass fed animal [56,57].

Dairy based powders will remain a leading source of protein for the foreseeable future. The number of recombined food products will continue to grow in emerging economies and will drive demand. Nutritional applications for different stages of life from infants to elderly will develop in parallel with population and economic development. The ability to easily process liquid milk from bovine origin will ensure it continues to provide safe, high-quality products with good microbiological and functional properties. The rate of technological advancement in dairy will continue to evolve and support production of high quality ingredients that will facilitate supply to meet global demand for protein.

4. New and Emerging Sources of Protein

4.1. Plant-Derived Protein (Pulses)

On a global basis, plant based protein is of immense importance and there is significant interest in its ability to meet growing demand for protein from non-meat sources. Thus, one of the key reasons for an increase in demand for plant-based protein is due to favourable comparisons with meat-based proteins. Plant-based protein is preferred to animal-based protein from an environmental perspective as it is associated with a lower land use requirement, and it is generally accepted that plant-based foods produce lower levels of greenhouse gases (GHG), which are associated with climate change, than animal-based foods [3]. Some plants offer unique advantages, e.g., pulses, being legumes, have the unique ability to fix nitrogen. Furthermore, due to the high cost and limited availability of animal proteins in several countries and consumer concerns over health benefits of animal-based proteins, increased attention has focused on the utilization of plant-based proteins as potential sources of low-cost dietary proteins for food use [58].

Among plants, pulses are considered an important source of dietary protein and other nutrients. In many regions of the world, pulses are the major source of protein in the diet, and often represent a necessary supplement to other protein sources. Indeed, pulses have a significant role in overcoming challenges associated with protein-energy malnutrition in developing and under-developed countries. For example, cowpea (*Vigna unguiculata*) is an important pulse grown and consumed in East and West African countries. Nutritionally pulses contain approximately 10% moisture, 21–25% crude protein, 1–1.5% lipids, 60–65% carbohydrates, and 2.5–4% ash. Chickpea is an exception as it contains about 4–5% lipids, and some pulses such as soy bean and lupin have been reported as having up to 45–50% protein. Proteins in pulses are found in the cotyledon and the embryonic axis of the seed with small amounts present in the seed coat. The cotyledons constitute the major portion of the pulse grain and hence make the highest contribution to protein content. Pulses like cereals show wide variation in protein content mainly due to genetic, environmental and agronomic factors.

There are some concerns that vegans, or diet containing plant proteins as the major source of protein, are nutrient deficient due to unbalanced protein sources [59]. This is because, unlike animal proteins, plant proteins may not contain all the essential amino acids in the required proportions. Out of the nine essential amino acids (histidine, leucine, isoleucine, valine, threonine, methionine, phenylalanine, tryptophan and lysine) required by humans, pulses are deficient in sulphur containing amino acids (e.g., methionine). Pulses may also contain many anti-nutritional compounds such as hydrolase inhibitors and lectins which constitute part of the defensive mechanism of the seed; these may inhibit several biological functions [60,61]. Products obtained by using a blend of cereal and pulses can balance the amino acids and may improve the nutritive value of the product. Furthermore, fortification and food supplements mean it is easier nowadays to adopt a vegan or vegetarian dietary regime. Table 2 shows the level of proteins in selected pulses consumed around the world.

Table 2. Proximate composition of different pulse grains (g/100 g dry weight).

Pulses	Protein Content	Pulses	Protein Content
Kidney bean	23.58 [62,63]	Navy beans	22.33 [62]
Chickpea	19.30 [62,64], 19.29 [65]	Gt. northern bean	21.80 [62]
Lentils	25.80 [64], 26.1 [66]	French beans	18.81 [62]
Mung bean	23.86 [62], 27.5 [67]	Winged beans	29.65 [62]
Mungo bean	25.21 [62], 26.22 [65]	Hyacinth beans	23.90 [62]
Pigeon pea	21.70 [62,65]	White beans	23.36 [62,67]
Peas	24.55 [62], 19.3 [63]	Horse gram	22.50 [68–70]
Adzuki bean	19.87 [62,71]	Cowpea	23.85 [62,66], 24.1 [70]
Black beans	21.60 [62], 23.6 [64]	Navy beans	22.33 [62]
Lima beans	21.46 [62]	Gt. Northern bean	21.86 [62]

Advances in production technologies, including technologies such as advanced monitoring systems (e.g., biomass monitoring and harvest monitoring) and remote sensing, and biological innovations in genomics and plant breeding will improve input efficiencies and help to anticipate and thus mitigate production risks (e.g., weather and pest related risks). New models of production are also emerging, e.g., soil-less growing and indoor farming. Production costs and energy requirements mean that such systems are currently confined to high value protected crops, e.g., salad crops, rather than field grown crops, with limited relevance for protein production currently. However, overall, in the foreseeable future, technological development will continue to position plant-based protein as a desirable option from a sustainability perspective.

4.2. Insect

Insect consumption (entomophagy), whereby eggs, larvae, pupae and adults of certain insects are consumed by humans, has occurred for thousands of years. Approximately 2000 species of insects have been used as food [71] and they are part of the traditional diets of at least 2 billion people [72], particularly in parts of Asia, Africa and South America where they provide important livelihood opportunities. Beetles are the most commonly consumed insects (31%), followed by caterpillars (18%); bees, wasps and ants (14%); grasshoppers, locusts and crickets (13%); cicadas, leafhoppers, planthoppers, scale insects and true bugs (10%); termites (3%); dragonflies (3); flies (2%); and others (5%) [72]. However, more recently, insects have been identified as an alternative source of protein for the Western world, not only as a delicacy or for emergency nutrition [73], supported by organizations such as the FAO [72] and the European Commission. Crickets, lesser mealworm and yellow mealworm are potential insects for application for food in the EU while black soldier fly, yellow mealworm and the common housefly have potential for use in feed products [13].

Proponents of entomophagy argue that it has a lower environmental impact compared to meat production [74]. Significantly, they argue that insects do not compete for land, require less water and emit lower levels of greenhouse gases and NH_3 than regular livestock. They can be reared on organic side-streams thus creating value from and reducing waste products. However, the environmental impact of insect production is significantly influenced by insect's diet, which in turn influences whether it can be used for food or feed purposes [75]. Another environmental advantage is that up to 80% of body weight is edible and digestible compared to 55% for chicken and 40% for cattle [35]. Being cold-blooded, they perform better in terms of feed conversion efficiency, and they reproduce more rapidly [76]. Many insects also have a favourable nutritional profile for humans, with most being highly digestible (77–98%), high in protein (crude protein 40–75% on a dry weight basis) [10] and a good source of essential amino acids, high in vitamins B_1, B_2 and B_3 and the minerals iron and zinc [77]. However, many insects are deficient in certain amino acids, including tryptophan and lysine, and those with chitin exoskeletons have lower levels of digestibility. Insects vary considerably in fat, and thus energy levels, with some insects recorded as having 77 g/100 g dry weight. Some research argues that insects pose less risk of transmitting zoonotic diseases to humans, compared to animals and birds [72].

Most insects are, however, still collected from their natural environment (usually forests), which restricts supply according to season and location [78]. Furthermore, some researchers question insect consumption from a food safety perspective, arguing that the safety of insects as food is under-researched [77]. While insects are similar to other animal-derived products in that they are rich in nutrients and moisture (providing a medium for growth of unwanted microorganisms in certain conditions), the fact that insects are phylogenetically far removed from mammals, birds and some aquatic species regularly consumed as food means that significant differences are expected when making comparisons between them in terms of risks [77]. Furthermore, many insects are consumed whole, which includes their gut microflora which may affect microbiological quality of the food product [10] and specific health implications associated with using organic feedstocks to produce the insects need to be assessed (for example (undesirable) substrate materials can be transferred into the protein products and thus into the human food chain).

Notwithstanding long traditions of domestication of bees and silk worms, and the practice of rearing insects for biological control, health and pollination [72] commercial insect farming for food purposes is only beginning to evolve, largely in proximity to concentrations of consumers. While insects are an established part of food culture in some countries and are eaten out of choice, there is evidence of some reluctance by Western consumers to accept insects as food as they are often considered as pests, and a source of contamination, and thus to be avoided. Van Huis and colleagues [72] relate the absence of a history of consuming insects to the difficulty of harvesting a proper meal of insects in temperate zones (insects tend to be less abundant, smaller and found less often in clumps in temperate zones). Nonetheless, food products are available on the market including cricket protein bars produced by British company Next Step Foods, and are on the shelves in European supermarkets including the Belgian supermarkets Delhaize and Carrefour. Widespread adoption of insects for food purposes will not however be achieved based on communicating their environmental and nutritional benefits. It will be dependent on addressing what Rozin and Fallon [79] identify as three important motives that lead to product rejection: negative sensory properties (distaste), harmful consequences (perceived danger) and "ideational" factors. Given that physical state is often used as a heuristic to evaluate food, processing insect protein to render it into an unrecognizable (e.g., as an ingredient in a familiar product) could be a productive strategy. Similarly, ensuring insects and insect-based products are acceptable from a sensory perspective and are safe to eat will be critical.

The Belgian food safety authority allowed the sale of ten different types of insects for food consumption in 2013 following a federal ruling. However, the EU novel food regulation does not yet allow insects for human consumption to be sold. Regulatory changes are expected but it is not clear when this will happen, which means companies are currently focusing on insects for animal feed and pet food rather than human consumption. Commercial production of insects for human consumption will require the establishment of new value chains. Given that they are most likely to be accepted as ingredients rather than whole, such a chain includes ensuring a safe and reliable feedstock for the insect, mass rearing of insect larvae, their processing into insect ingredients, and application of these ingredients in final food products. All of these activities need to be developed simultaneously, resulting in costs and risks. Some organizations and commercial companies are setting up production on an industrial scale, e.g., the FAO established an insect farming project in the Philippines in 2010 [80], Swiss company Bühler is setting up pilot facility in China to process fly larvae and mealworms, and there are a few industrial scale enterprises in various stages of start-up within Europe for raising insects such as black soldier flies. The establishment of a trade group called International Platform for Insects for Food and Feed (IPIFF) is also significant. Nonetheless, current primary production systems remain expensive, with a need to develop automation processes to make insect production economically competitive; extraction processing systems are also too costly and need further development [72].

4.3. *Algae*

Marine plants such as seaweeds and microalgae represent a promising and novel future protein source. Both are collectively termed algae, however, seaweeds are complex multicellular organisms that grow in salt water or a marine environment, whereas microalgae are single celled organisms that can grow in a range of environmental conditions. Already €6.4 billion or 24 million tonnes of algae (mainly seaweed) are farmed globally [35] and protein-rich micro-algae are seen as a forerunner resource to close the so-called "protein gap". Microalgae used for human consumption include *Arthrospira* spp. (*Spirulina* spp., namely, cyanobacteria), *Chlorella* spp. and *Dunaliella salina*. Only two microalgae are used primarily in the EU—*Spirulina* and *Chlorella* spp. Japan and China also supply these microalgae but there are some concerns regarding toxins, microbial load and other sanitary issues concerning their production. Microalgae are ordinarily produced in open, outdoor raceway ponds [81].

Nutritionally they are comparable to vegetable proteins but their development has been hampered by high production costs, technical difficulties in extraction and refining and sensory and palatability issues concerning their incorporation into food products. At present, microalgae are largely targeted for their EPA/DHA content and are sold as health foods, as cosmetics or as animal feed [82]. Approximately 30% of global algal production is sold for animal feed [13] with potential for further increases as dried, defatted algae could compete with soybean in pig and chicken feed, potentially replacing up to one third of soybean meal in their diets [31]. A US based company Solazyme is currently producing AlgaVia a whole algal product with 65% protein for sale as a food additive. They have two full-scale production lines in the US, with another in Brazil (capacity of 100,000 mt/year) [35].

Seaweeds can be harvested from the sea but are also increasingly cultivated. They, especially the red and green varieties, are very rich in protein (up to 47% dry weight in the red seaweed *Porphrya* sp.), are already considered as sea vegetables in countries such as France, and have consumer acceptance. The protein content of seaweeds differs according to species. Generally, the protein fraction of brown seaweeds is low (3 ± 15% of the dry weight) compared with that of the green or red seaweeds (10 ± 47% of the dry weight) [83]. The protein content of seaweeds also varies depending on the geographical location of harvest and the season of harvest. The amino acid content of seaweeds is comparable to protein sources such as eggs or soybean. The amino acids aspartic and glutamic acids constitute a large proportion of the amino acids of brown seaweeds (22–44% of total amino acids in some brown algal species). Studies carried out previously suggest that the method of extraction has an impact on digestibility of protein but that seaweed derived proteins are for the most part digestible and comparable to casein but the carbohydrate fraction and phlorotannins found in brown seaweeds can affect digestibility [83]. Some safety hazards include potential accumulation of heavy metals, high levels of iodine, and contaminants such as dioxins and pesticides [13].

Seaweeds could also be used in aquaculture diets or as animal feeds. A recent review by García-Vaquero [84] and others [83] suggest that the red seaweed *Porphyra* sp., has potential for use as feed for sea bream due to the high protein content and excellent amino acid profile of this species. Brown seaweeds including *Laminaria* sp., *Fucus* sp., and others, however, contain phlorotannins that can be viewed as anti-nutritional due to their protein binding capacity [85].

In addition to use in animal feed, seaweed protein and peptides can impart taste to foods or when used as food ingredients. Recently, a group in Oregon developed red seaweed, vegan bacon-tasting alternative to bacon. The taste of seaweeds is closely related to the fifth taste—umami. Umami taste is linked to the substance monosodium glutamate (MSG), the sodium salt of glutamic acid, an amino acid found in great abundance in seaweeds. Brown algae such as *Saccharina japonica* or *Laminaria jabonica* (common name is Konbu) have a particularly high MSG content which is released when the seaweed is heated or softened. In addition, red seaweeds including *Porphyra* sp, (common name Nori) also impart umami taste due to nucleic acids present within their cells. These compounds that cause umami taste in seaweeds are present due to the environment in which seaweeds grow which are often hostile. Seaweed ash salt was used in the past in regions where salt was not plentiful in countries such as Denmark and Brittany. Seaweed ash salt can impart a smoky taste to foods.

Regulatory restrictions on novel food and novel food ingredients, food safety, nutrition and food health claims can delay the pace of commercialization of algae [86]. Furthermore, the lack of publicly available economic and market information make it difficult to evaluate their industrial potential which is a hindrance in terms of seeking funding for research and development and for influencing policy.

5. Muscle protein sources

5.1. Fish

Seafood consumption is generally increasing in many parts of the World, particularly for coastal communities. In many places, fish and shellfish are the only readily available sources of protein that people can self-harvest, often throughout the year.

Aquaculture supplies 50% of all fish consumed globally today and by 2030 it is predicted that aquaculture will be the prime source of fish due firstly to demands from consumers, and secondly due to depletion of wild capture fisheries [87]. However, aquaculture needs to become more sustainable. Currently, it is over-dependent on a supply of fishmeal and fish oil, and there are concerns surrounding pollution and water quality, use of soy and chemicals in aquaculture feed, and habitat destruction. Negative social outcomes are also associated with aquaculture in countries where there are weak regulatory frameworks and there is concern about emerging diseases and disease transmission as a result of increased intensification and globalisation [35]. Companies such as Findus are however progressing towards more sustainable aquaculture systems, e.g., they now source shrimp farmed in Sweden that are produced using the "Biofloc" method [88]. This involves feeding shrimp excrement to microorganisms which in turn are consumed by the shrimp as a source of protein. Biofloc can produce up to 40 times more shrimp that conventional shrimp farming [88]. Thus, aquaculture has the potential to be a key element of a circular food system.

Reform of the European Common Fisheries Policy (CFP) in 2015 (Commission Delegated Regulation (EU) No. 1394/2014 further supports the marine sector being a key source of protein. The CFP aims to gradually eliminate the practice of discarding undersized fish as by-products at sea. Landing obligations will be phased in from 2015 to 2019 and will result in greater quantities of catch being landed at port. Undersized fish or under-utilised species such as Blue Whiting are not aesthetically pleasing to consumers and traditionally were thrown back to sea. However, there now exists an exciting opportunity for processors to convert this by-catch or marine "rest raw material" into value added ingredients due to the high protein and oil content of this by-catch. Several recent publications highlight the potential of this resource as a source of bioactive peptides [89], lipids [90] and small molecules/nutrients including vitamins and minerals [91].

Regular fish consumption is widely promoted as part of a healthy diet. There is strong scientific evidence for this. Fish have a high protein content compared to terrestrial animal meat and have a lower feed conversion rate (FCR) than land animals and more protein can be produced using this lower feed rate from fish [83]. Furthermore, fish protein is highly digestible and rich in essential amino acids that are limited in animal sourced protein including methionine (6.5% compared to 5.7% of total essential amino acids in fish compared to animal meat) and lysine (19.6% compared to 19.0% of total essential amino acids in fish compared to animal meat). Fish and shellfish consumption has several reported health effects including decreased risk of heart diseases, inflammation and arthritis. The health benefits of fish are mainly linked to the presence of long chain omega 3 (n-3) polyunsaturated fatty acids (PUFA) [83]. However, fish protein is also a rich source of bioactive peptides. These can have a myriad of positive health effects if consumed in appropriate concentrations and bioavailable within the human body. Health effects include control of blood pressure through inhibition of enzymes within the renin-angiotensin aldosterone system (RAAS), maintenance of bone health, control of inflammation (antioxidant peptides), mental health through the action of opioid peptides and platelet activating factor acetyl-hydrolase inhibitory (PAF-AH) peptides) and several other bioactivities [92].

While fish is seen as a healthy food by consumers, food safety risks such as heavy metal content could represent a potential barrier to consumption frequency, particularly concerning contamination of wild fish [89]. It is well known for example that the Baltic sea is heavily contaminated with heavy metals due to the presence of heavy industry in the landmass surrounding it that have used the sea as a "sink" for industrial waste for many years. Such chemical contamination, e.g., by heavy metals such as arsenic, mercury and cadmium and polychlorinated biphenyls (PCBs), can result in carcinogenic and toxicological impacts, as well as potentially mitigating the beneficial impact of omega 3 [93].

5.2. In Vitro Meat

In vitro meat, also referred to as cultured meat, cell-cultured meat or clean meat, is an animal product produced following cell isolation and identification, cell culture and tissue engineering protocols [94]. Several sources of cell can be used including embryonic stem cells from pre-implantation embryos or adult stem cells. Proponents argue that an in vitro meat bioreactor the size of a swimming pool could feed 40,000 people for a year, and that it would use 99% less land than the average for farmed beef [35]. The first cultured hamburger was created by Mark Post of Maastricht University in 2013, cost £200,000 and took two years to create [35]. Mosa Meat, a Dutch start-up, continues to develop it, focusing on improving flavor through co-culturing of fat cells, and cost reduction. Other companies, in the US and elsewhere, are also working towards similar goals. Current costs are estimated to be in the region of $11. Significant media attention and innovative research funding mechanisms, e.g., crowd funding, will support further technical development.

In addition to economic and sensory factors, environmental, ethical and societal factors are at play. The possibility of producing meat without slaughtering animals carries an obvious benefit, as does the availability of a meat product with a much-reduced environmental impact. However, while on the positive side, ruminant greenhouse gases are lower by 96% [35], cultured meat will require more industrial energy than is required in livestock production. Furthermore, as cultured cells do not have an immune system, sterilization is a perquisite. Improvements in energy efficient processes may help circumvent this significant energy demand. The control afforded from the process means that consumers can be confident that they are being offered a standard product. However, the production process requires the use of chemicals e.g., hormones, nutrients, which while being of food grade, may not be attractive to consumer segments that value natural production systems. In a wider societal context, it is necessary to consider the impact on farmers' livelihoods, in particular smaller operators and those in under-developed economies, who rely on livestock production for income and wealth.

Technical challenges remain with the production of in vitro meat. Co-culturing of cells, e.g., muscle cells and adipose cells, and potential genetic instability due to speed of growth leading to cancerous cell formation, are two of particular note. The composition of the culture medium needs to meet the needs of growing cells and be considered safe for consumption. Further developments in industrial scale bioreactors are necessary. A big issue with the development of in vitro meat products is the complexity of factors which contribute to the final eating quality of real meat. While cell culture technology may be able to work towards providing the range of necessary micro and macro molecules which are reflective of raw muscle, mimicking the influence of environmental factors will remain very challenging. After slaughter, there are a series of events including glycolysis, calcium release, energy utilization, rigor onset, oxidation and denaturation, and enzyme effects, which all impact on the ultimate eating quality. These in turn are influenced by the environment in the abattoir such as chilling regime, hanging method and electrical inputs. Similarly, on-farm factors, in particular diet, can influence the flavour. When we consider meat production, while meat is the main product there are a wide variety of side products or co-products which are also produced. A systemic environmental analysis of in vitro meat is called for and must take this aspect into consideration [95].

Consumer attitudes will, obviously, have a great influence on the level of adoption of such a novel food. There is likely to be some public reluctance—a "yuck" factor—however in vitro meat offers some potential to respond to health-focused meat reducers and ethically motivated consumers who see environmental and animal welfare benefits. Interestingly, Verbeke et al. [96] suggest that vegetarians may not be the ideal primary target group for this product, given that the starting material has come from an animal. In a study of French consumers, Hocquette et al. [97] found that over half of the consumers surveyed accepted the feasibility of the concept but most were sceptical about it from a taste and health perspective, and only a minority would recommend or accept eating it instead of real meat. Taste tests with food critics in the UK actually reported that its taste was acceptable, albeit slightly bland. Willingness to try the new product but reluctance to incorporate into one's daily diet was evident in US [98].

6. Discussion and Conclusions

Many scenarios are possible regarding future protein demand. While it is possible to adequately meet the nutritional requirements of a larger global population based on current supply volumes, it is highly likely the balance of driving and opposing forces, and a reluctance of policy makers to impose austere diets on citizens, will result in a world which will demand significantly more protein in the future. Furthermore, changing consumption patterns is a slow cultural process so, while demand size initiatives may be part of the solution in the longer term, for the short to medium term, more rapid changes are likely to originate from the supply side. While existing protein sources have been subject to various criticisms, and new concerns are emerging including the growth in antibiotic resistance associated with intensive animal production, they will have to continue to be significant sources of supply. This is for pragmatic reasons including the fact that they have established production and supply systems, are important economic sectors in rural areas and provide additional economic benefits in developing countries, their use is well established in consumers' routines and practices and animal-based protein, in particular, has a significant role in diets, lifestyles and culture. However, it is also because we have not yet definitely established which diets are more sustainable; recent research makes a strong case that animal-based protein is an essential component of a sustainable diet. Van Kernebeek et al. [50] conclude that moderate consumption of meat is better for the environment compared to a vegan and vegetarian diet. Their results, which are based on a land use optimization model, contradict the results of previous life cycle assessment studies (which indicate that vegan diets use the least land followed by vegetarian diets) as "they do not consider the unsuitability of marginal lands to grow crops, the suitability of animals to use human-inedible products, and the co-production of meat and milk" [50] (p. 685). This leads them to argue that large populations can only be sustained if animal protein is consumed. Their contention is that land unsuitable for crop production is necessary to meet the dietary requirements of large populations and that this provides animal protein without competing with land for crops. A similar case is made by van Zanten et al. [4] when they assessed the land use efficiency of different livestock systems. They conclude that certain livestock systems can produce human digestible protein more efficiently than crops and that therefore livestock systems have a role in future food supply and contribute to food security. This research supports a case for the transformation of the food system, whereby livestock become vehicles to use resources (i.e., grass and food by-products) that cannot otherwise be used for food production, rather than being sources of high-quality protein [99]. Other recent research, linking GHG emissions to nutrition, further challenges conventional wisdom by indicating that dietary patterns with high levels of meat consumption do not necessarily result in the highest levels of GHG emissions [100].

Notwithstanding the argument to maintain existing protein sources in the diet, it is clear that the way we produce and consume protein has significant impacts on the environment and on human and animal health which necessitate action. Efforts have been underway since the 1960s to reduce the environmental impact of existing protein sources, particularly animal-based protein, in terms of GHG emissions, land, water and energy use, and biodiversity, as well as their social and public health impacts. Breeding, management, and nutrition efforts have led to significant increases in feed conversion efficiency, per-animal yields and decreases in GHG emissions per kg of animal product [92]. However there is significant potential for further efficiency gains because while technologies and practices that reduce emissions for example are in existence, they are not widely adopted despite most mitigation interventions offering economic as well as environmental benefits [37]. Such efficiency strategies are reported to have the potential to result in a reduction of up to 30% in GHG emissions from the livestock sector, if applied globally [99]. In this increased efficiency strategic perspective, the livestock sector (and its sectoral organizations) is identified as an important stakeholder in delivering on the mitigation efforts necessary to reduce GHG emissions and to improve its environmental footprint [37] with public policy to support on-farm investment and reduce risk associated with the adoption of new technologies, and extension efforts that disseminate best practice and highlight success stories, being important approaches.

As a complementary strategy, the UN, largely through the FAO, the EU and many NGOs, is driving the production of alternative proteins. Academics also are having an influential role. Halloran, Roos, Eilenberg, Cerutti and Bruun (2016) [101] identify the publication of the book Edible insects: Future prospects for food and feed security [72] as a "notable landmark" and state that it "captivated a multitude of stakeholders and successfully drew significant attention to the area" [72] (p. 2). However, as highlighted above, such proteins cannot be assumed to be safe or sustainable. They require thorough evaluation including holistic assessment, and appropriate regulatory frameworks. Their success will depend on "proving food safety, production costs, nutritional qualities, scalability and consumer acceptance" [35]. Consumer acceptance is a key hurdle and will require a significant shift for real growth to occur, whereby consumers incorporate such products into their habitual shopping behaviours as opposed to treating such products as novelties. Furthermore, their realization will require the development of new value chains and will require new initiatives to bring actors together who may not have traditionally worked together. An example of this is provided by the Forum for the Future, a project which brought the World Wide Fund for Nature (WWF), the Global Alliance for Improved Nutrition (GAIN), dairy firm Volac, flavour firm Firmenich, confectionery manufacturer Hershey and the meat-free brand Quorn together under the project "the Protein Challenge 2040". This project aims to increase plant-based protein consumption, scale up sustainable feed and reduce protein waste.

When comparing existing and novel sources of proteins and developing an overall strategy to address protein demand and thus to reflect prioritized sources, it is important that a balanced and holistic perspective is adopted as the story with respect to individual protein sources is equivocal. For example, while livestock production results in greenhouse gas emissions, livestock production can also result in a positive impact on biodiversity, e.g., through the maintenance of semi-natural grassland habitats that have high biodiversity levels [33] and nutrient recycling [10]. The same is true from a health perspective, e.g., while meat is an important source of nutrition and consuming meat has positive effects on health in developing countries [10], over-consumption of meat in developed countries has negative health impacts. The identification of appropriate comparison boundaries for novel and existing protein sources is also important as highlighted by Smetana et al. [75] in relation to insect-based and traditional products. These authors also cautioned against using results from one modelling scenario to approximate other models. Moreover, the value of many novel protein sources may extend beyond food. As highlighted above, microalgae have a value as ingredients for cosmetic, pharmaceutical and functional food applications as well as in polymer manufacture and in industry. Indeed, marine-derived bioactive peptide hydrolysates could return higher value than isolated proteins depending on their health benefits, which must be proven in accordance with the European Food Safety Authority legislation in Europe and the FDA in America and elsewhere. Quantifying their (higher) aggregate value may support more widespread adoption. Overall, this discussion points to the need for a change in the narrative from a discussion around "good" and "bad" sources of protein towards a better balance of sustainable protein [29].

There are many strategies are available to respond to growing global protein demand which span technological advancements, changes in the agri-food system and shifts in consumption patterns. Technology advancements include advances in data-enabled technologies as well as improvements in production and processing technologies. While there are concerns in some countries that the agrifood system is becoming more opaque [80], the future agri-food system "will need to play an active role in helping consumers make healthier and more sustainable food choices" [31]. Indeed, many stakeholders are needed to contribute to a "future food system which provides enough safe, authentic food for us all to have healthy lives now and in the future" [81]. Multi-stakeholder action is required at global level to be effective and fair [37]. Overall, this is quite a significant challenge, as, while it is clear that people need to start thinking globally about food, this is new and challenging and not something they are not yet used to doing (FSA). Active policy involvement is required to prevent irreversible consequences of dietary changes [5] and to ensure food security.

While the drive towards diversified protein sources could be considered a threat by companies in meat and dairy sectors, for example, there are also significant opportunities for such companies to create new, high value by-products from undervalued waste streams as well as becoming directly involved in the development of new products from diversified protein sources. Policy makers have a key role to play in this regard, as well as in supporting opportunities to realize efficiencies and reduce food waste [35]. The WRAP report states that "ensuring the UK has a diversified, sustainable and healthy supply of protein will be one of the defining challenges of the coming decades". This challenge is truly scalable, i.e., it will be one of the defining challenges for individual nation stages, regions and indeed the globe in the coming decades.

Acknowledgments: This work forms part of the ReValueProtein Research Project (Exploration of Irish Meat Processing Streams for Recovery of High Value Protein Based Ingredients for Food and Non-food Uses, Grant Award No. 11/F/043) supported by the Department of Agriculture, Food and the Marine (DAFM) under the National Development Plan2007–2013 funded by the Irish Government.

Author Contributions: Maeve Henchion initiated the paper, developed the framework and wrote Sections 1, 2, 4.2 and 5. Maria Hayes wrote Sections 3.1, 4.3 and 5.1. Anne Maria Mullen wrote Section 3.2. Brijesh Tiwari wrote Section 4.1. Mark Fenelon wrote Section 3.3. All authors read and approved the final version of the paper.

Conflicts of Interest: The authors declare no conflict of interest.

References

1. United Nations. 2015 Revision of World Population Prospects, United Nations. Available online: https://esa.un.org/unpd/wpp/publications/files/keyfindingswpp2015.pdf (accessed on 25 April 2016).
2. Westhoek and Colleagues. Available online: http://www.fao.org/fileadmin/user_upload/animalwelfare/Protein_Puzzle_web_1.pdf (accessed on 17 July 2017).
3. Tilman, D.; Clark, M. Global diets link environmental sustainability and human health. *Nature* **2014**, *515*, 518–522. [CrossRef] [PubMed]
4. Van Zanten, H.H.E.; Mollenhorst, H.; Klootwijk, C.W.; van Middelaar, C.E.; de Boer, I.J.M. Global food supply: Land use efficiency of livestock systems. *Int. J. Life Cycle Assess.* **2016**, *21*, 747–758. [CrossRef]
5. Delgado, C.L. Rising consumption of meat and milk in developing countries has created a new food revolution. *J. Nutr.* **2003**, *133*, 3907S–3910S. [PubMed]
6. Popkin, B.M.; Adair, L.S.; Nq, S.W. Global Nutrition and the pandemic of obesity in developing countries. *Nutr. Rev.* **2012**, *70*, 3–21. [CrossRef] [PubMed]
7. Kearney, J. Food consumption trends and driver. *Philos. Trans. R. Soc.* **2014**, *365*, 2793–2807. [CrossRef] [PubMed]
8. FAO. *Mapping Supply and Demand for Animal-Source Foods to 2030, Animal Production and Health Working Paper*; FAO: Rome, Italy, 2011.
9. Veldhorst, M.; Smeets, A.; Soenen, S.; Hochstenbach-Waelen, A.; Hursel, R.; Diepvens, K.; Lejeune, M.; Luscombe-Marsh, N.; Westerterp-Plantenga, M. Protein-induced satiety: Effects and mechanisms of different proteins. *Physiol. Behav.* **2008**, *94*, 300–307. [CrossRef] [PubMed]
10. Klunder, H.C.; Wolkers-Rooijackers, J.C.M.; Korpela, J.M.; Nout, M.J.R. Microbiological aspects of processing and storage of edible insects. *Food Control* **2012**, *26*, 628–631. [CrossRef]
11. Bühler. Background Report. Available online: https://www.buhlergroup.com/global/en/downloads/Background_Report_Proteins.pdf (accessed on 2 March 2017).

12. FAO. *The State of Food Insecurity in the World, Addressing Food Insecurity in Protracted Crises*; FAO: Rome, Italy, 2010.
13. Van der Spiegel, M.; Noordam, M.Y.; van der Fels-Klerx, H.J. Safety of novel protein sources (insects, microalgae, seaweed, duckweed and rapeseed) and legislative aspects for application in food and feed production. *Compr. Rev. Food Sci. Food Saf.* **2013**, *12*, 662–678. [CrossRef]
14. Shewry, P.R.; Halford, N.G. Cereal seed storage proteins: Structures, properties and role in grain utilization. *J. Exp. Bot.* **2002**, *53*, 947–958. [CrossRef] [PubMed]
15. Cunsolo, V.; Muccilli, V.; Saletti, R.; Foti, S. Mass spectrometry in the proteome analysis of mature cereal kernels. *Mass Spectrom. Rev.* **2012**, *31*, 448–465. [CrossRef] [PubMed]
16. Jansen, G.R.; DiMaio, L.R.; Hause, N.L. Cereal proteins: Amino acid composition and lysine supplementation of TEFF. *J. Agric. Food Chem.* **1962**, *10*, 62–64. [CrossRef]
17. Shiferaw, B.; Prasanna, B.M.; Hellin, J.; Bánziger, M. Crops that feed the world 6. Past successes and future challenges to the role played by maize in global food security. *Food Secur.* **2011**, *3*, 307. [CrossRef]
18. Cavazos, A.; Gonzalez de Mejia, E. Identification of bioactive peptides from cereal storage proteins and their potential role in prevention of chronic diseases. *Compr. Rev. Food Sci. Food Saf.* **2013**, *12*, 364–380. [CrossRef]
19. Klose, C.; Schehl, B.D.; Arendt, E.K. Fundamental study on protein changes taking place during malting of oats. *J. Cereal Sci.* **2009**, *49*, 83–91. [CrossRef]
20. Shih, F.; Daigle, K. Preparation and characterization of rice protein isolates. *J. Am. Oil Chem. Soc.* **2000**, *77*, 885–889. [CrossRef]
21. Sylvester-Bradley, R.; Folkes, B.F. Cereal grains: Their protein components and nutritional quality. *Sci. Prog. Oxf.* **1976**, *63*, 241–263.
22. Hernández-Ledesma, B.; Del Mar Contreras, M.; Recio, I. Antihypertensive peptides: Production, bioavailability and incorporation into foods. *Adv. Colloid Interface Sci.* **2011**, *165*, 23–35. [CrossRef] [PubMed]
23. Gobbetti, M.; Minervini, F.; Rizzello, C.G. Angiotensin I-converting enzyme-inhibitory and antimicrobial bioactive peptides. *Int. J. Dairy Technol.* **2004**, *57*, 173–188. [CrossRef]
24. Udenigwe, C.C.; Aluko, R.E. Food protein-derived bioactive peptides: Production, processing, and potential health benefits. *J. Food Sci.* **2012**, *77*, R11–R24. [CrossRef] [PubMed]
25. Iwaniak, A.; Dziuba, J. Animal and plant proteins as precursors of peptides with ACE Inhibitory Activity—An in silico strategy of protein evaluation. *Food Technol. Biotechnol.* **2009**, *47*, 441–449.
26. Malaguti, M.; Dinelli, G.; Leoncini, E.; Bregola, V.; Bosi, S.; Cicero, A.F.G.; Hrelia, S. Bioactive peptides in cereals and legumes: Agronomical, biochemical and clinical aspects. *Int. J. Mol. Sci.* **2014**, *15*, 21120–21135. [CrossRef] [PubMed]
27. Pihlanto, A.; Mäkinen, S. Antihypertensive properties of plant protein derived peptides. In *Bioactive Food Peptides in Health and Disease*; Intech: Rijeka, Croatia, 2013.
28. Gangopadhyay, N.; Wynne, K.; O'Connor, P.; Gallagher, E.; Brunton, N.P.; Rai, D.K.; Hayes, M. In silico and in vitro analysis of the angiotensin-I-converting enzyme inhibitory activity of hydrolysates generated from crude barley (*Hordeum. vulgare*) protein concentrates. *Food Chem.* **2016**, *203*, 367–374. [CrossRef] [PubMed]
29. Forum for the Future. The Future of Protein: The Protein Challenge 2040. Shaping the Future of Food. Available online: https://www.forumforthefuture.org/sites/default/files/TheProteinChallenge2040SummaryReport.pdf (accessed on 3 May 2017).
30. Henchion, M.; McCarthy, M.; Resconi, V.; Troy, D. Meat consumption: Trends and quality matters. *Meat Sci.* **2014**, *98*, 561–568. [CrossRef] [PubMed]
31. Rosegrant, M.W.; Paisner, M.S.; Meijer, S.; Witcover, J. *Global Food Projections to 2020: Emerging Trends and Alternative Futures*; IFPRI: Washington, DC, USA, 2001; Volume xvi, p. 206.
32. OECD/FAO 2013. OECD-FAO Agricultural Outlook 2013–2022. Available online: http://www.oecd.org/berlin/OECD-FAO%20Highlights_FINAL_with_Covers%20(3).pdf (accessed on 17 July 2017).
33. Henchion, M.; De Backer, C.; Hudders, L. Ethical and sustainable aspects of meat production; Consumer perceptions and system credibility. In *Meat Quality Aspects: From Genes to Ethics*; Purslow, P., Ed.; Elsevier: Cambridge, MA, USA, 2017.
34. Kanerva, M. *Meat Consumption in Europe: Issues, Trends and Debates*; Artec-Paper 187; Universität Bremen: Bremen, Germany, 2013; p. 58, ISSN 1613-4907.
35. WRAP. Food Futures: From Business as Usual to Business Unusual. Available online: http://www.wrap.org.uk/content/food-futures (accessed on 12 March 2017).

36. Uauy, R.; Waage, J. Feeding the world healthily: The challenge of measuring the effects of agriculture on health. *Philos. Trans. R. Soc. B* **2010**, *365*, 3083–3097.
37. Gerber, P.J.; Mottet, A.; Opio, C.I.; Falcucci, A.; Teillard, F. Environmental impacts of beef production: Review of challenges and perspectives for durability. *Meat Sci.* **2015**, *109*, 2–12. [CrossRef] [PubMed]
38. Bax, M.-L.; Buffière, C.; Hafnaoui, N.; Gaudichon, C.; Savary-Auzeloux, I.; Dardevet, D.; Santé-Lhoutellier, V.; Rémond, D. Effects of meat cooking, and of ingested amount, on protein digestion speed and entry of residual proteins into the colon: A study in minipigs. *PLoS ONE* **2013**, *8*, e61252. [CrossRef] [PubMed]
39. Berner, L.K.; Miller, D.D.; VanCampen, D. Availability to rats of iron in ferric hydroxide polymers. *J. Nutr.* **1985**, *115*, 1042. [PubMed]
40. Hurrell, R.; Egli, I. Iron bioavailability and dietary reference values. *Am. J. Clin. Nutr.* **2006**, *91*, 1461S–1467S. [CrossRef] [PubMed]
41. International Agency for Research on Cancer. *Consumption of Red Meat and Processed Meat*; IARC Working Group: Lyon, France, 2015.
42. De Smet, S.; Voosen, E. Meat: The balance between nutrition and health. A review. *Meat Sci.* **2016**, *120*, 145–156. [CrossRef] [PubMed]
43. O'Connor, L.E.; Kim, J.E.; Campbell, W.W. Total red meat intake >0.5 servings/d does not negatively influence cardiovascular disease risk factors: A systemically searched meta-analysis of randomized controlled trials. *Am. J. Clin. Nutr.* **2016**, *105*, 57–69. [CrossRef]
44. Wu, G.Y.; Fanzo, J.; Miller, D.D.; Pingali, P.; Post, M.J.; Steiner, J.L.; Thalacker-Mercer, A.E. Production and supply of high-quality food protein for human consumption: Sustainability, challenges, and innovations. *Ann. N. Y. Acad. Sci.* **2014**, *1321*, 1–19. [CrossRef] [PubMed]
45. Agarwal, S.; Beausire, R.L.W.; Patel, S.; Patel, H. Innovative uses of milk protein concentrates in product development. *J. Food Sci.* **2015**, *80*, A23–A29. [CrossRef] [PubMed]
46. Miller, P.E.; Alexander, D.D.; Perez, V. Effects of whey protein and resistance exercise on body composition: A meta-analysis of randomized controlled trials. *J. Am. Coll. Nutr.* **2014**, *33*, 163–175. [CrossRef] [PubMed]
47. Coker, R.H.; Miller, S.; Schutzler, S.; Deutz, N.; Wolfe, R.R. Whey protein and essential amino acids promote the reduction of adipose tissue and increased muscle protein synthesis during caloric restriction-induced weight loss in elderly, obese individuals. *J. Nutr.* **2012**, *11*, 105. [CrossRef] [PubMed]
48. Markus, C.R.; Olivier, B.; Panhuysen, G.E.; Van der Gugten, J.; Alles, M.S.; Tuiten, A.; Westenberg, H.E.; Fekkes, D.; Koppeschaar, H.F.; de Haan, E.E. The bovine protein a-lactalbumin increases the plasma ratio of tryptophan to the other large neutral amino acids, and in vulnerable subjects raises brain serotonin activity, reduces cortisol concentration, and improves mood under stress. *Am. J. Clin. Nutr.* **2000**, *71*, 1536–1544. [PubMed]
49. Rutherfurd, S.M.; Moughan, P.J. Digestible reactive lysine in selected milk-based products. *J. Dairy Sci.* **2005**, *88*, 40–48.
50. Van Kernebeek, M.R.J.; Oosting, S.J.; Van Ittersum, M.K.; Bikker, P.; De Boer, I.J.M. Saving land to feed a growing population: Consequences for consumption of crop and livestock products. *Int. J. Life Cycle Assess.* **2016**, *21*, 677–687. [CrossRef]
51. Capper, J.L.; Cady, R.A.; Bauman, D.E. The environmental impact of dairy production: 1944 compared to 2007. *J. Anim. Sci.* **2009**, *87*, 2160–2167. [CrossRef] [PubMed]
52. Gerber, P.; Steinfeld, H.; Henderson, B.; Mottet, A.; Opio, C.; Dijkman, J.; Falcucci, A.; Tempio, G. *Tackling Climate Change through Livestock—A Global Assessment of Emissions and Mitigation Opportunities*; Food and Agriculture Organization: Rome, Italy, 2013.
53. Teagasc. Teagasc Technology Foresight 2035: Technology Transforming Irish Agri-Food and Bioeconomy. Available online: https://www.teagasc.ie/media/website/publications/2016/Teagasc-Technology-Foresight-Report-2035.pdf (accessed on 17 July 2017).
54. Flysjö, A.; Thrane, M.; Hermansen, J.E. Method to assess the carbon footprint at product level in the dairy industry. *Int. Dairy J.* **2014**, *34*, 86–92. [CrossRef]
55. Vergé, X.P.C.; Maxime, D.; Dyer, J.A.; Desjardins, R.L.; Arcand, Y.; Vanderzaag, A. Carbon footprint of Canadian dairy products: Calculations and issues. *J. Dairy Sci.* **2013**, *96*, 6091–6104. [CrossRef] [PubMed]
56. O'Callaghan, T.F.; Hennessy, D.; McAuliffe, S.; Kilcawley, K.N.; O'Donovan, M.; Dillon, P.; Ross, R.P.; Stanton, C. Effect of pasture versus indoor feeding systems on raw milk composition and quality over an entire lactation. *J. Dairy Sci.* **2016**, *99*, 9440. [CrossRef] [PubMed]

57. O'Callaghan, T.F.; Faulkner, H.; McAuliffe, S.; O'Sullivan, M.G.; Hennessy, D.; Dillon, P.; Kilcawley, K.N.; Stanton, C.; Ross, R.P. Quality characteristics, chemical composition, and sensory properties of butter from cows on pasture versus indoor feeding systems. *J. Dairy Sci.* **2016**, *99*, 9441–9460. [CrossRef] [PubMed]
58. Wang, B.; Li, Y.; Wu, N.; Lan, C.Q. CO_2 bio-mitigation using microalgae. *Appl. Microbiol. Biotechnol.* **2008**, *79*, 707–718. [CrossRef] [PubMed]
59. Elorinne, A.-L.; Alfthan, G.; Erlund, I.; Kivimäki, H.; Paju, A.; Salminen, I.; Turpeinen, U.; Voutilainen, S.; Laakso, J. Food and nutrient intake and nutritional status of Finnish vegans and non-vegetarians. *PLoS ONE* **2016**, *11*, e0148235. [CrossRef] [PubMed]
60. Campos-Vega, R.; Loarca-Piña, G.; Oomah, B.D. Minor components of pulses and their potential impact on human health. *Food Res. Int.* **2010**, *43*, 461–482. [CrossRef]
61. Thompson, L.U. Potential health benefits and problems associated with antinutrients in foods. *Food Res. Int.* **1993**, *26*, 131–149. [CrossRef]
62. USDA National Nutrient Dabase For Standard Reference for What We Eat in America, NHANES (Survey-SR). Available online: https://www.ars.usda.gov/northeast-area/beltsville-md/beltsville-human-nutrition-research-center/nutrient-data-laboratory/docs/usda-national-nutrient-database-for-standard-reference-dataset-for-what-we-eat-in-america-nhanes-survey-sr/ (accessed on 27 March 2017).
63. Alonso, R.; Aguirre, A.; Marzo, F. Effects of extrusion and traditional processing methods on antinutrients and in vitro digestibility of protein and starch in faba and kidney beans. *Food Chem.* **2000**, *68*, 159–165. [CrossRef]
64. Singh, U.; Singh, B. Tropical grain legumes as important human foods. *Econ. Bot.* **1992**, *46*, 310–321. [CrossRef]
65. Rossi-Fanelli, A.; Antonini, E.; Brunori, M.; Bruzzesi, M.; Caputo, A.; Satrani, K. Isolation of a monodisperse fraction from cottonseeds. *Biochem. Biophys. Res. Commun.* **1964**, *15*, 110. [CrossRef]
66. Iqbal, A.M.; Wani, S.A.; Lone, A.A.; Nar, Z.A. Breeding for Quality Traits in Grain Legumes. 2006; pp. 1–20. Available online: https://www.researchgate.net/profile/Ajaz_Lone/publication/257645622_Breeding_for_Quality_Traits_in_Grain_Legumes/links/562b225408ae22b17031f063.pdf (accessed on 18 July 2017).
67. Mubarak, A.E. Nutritional composition and antinutritional factors of mung bean seeds (*Phaseolus aureus*) as affected by some home traditional processes. *Food Chem.* **2005**, *89*, 489–495. [CrossRef]
68. Bhat, R.; Karim, A.A. Exploring the nutritional potential of wild and underutilised legumes. *Compr. Rev. Food Sci. Food Saf.* **2009**, *8*, 305–331. [CrossRef]
69. Tresina, P.S.; Mohan, V.R. Assessment of nutritional and antinutritional potential of underutilised legumes of the genus MUCANA. *Trop. Subtrop. Agroecosyst.* **2013**, *16*, 155–169.
70. Sreerama, Y.; Seshikala, V.; Pratape, V.; Singh, V. Nutrients and antinutrients in cowpea and horse gram flour in comparison to chickpea flour. Evaluation of their flour functionality. *Food Chem.* **2012**, *131*, 462–468. [CrossRef]
71. Jongema, Y. World List of Edible Insects 2015. Wageningen Laboratory of Entomology, Wageningen University. Available online: https://www.wur.nl/upload_mm/7/4/1/ca8baa25-b035--4bd2--9fdc-a7df1405519aWORLD%20LIST%20EDIBLE%20INSECTS%202015.pdf (accessed on 2 March 2017).
72. Van Huis, A.; Van Itterbeeck, J.; Klunder, H.; Mertens, E.; Halloran, A.; Muir, G.; Vantomme, P. *Edible Insects: Future Prospects for Food and Food*; Food and Agriculture Organisation of the United Nations: Rome, Italy, 2013.
73. Borello, M.; Caracciolo, F.; Lombardi, A.; Pascucci, S.; Cembalo, L. Consumer perspective on circular economy strategy for reducing waste. *Sustainability* **2017**, *9*, 141. [CrossRef]
74. Oonicx, D.G.A.B.; de Boer, I.J.M. Environmental impact of the production of mealworms as a protein source for humans—A life cycle assessment. *PLoS ONE* **2012**, *7*, e51145.
75. Smetana, S.; Palanisamy, M.; Mathys, A.; Heinz, V. Sustainability of insect use for feed and food: Life Cycle Assessment perspective. *J. Clean. Prod.* **2016**, *137*, 741–751. [CrossRef]
76. Illgner, P.; Nel, E. The Geography of Edible Insects in Sub-Saharan Africa: A study of the Mopane Caterpillar. *Geogr. J.* **2000**, *166*, 336–351. [CrossRef]
77. Belluco, S.; Losasso, C.; Maggioletti, M.; Alonzi, C.C.; Paoletti, M.G.; Ricci, A. Edible insects in a food safety and nutritional perspective: A critical review. *Compr. Rev. Food Sci. Food Saf.* **2013**, *12*, 296–313. [CrossRef]
78. Nonaka, K. Feasting on insects, invited review. *Entomol. Res.* **2009**, *39*, 304–312. [CrossRef]
79. Rozin, P.; Fallon, A.E. A perspective on disgust. *Psychol. Rev.* **1987**, *94*, 23–41. [CrossRef] [PubMed]

80. Carrington, D. Insects Could be the Key to Meeting Food Needs of Growing Global Population. Available online: https://www.theguardian.com/environment/2010/aug/01/insects-food-emissions (accessed on 14 July 2017).
81. Becker, E.W. Microalgae as a source of protein. *Biotechnol. Adv.* **2007**, *25*, 207–210. [CrossRef] [PubMed]
82. Fleurence, J. Seaweed proteins; biochemical, nutritional aspects and potential uses. *Trends Food Sci. Technol.* **1999**, *10*, 25–28. [CrossRef]
83. Sampels, S. Towards a more sustainable production of fish as an important source for human nutrition. *J. Fish. Livest. Prod.* **2014**, *2*, 119. [CrossRef]
84. García-Vaquero, M.; Hayes, M. Red and green macroalgae for fish, animal feed and human functional food development. *Food Rev. Int.* **2016**, *32*. [CrossRef]
85. Target, V.M.; Arnold, T.M. Mini review—Predicting the effect of brown algal phlorotannins on marine herbivores in tropical and temperate oceans. *J. Phycol.* **1998**, *34*, 195–205. [CrossRef]
86. Vigani, M.; Parisi, C.; Rodriguez-Cerezo, E.; Barbosa, M.J.; Sijtsma, L.; Ploeg, M.; Enzing, C. Food and feed products from micro-algae: Market opportunities and challenges for the EU. *Trends Food Sci. Technol.* **2015**, *42*, 81–92. [CrossRef]
87. FAO. *The State of World Fisheries and Aquaculture Contributing to Food Security and Nutrition for All*; FAO: Rome, Italy, 2016.
88. Hanson, T. Economic analysis project rising returns for intensive biofloc shrimp system. *Glob. Aquac. Advocate* **2013**, *16*, 24–26.
89. Hayes, M.; Mora, L.; Hussey, K.; Aluko, R.E. Boarfish protein recovery using the pH shift process and generation of protein hydrolysates with ACE-I and antihypertensive bioactivities in Spontaneously Hypertensive Rats (SHRs). *Innov. Food Sci. Emerg. Technol.* **2016**, *37*, 253–260. [CrossRef]
90. Adeoti, I.A.; Hawboldt, K. A review of lipid extraction from fish processing by-product for use as biofuel. *Biomass Bioenerg.* **2014**, *63*, 330–340. [CrossRef]
91. Abbey, L.; Glover-Amengor, M.; Atikpo, M.O.; Atter, A.; Toppe, J. Nutrient content of fish powder from low value fish and fish by-products. *Food Sci. Nutr.* **2016**, *5*, 374–379. [CrossRef] [PubMed]
92. Hayes, M. Bioactive peptides and their potential use for the prevention of diseases associated with Alzheimers' disease and mental health disorders: Food for Thought? *Ann. Psychiatry Ment. Health* **2014**, *2*, 1017.
93. Sidhu, K.S. Health benefits and potential risks related to consumption of fish or fish oil. *Regul. Toxicol. Pharmacol.* **2003**, *38*, 336–344. [CrossRef] [PubMed]
94. Langelaan, M.L.P.; Boonen, K.J.M.; Polak, R.B.; Baaijens, F.P.T.; Post, M.J.; van der Schaft, D.W.J. Meet the new meat: Tissue engineered skeletal muscle. *Trends Food Sci. Technol.* **2010**, *21*, 59–66. [CrossRef]
95. Mattick, C.S.; Landis, A.E.; Allenby, B.R. A case for systemic environmental analysis of cultured meat. *J. Integr. Agric.* **2015**, *14*, 249–254. [CrossRef]
96. Verbeke, W.; Sans, P.; Van Loo, E.J. Challenges and prospects for consumer acceptance of cultured meat. *J. Integr. Agric.* **2015**, *14*, 285–294. [CrossRef]
97. Hocquette, J.-F. Is in vitro meat the solution for the future? *Meat Sci.* **2016**, *120*, 167–176. [CrossRef] [PubMed]
98. Wilks, M.; Phillips, C.J.C. Attitudes to In Vitro meat: A survey of potential consumers in the United States. *PLoS ONE* **2017**, *12*, e0171904. [CrossRef] [PubMed]
99. Schader, C.; Muller, A.; Sciaballa, N.E.-H.; Hecht, J.; Isensee, A.; Erb, K.-H.; Smith, P.; Makkar, H.P.S.; Klocke, P.; Leiber, F.; et al. Impacts of feeding less food-competing feedstuffs to livestock on global food system sustainability. *J. R. Soc. Interface* **2015**, *12*. [CrossRef] [PubMed]
100. Hyland, J.J.; McCarthy, M.B.; Henchion, M.; McCarthy, S.N. Dietary emissions patterns and their effect on the overall climate impact of food consumption. *Int. J. Food Sci. Technol.* **2017**. [CrossRef]
101. Halloran, A.; Roos, N.; Eilenberg, J.; Cerutti, A.; Bruun, S. Life cycle assessment of edible insects for food protein: a review. *Agron. Sustain. Dev.* **2016**, *36*, 57. [CrossRef]

MDPI
St. Alban-Anlage 66
4052 Basel, Switzerland
Tel. +41 61 683 77 34
Fax +41 61 302 89 18
http://www.mdpi.com

Foods Editorial Office
E-mail: foods@mdpi.com
http://www.mdpi.com/journal/foods

www.ingramcontent.com/pod-product-compliance
Lightning Source LLC
Chambersburg PA
CBHW051858210326
41597CB00033B/5940
* 9 7 8 3 0 3 8 4 2 8 6 3 3 *